STEAM IN THE BLOOD

A Railwayman's Journey 1941-1982

STEAM IN THE BLOOD

A Railwayman's Journey 1941-1982

RHN Hardy

A Goodall paperback
from
Crécy Publishing Limited

STEAM IN THE BLOOD
Copyright © R.H.N. Hardy 1971

First published 1971 by Ian Allan Publishing

RAILWAYS IN THE BLOOD
© R.H.N. Hardy 1985

First published 1985 by Ian Allan Publishing

This combined paperback edition
published by Crécy Publishing 2022

ISBN 9781800351455

Front cover:
In August 1959, A4 Pacific No 60022 Mallard arrives at King's Cross with a train
consisting of former LNER carriages in both the BR maroon and carmine and cream
liveries. *LRTA (London Area) Collection/Online Transport Archive*
Back cover:
Designated Class G by the Metropolitan and Class M2 by the LNER, No 96 Charles
Jones was withdrawn by BR in 1948 (as LNER No 9077).
J. Joyce Collection/Online Transport Archive

Printed in Bulgaria by Multiprint

A Goodall paperback
published by

Crécy Publishing Limited
1a Ringway Trading Estate, Shadowmoss Road, Manchester M22 5LH
www.crecy.co.uk

STEAM IN THE BLOOD

RHN Hardy

Contents

Acknowledgements

To Cecil J. Allen, who has done so much since I was a child by his writing and later by his friendship to foster in me the railway spirit, and who has shown me that there is a great deal more in writing a book than I thought there ever could be.

To those who have checked some of the chapters and to those who have succeeded in reading and then typing my illegible script.

And lastly, to my wife and our family, who have always been an inspiration to me in everything that I have had to tackle and who have encouraged me and indeed put up with me whilst I have wrestled with this unaccustomed task.

Preface

THIS BOOK has been written at the request of Ian Allan Ltd, the publishers, who have been good enough to print, now and again, such articles as I have had time to write over the past years under the pen-name of "Balmore".

I have been lucky enough to choose a great profession, great in history and tradition despite the criticism often heaped on us, great in present-day achievement and great, too, as to the future, make no mistake about that. But this book deals with the past. In it I have recalled incidents and experiences in my railway life up to 1962, some interesting, some amusing, and some just ordinary, to give an insight into the kind of existence I have enjoyed during the first twenty years of my railway career.

But I find that, once one becomes a manager, particularly a manager of men, it is extremely difficult to portray the really true picture of one's work. Routine does not necessarily make interesting reading, whilst so many other things which are not routine must go unrecorded because the feelings of men still working, or living in retirement, must be respected. The earlier chapters were written, not with the hindsight and experience of years, but as seen through the eyes of youth and inexperience. I think it has been better and more natural that way, for such a method is more likely to convey to the reader the effect on a young man of the thrill and excitement of being involved in railways.

By now I suppose I must know several thousands of railway employees. One's training and responsibilities must breed interest in people as well as in machines, in the organisation, and in one's own ambitions. I have had to draw on my memory to write this book, for I have never kept notes or a diary of any sort, and if the odd discrepancy in dimension, date or figure creeps in, it must be put down to that lack. But there will be no discrepancy when it comes to people, for one rarely forgets the men one works with. Indeed, it would be ungrateful to do so, for it is they who have helped one to climb the ladder of promotion, and it is they who make one's work alive.

And so this is my chance to express my gratitude to those thousands, so few of whom can be mentioned in these pages, for what they have done for me. Some helped me as a young man, some have been my colleagues through rough and smooth, some have done battle with me. All have made my life worth living.

R. H. N. Hardy
Burton-in-Wirral, Cheshire.

I

The Irresistible Urge

IT WAS at a very tender age that I became interested in railways. I was given the January 1929 *Railway Magazine* and was enthralled by it; I read all sorts of books on railways, many of which I could not understand. Before I reached double figures in age, I could bore (or fascinate) people by quoting Cecil J. Allen at some length (I'm not sure if that is a very tactful remark), but the locomotive, live and wonderful, was my real love; the nearer I could get to these romantic machines, the better it was for me. And if I was slipped into the cab and taken for a spot of shunting in Leatherhead goods yard, I was no more trouble to my parents for days on end.

I never jotted down the numbers of engines that I had seen but I did draw them. Extraordinary though it may seem, I won a gold medal at a children's exhibition organised by the Royal Drawing Society when I was just under five years old – the engine was an LNER A1 Pacific, and you could see not only the smokebox door, but the full side view at the same time.

In March 1931 my governess, a wonderful person whom I called Mitts, offered to take me for a fortnight to her home at Mexborough in Yorkshire. Her father was a coal merchant, and her uncle, Willie Crossley, was the chief clerk at Doncaster Locomotive Depot, whence he had gone from Mexborough some time after the Grouping. I was seven-and-a-half years old and the holiday still stands out as one of the most marvellous in my life. Apart from delivering coal with a character called Reuben who drove a Ford truck, going to Manvers Colliery, seeing a glass-blowing factory, sailing on the Don in a barge, and visiting the "trackless tram" depot, I spent a day in Doncaster locomotive sheds, a day at York, and endless hours on Mexborough station. But I never thought that 10 years later I should be on the same station as a railwayman, with a seniority of one day!

In 1934, my father, who had been a tea planter, and then, after the first war, in the tea business, retired and we moved from Surrey to Amersham in Buckinghamshire. The railway that passed through Amersham was of particular interest in that it was jointly owned by the Metropolitan and London and North Eastern Railway Companies, and it was at this stage that railways began to get really deeply into my blood. I was fascinated by the ex-Great Central locomotives as well as by the beautifully kept Metropolitan tanks, and by the time I was 12 I had got to know quite a few footplatemen, who seemed to look forward to meeting me at the end of the platform for a few seconds before they were on their way to the next station with a Metropolitan stopping train.

Nor was my interest confined to locomotives, for elementary signal box work attracted me through the kindness of one of the signalmen, Hector Ratcliffe, and I was also initiated into the secrets of a booking office by two of the clerks. The stationmaster, however, had no use for me at all, though he did not actively discourage me, and I think I learned not to overdo things.

Now came an interruption in this railway contact, though certainly not in my interest in railways, caused by the years that I spent at Marlborough College. Here, it was a long time before I found a kindred spirit of my own age who was a lover of the railway, although work and games provided plenty for me to get on with. Now and again I used to go up to the station at Marlborough on the Midland and South Western Junction line from Cheltenham to Southampton, but a much greater attraction was Savernake, five miles away, on the Great Western main line, where I could see a great deal of the GWR locomotives about which I had read so much. Here I got to know two of the signalmen, and spent hours in their boxes on Saturday afternoons.

All the different classes of Great Western express engine were in evidence, and then, at 5.48pm, a little ex-M&SW J 2-4-0 would arrive with a stopping train from Reading, and would run round the coaches before returning at 6.13 pm to Newbury. Occasionally, and always to my great delight, an ex-ROD Great Central 2-8-0 would rattle through on a freight train, while into the adjacent M&SW J station a Great Western "Duke" or "Bulldog" 4-4-0 would run quietly on its way to Swindon or Southampton. So there was plenty of variety to occupy my attention.

Occasionally I used to travel over the M&SW J line as far as Swindon, and on one occasion we were allowed round the Swindon Works. But my greatest joy, apart from the Savernake and Woodborough signalboxes, was to have a quiet hour in the cab of a 45 2-6-2 tank engine with my friend Arthur Wilkins from Andover Junction. He was a very interesting man, and gave me a lot of instruction about locomotives, as well as a scrapbook of photographs of American locomotives which I treasured for many years. Much of his work was on these 2-6-2 tanks, though he also covered the M&SW J main line, changing footplates on the single line at Swindon or Chiseldon or some such place.

Arthur was always pleased to see me except when he had *Trefusis*, which was an old Great Western "Duke" 4-4-0, loved by nobody very much except myself, for she had a big dome, and the sight of her outside cranks whirling round I thought quite marvellous. But she had one foot in the grave, and was shy of steam, whereas the "Bulldogs" and the 2-6-0s were masters of their work. Then one day, to my amazement, Arthur went by on a Drummond "Greyhound" 4-4-0 of the Southern, which he was driving as Conductor on a double-headed special train.

We did not have a Railway Society at Marlborough, but our house ran a Casuals' cricket team which from time to time played the surrounding villages. One of these was Burbage, just west of Savernake, which I well remember because that game drew me even closer to the railway. The Burbage captain and wicket-keeper was a ganger at Savernake, and the fast bowler, K. Fear (and his name fitted his bowling!) was a porter-signalman at Savernake Low Level, who from that day on always looked after me whenever we met.

But much as I admired the Great Western engines, it was still the London and North Eastern, and above all the Great Central engines, that took my fancy, and there was nothing else to equal getting back home from school, travelling behind an A5 4-6-2 tank, taking my place once again at the end of Amersham platform, and waiting for a full-blooded Robinson Atlantic or 4-4-0 or 4-6-0 to come roaring up Amersham bank. *Valour* or *Butler-Henderson* would come under the bridge, accelerating gradually where the gradient eased and then thundering by, with the white feather from the safety-valves standing out clearly against the background of black smoke shooting skywards, and the driver, nearly always a solid elderly moustachioed men, sitting comfortably and relaxed in his seat. Sometimes these great man winked or even waved as they passed by, and I could live on that acknowledgement for days afterwards.

In 1938 my father died. His death created somewhat of a financial problem, but my mother was able to keep me at Marlborough until the end of 1940.

Soon afterwards I decided that I must become a railwayman, although I was not clear how or where. Lack of funds dictated that there could be no going to a university, and so by the beginning of 1941 I took a step that I have never regretted, and joined the London and North Eastern Railway as a premium apprentice. But I am moving ahead a little too fast, because between the years 1938 and 1940 a gradual change had come over my feelings towards railways.

This change is difficult to put into words. It was that instead of a straightforward dedication to a machine, my interest widened to include people, their work, their way of life. Whereas in earlier years I had saved every penny I could to pay for a holiday journey from Marylebone to Leicester and back behind a "Director" 4-4-0, now I was content to stay at Amersham throughout my holidays, meeting as many engine crews as I could and hoping that they would invite me to travel with them to the next station. Sometimes they did, and I was unbelievably happy, because I was beginning to see for myself the art of enginemanship – just beginning, but that was enough.

The enginemen whom I knew belonged to Neasden depot. They were either ex-Great Central or ex-Metropolitan men. The Met men came across to the LNER depot a little while before the war but remained in their own links on Metropolitan work, retaining their conditions of service, which included a different type of overall and a distinctive cap.

The senior Great Central drivers were men of long experience of main line work between Marylebone, Leicester, Sheffield and Manchester, a road which demanded a considerable degree of skill and knowledge. George Parks and Ted Simpson had been firemen when the London extension was opened in 1899 and had come south to fill vacancies at the new Neasden depot.

Ted Simpson, if my memory serves me correctly, came from Liverpool, having been a cleaner at Brunswick Cheshire Lines depot; he was a fine and economical engineman. As a GC man, I expected him to be passionately devoted to John G. Robinson and all his works and to have no love for Gresley Pacifics or "Sandringham" 4-6-0s. But although he had known and deeply respected Robinson, he was able to appreciate a remarkable locomotive and the A1 Pacifics which came to Neasden and Gorton before the war were in fine fettle and were welcomed accordingly.

Simpson retired in April 1941, soon after I started work, but not before I had learnt much from him and been to his home in Gresham Road, Neasden, to listen for hours on end to him talking about locomotives and railwaymen. I still treasure the letter that he wrote to me at Marlborough telling me that, if my mother agreed, he would take me from Marylebone to Aylesbury on the footplate of the locomotive working the 10.0pm Mail. That letter burnt my pockets day after day and was dogeared long before the end of the term. I longed for a class B3 4-6-0 with Caprotti valves such as *Earl Haig* or *Lord Stuart of Wortley.* Night after night, I had heard these old engines roar through Amersham on the Mail, more often it seemed than the Pacifics. But on the night that I was to ride, as I stood waiting on No 4 platform under the bridge at Marylebone, it was a high Pacific tender that backed down in the blackout on to the heavy 12-coach train.

The fireman was Sam Oldknow and the engine was 4-6-2 No 2554 *Woolwinder.* We made a calm and purposeful run to Aylesbury, quietly climbing the long Amersham bank at about 30mph, and arriving exactly on time at 10.54. I had six minutes to catch the last Met train home. Sam saw me across from the engine on to the up platform. In the dark I dashed along and fell headlong over a pile of mailbags; but bruises and lack of wind made no difference to me. It had been a wonderful evening!

Sam Oldknow lived at High Wycombe and finished up as Running Foreman at Marylebone station; Ted Simpson battled with wartime conditions until he retired. One of his last journeys was with the 10.0am from Marylebone to Leicester. His Pacific was stopped at the last moment and he was given an ex-GC C4 Class 4-4-2, No 5363. This old dear had seen better days; her sands were not working and she was light on her feet. They backed on to 11 bogies at Marylebone; the normal load was 10 but an extra coach full of expectant mothers had been added. Where they were going and why, history does not relate.

It took Ted Simpson and No 5363 about 20 minutes to get out of Marylebone and they felt their way through the first and second tunnels satisfactorily. But when they struck the 1 in 100 in the third tunnel, the old engine slipped and slipped and slipped. The cab was full of smoke and steam, and the train came gently to a stand, with the expectant mothers' coach standing on a set of catch-points at the London end of the tunnel. Gently, the train moved back, with No 5363's driving wheels still spinning furiously; and equally gently the last coach came to a stand on the ballast near the bridge over the LMS main line.

Pandemonium reigned, but after a while driver and guard got things sorted out, and the mothers with their baggage were taken through to seats further up the train. No 5363 had another go with the lightened load, got her train under control, and eventually made Woodford, where she and Ted came off, she to go to the shed in disgrace, he to pick up the train he should have worked from Leicester. In after years I used to enjoy hearing this story. I think this happened during his last week's work, but his last days on the road were completed without further incident.

Jack Proctor was another of the top link drivers, and Fisher made up the "Big Four". The next link was called the Piped Goods Link and of these drivers I knew Jack Kitching particularly well. He came from Barnsley to Neasden and in later years, after I joined the railway, he gave me very clear and practical instruction and experience in the handling of the vacuum brake on stopping trains, with "Director" 4-4-0s, M 4-6-2 tank engines, B3 and B7 4-cylinder 4-6-0s, and, in fact, any old thing that came to hand. I last saw him in 1950 when I was shedmaster at Woodford 20 years ago, but he and all the others are as clear in my mind as if it were yesterday. Youth is very impressionable and one never forgets the kindness and generosity of these people.

Then, in the Local Train Link, there was Bill Collins; his mate was Bill Palfreyman, killed by a moving train not so long ago near Neasden. Bill took me up on his tank engines, and, as they also worked a wartime train to Woodford, leaving London at 1.50pm, this was a favourite for me as sometimes locomotives strange to London used to come south on the outward working and I lived for days on the strength of a journey on one of the original Robinson 4-6-0s, Class Br, No 5196. Bill is still alive; he lives at Marlow and we correspond from time to time.

Arthur Ross, a top link fireman in 1939, yet another north countryman from Keadby, near Frodingham, and now a driver at Marylebone, came to see me not so long ago at King's Cross. He used to write to me at school: long letters which must have taken him ages to compose, on locomotive technicalities. I wonder if he realised how much pleasure he was giving a young man on the threshold of a railway career?

14

The Metropolitan men had been reared on the G and H Class tank engines for passenger work and the K Class for freight work, 0-6-4, 4-4-4 and 2-6-4 tank engines respectively, of which I had gained a little experience. Among Met drivers I had known Len Hyde since he had been on goods work in 1936. I used to get up very early in the morning and travel to Aylesbury and back with Len, being put off on to the platform in a cloud of steam from the injector steam valve. The K Class locomotives were of a design based on Maunsell practice on the Southern Railway.

They were strong and rugged engines and gave the impression of being much larger than their Southern N Class counterpart. After the Metropolitan locomotives had been absorbed into the LNER stock and transferred to Neasden GC shed, these locomotives got on to passenger trains. This had been virtually unknown in Met days, but they were soon working turn-and-turn-about with the LNER A5 4-6-2 tanks and putting up excellent performances.

Some of the Met men were very hard runners, and were meticulous in their timekeeping. All of them could do a great deal with the Metropolitan engines until their condition began to deteriorate under wartime conditions.

So in those years I got to grips with railways and railwaymen; the fellow feeling began to go deep into my heart. And yet I was still undecided where I should go for my training. Just before the war, a great friend of mine at Marlborough, Dick Lawrence, had left school and gone to Derby Works to be trained as a locomotive engineer. My boyish love was for the LNER, because I lived on that railway, knew its locomotives and just a few of the employees. And yet I felt I ought to follow Dick.

However, when I did make my application to the LMS it did not evoke any particularly enthusiastic reply, and eventually my housemaster arranged an interview with a retired Army officer, who at that time was advising public schoolboys on their careers. I remember the interview vividly. It lasted about two minutes and it changed my life. I was 16½ at the time, easily led and impressionable: the interview went something like this:

CoL P.T.	"Now, boy, what are you going to do as a career?"
R.H.	"I'm going on the railway, sir."
CoL P.T.	"Where, boy, where?"
R.H.	"I think I'm going to follow a friend of mine to Derby on the LMSR as a locomotive engineering apprentice, sir."
COL P.T.	"Don't know anything about the LMS, boy, know the LNER. Gentlemen at the top. Go there."
R.H.	"Yes, sir."

And so I did! The old gentleman's methods were unsophisticated by present day standards, but I have had no reason whatsoever for regretting taking his advice. And so, with visions of becoming one day a Sir Nigel Gresley, and imagining that locomotives were designed on a sheet or two of paper, I wrote to Doncaster. By return, I received a letter signed by Edward Thompson, the Mechanical Engineer, Doncaster. There were vacancies: my name would be entered but he advised me to start work as near the maximum age of 17½ as possible. I was to attend his office for an interview.

I remember so clearly my first impression of Mr Thompson. He was tall, elegant, and well dressed; he welcomed me kindly and interviewed me at some length. What were my ambitions, my home life, school, sport, interests, the lot? Then followed a short trip into the town in search of digs and away home again. This was in the summer of 1940 just before the Battle of Britain began. In October, 1940 I was seventeen years old; so I went back to school and counted the days until my railway life was to begin.

II

Doncaster Works

I COULD SAY that I am one of Gresley's men; but to me he is little more than a legend, for he died very shortly after I went to Doncaster. And so Edward Thompson was the man who mattered and of that I am proud.

It was on January 16, 1941, that I made my way to Doncaster in readiness to start the next day, a Friday. It was a very dark and bitterly cold morning, but Miss Marsh, my ageless landlady, who was to be a second mother to me for the next four-and-a-half years, got me off to work, and I was soon one of a great crowd of men and women swarming into the works – a bit apprehensive and a bit overawed, but not unduly so. I was soon sent to B shop, and put on a turret lathe with a lad called Denis Brandon. He was on the early turn and was relieved at 3.0pm by John Hyde, who had risen to be a senior foreman in the Doncaster plant when I last saw him some years ago. They were both about 15½ years old but seemed over 20 to my untutored eyes, as indeed did most of the other lads in the shop, all South Yorkshire born and bred except for a Pole called Stan who claimed aristocratic connections. They were very kind to me, spared me most of the traditional tricks played on an apprentice, but seemed to be fascinated by the way in which I spoke. This had always seemed natural enough to me but now I began to have grave doubts.

That first day seemed endless; the work was repetitive, turning out pins large and small for I knew not what – the war was on with a vengeance and we were driven hard. But there was just time to go to the adjacent blacksmith's shop to meet Harry Worthington, whose brother was a fireman at Neasden and whom by great good fortune, John Hyde knew. This made my day, for despite the kindness already shown me I was feeling a bit lost and in need of finding somebody who had some contact with the world from which I had come. It was dark again long before I finished work but as I went over the bridge above the station on my way home, I saw beneath it two Great Central engines, a "Fish" 4-6-0 and an Atlantic; this sight made me realise that I should see plenty of my beloved GC locomotives outside the works, even if I never set eyes on them inside.

On the Saturday, work over, I went home to lunch and got back to Doncaster station as quickly as I could, hoping to catch a train to Mexborough. But I just missed the 2.15, and so had to hang around, cold but fascinated, for the next, some two-and-a-half hours later. During this long wait I saw an endless stream of locomotives, LNER, ex GN, ex GC, ex NE, many of which I had only seen in photographs. About a quarter past four, the express from York to the Eastern Counties ran in. This was a Lincoln turn and it was almost always worked by a GC engine, either a B2

"City" Class 4-6-0 or, as on this occasion, a B4 4-6-0, No 6097 *Immingham* herself. It was the first time I had seen her or indeed any of her breed, for they had rarely got as far south as Marylebone except on special workings.

Alongside the cab I stood sheltering from the snow, which had begun to come down really hard. I must have looked pathetic enough to the driver, standing there with my coat collar up and my hair on end, but when the great man spoke and invited me on to the footplate, I imagined myself already on the way to Lincoln. This was stretching it a bit far, however, and I was put back on to the platform when the *rightaway* was given. Herbert Harrison, the driver, became a good friend to me. I saw him many times, indeed whenever he came to Doncaster, even if only for a few minutes; but always he had something to teach me or show me or explain to me.

He was Herbert Harrison of Lincoln GC (his brother was a GN man but not he) for in his eyes nothing good ever came out of Doncaster, certainly not the unmentionable "Klondyke" 990s or the later Ivatt big-boilered Atlantics or the impossible little W Class 4-4-0s that he and his mate had to thrash about Lincolnshire, setting fire to the crops. But a GC compound Atlantic – now that *was* an engine! He was a wise old man. His son, Frank, wanted to follow him on to the footplate but he was told he was going to be a "clurk." And so he did and now, having been a Divisional Manager in the Southern Region, he has risen to a high position in Industrial Relations. The old gentleman looked the part that day – he was typical of the generation of men who had started work around the turn of the century. Dedicated, a strong personality, he was a disciplinarian on the job. A big man, he was not too good on his feet; this he put down to the grouping of the railways, to the fact that Gresley had become Chief Mechanical Engineer and not John G. Robinson, and because you could not work on a Great Northern engine unless you had flat feet!

Soon after No 6097 had disappeared into the blizzard, my train was set back into the bay platform and away I went to Mexborough to relive my 1931 holiday and to surprise one or two of the people I had met ten years before. It all seemed wonderful!

The early days, indeed months, of my railway life in Doncaster "Plant" are clear in my mind. The people I worked with, many of whom must now be dead and certainly retired, imprinted themselves on my memory. I learnt very early not to fool around with fast moving steel pins, one of which took a big chunk out of the first finger of my right hand, and necessitated a quick trip to the Infirmary to be stitched up; I learnt to keep away from the all-seeing eyes of the foreman, whose bowler or Homburg hat struck fear into my soul; and I learnt that clogs were the warmest and most comfortable footwear for the job in hand.

I had started off with a pair of school OTC boots which were too small and hurt like the devil. After a week or two, somebody said "Why not buy a pair of clogs?" and this I did for 8s 6d a pair (1 coupon). These were a godsend in more ways than one. With a weekly wage of 16s 2d and digs costing 30s a week, such economies were valuable. Apart from that the clogs were warm and comfortable, once one had mastered the technique of walking with a solid wooden sole. Most important of all, when in 1945 I caught my foot in between the fall-plate and the tender end of a GN Atlantic in York yard, the wooden sole of the old clogs took the squeeze, not my foot, for which I was everlastingly thankful.

Life in a locomotive works has been described by more experienced authors than I; one is very much like another, so there is no need for a detailed account of my three years in Doncaster Works. One was moved from gang to gang in the machine and fitting shops and so, to some extent in the first year at least, one was a bird of passage, but once one got to the Crimpsall, which was the main repair shop, and so to grips with locomotives, life began in earnest.

I had eleven months erecting on Bill Umpleby's pit as mate to an erector called Edgar Elvidge. Had I needed deflating, he would have seen to that, but he was good to me. He was a very hard worker, perhaps the hardest in this particular gang: he continually complained that as a gang we were not going hard enough – not earning enough money maybe – although Bill Umpleby with his piecework record at the end of each week was too honest. I didn't know nor care; I was happy working with a man who, though slight of build and short in stature, could lift the end of a heavy rod, or a strap, or a piston while I was still looking at it; he taught me all he knew ("and now you know nowt, Dick" – was one of his milder remarks in times of stress). He worked me hard, gave me plenty of barrow-pushing and the rougher menial work which, quite frankly, I enjoyed; and generally he made it clear that an apprentice, though certainly to be encouraged and taught, was really of rather little consequence.

This, perhaps, was one of the reasons why an engineering apprenticeship training is an education, not only in engineering but in human relations. It is no bad thing, in your late teens, to be of little consequence in a tough and hard workshop, particularly if you have had the advantages of education and have lived an easier life than those who are responsible for your training. How you get on with men, old enough to be your father and even your grandfather, rests entirely with you. Most of such men at Doncaster were Yorkshiremen who had spent a lifetime in the Plant, although there were also the odd Londoner, Geordie, Lancastrian or Norfolkman. All of them would help you – provided you were natural, provided you were interested, provided you worked hard and provided you were one of them; you were judged on your merits as a person and that is how it should be.

Edgar and I had many of the smaller engines through our hands; they included Atlantics, ex Great Northern tank engines, K2 and K3 2-6-0s,

J1 and J6s 0-6-0s, and, of course, the Jas and Jos, including No 091, a GN type 0-6-0 which had been built for the Midland and Great Northern Joint Line. I never worked on the Pacifics, for I had developed a devotion to Edgar and so asked to spend all my period of Crimpsall training with him, a decision I never regretted. Our Shop Foreman was George Andrews, who came from Tuxford on the Lancashire, Derbyshire and East Coast Railway, and the Chief Foreman, under the Superintendent, over the whole Crimpsall, was the terrifying Bob Whittaker. I was told that he was a singer and interested in light opera, but there was no evidence of this when he caught me in the bicycle shed trying to be first out the gate at lunch time – a dodge known as "getting a flyer." There followed a painful interview with the Superintendent, but mercifully this was no longer the legendary Rupert Vereker under whom I was to serve in later years.

Nevertheless, Marlborough had prepared me for a hard and spartan existence, so there was no real shock for me as far as my work in the Plant was concerned. Without a qualm I took to the life, to the seemingly autocratic attitude of many of the foremen and chargehands, and to the authority of the fitters and erectors with whom I worked. But it was a different matter at Doncaster Technical College, where I went to day and night classes. This *did* pin me down to size. I was told to go to see the Principal during my first week at the Plant, to acquaint him of my scholastic standards and achievements. These had amounted to Matriculation, seven credits at school certificate level, and Science. Upper V at school – not too bad, I thought.

But this in no way impressed Mr Clewer, the Principal; he looked at me whimsically, standing there in my overalls, and simply remarked "We had a young 'un here from Marlborough, called Dirty Neck Smith; he was a bit dumb, too"! I took all this very seriously, though I was not quite sure whether I liked the Principal! However, Doncaster "Tech" did me good. I had never found work so hard; maybe I had never tried so hard, for I am neither a good mathematician nor does the theoretical side – or indeed the practical side of engineering either – come easily to me. I started halfway through the first year of the five-year course which would culminate in the Higher National Certificate in Mechanical Engineering.

Many of my colleagues at the Tech. had left school at 14, and had grafted their way to day school; some of these worked on the railway, some at Denaby or Cadeby Collieries, some at Peglers or British Ropes; some were premium apprentices like myself who lived locally and had school certificates. All were Yorkshiremen and all were streets ahead of me in every subject. In the end I caught up, but it was hard going and was no more than a prelude to the next four hard years, which culminated for me in July, 1945, with the Higher National Engineering Certificate and blessed relief from examinations. Apart from the endorsements which I needed to become a Graduate of the Mechanicals, I swore I would finish with exams for ever. How easy it is to take that sort of decision at 21.

My next door neighbour in class for the last two years of the course was John Stephenson, now in a responsible position in the Design Office at Derby. John was gifted; it all fell into place with him and he was a great encouragement to me. He was patient; he helped me during class, with homework; and all I could ever do in return was to draw for him – or doodle – little engines of all shapes and sizes when I was being driven to insanity by our pure maths teacher, a brilliant man whose reasoning I could never follow. John lived just outside the works and spent much of his time in R shop on Millwright's work, ultimately going into the Drawing Office, where he rose to Chief Draughtsman within the comparatively short time of 25 years.

At Marlborough I had never been particularly conscious of the fact that the masters had had university backgrounds, or that most of them were academically very good indeed. One simply accepted them as Marlborough Masters. Maybe the biggest shock of all when starting at Doncaster "Tech" was that some of the masters spoke broad Yorkshire, that their methods were very different, that they allowed smoking between lessons, and that they were extraordinarily friendly with us boys, in a way I had but rarely known before. Secretly, I felt indignant over this, but I soon got used to it; most of them were very able men and excellent and patient teachers as indeed they had to be with some of us. Looking back I really enjoyed myself on my weekly visits, hard though the task seemed, not to mention the evening visits that we had to make for the less onerous additional subjects.

I seemed to need very little sleep in those days. But Miss Marsh, at 43 St Marys Road – my digs recommended by the Works – saw to it that I had enough. I may have been strongly disciplined at work, and I was certainly disciplined when away from it. My off-duty activities were subject to microscopic examination. In those days I was not very interested in girls; MM knew this and indeed encouraged the lack of interest. Nevertheless, if I came in at 11 o'clock at night in ordinary clothes, I had to give chapter and verse as to what I had been up to, whereas I could come in at any hour provided I was black and in overalls. How she made a profit in 1941 on providing full board at 30s a week, I do not know. We lived communally and she made the rations go round, so that we all had enough, bolstered up from time to time by fish and chips from down the road. A vast number of lodgers came and went; some stayed quite a while but most of them were in Doncaster for a few weeks only, and any who did not measure up to MM's standards were firmly told to go.

The first visit of my mother was memorable. It was the first time that she had ever tasted real Yorkshire pudding, but not only for this reason. For she found that in a month I had got so much Yorkshire in my speech that she could barely understand me; also she fell for clogs, and bought from a shop in Hexthorpe a fancy women's pair with red beading which she loved and used in the garden at home until she sold up and left Amersham in 1948. She and MM became great "buddies".

But there was another reason why I shall never forget this first visit of my mother. She was arriving at about 5 o'clock on a Friday evening and going straight from the station to my digs. About 4.30pm I was sent for by the foreman, who told me to report to Edward Thompson's office. He was still, of course, Mechanical Engineer, Doncaster. I was ushered into the presence, overalls, clogs and all; the great man simply said: "Go and meet your mother and bring her here to tea and you come too". So off to the Western platform I went, met the 1.10pm from King's Cross, and after my mother had finally recognised me under my grime I took her to his office, where we duly had tea, me still in my overalls and clogs. How Edward Thompson discovered that my Mother was coming to see me I shall never know: I couldn't ask him and I never tried to find out, for it didn't do to advertise the fact an apprentice had had tea in the boss's office. But it was very typical of the Thompson that I knew, and typical of him too in that he made me feel it was quite the normal thing for a 17-year-old boy to be eating sandwiches and drinking china tea in overalls in a very luxurious office

In August 1941 a young man, who was destined to become my closest friend, joined me at No 43: Basil de Iongh, now Group Captain, RAF, whose home was at Welwyn Garden City, the eldest son of a remarkable Dutchman and an even more remarkable Englishwoman. Where I was untidy, slovenly, very tolerant (except of Basil) and quite unsophisticated, Basil was scrupulously tidy, perfectly dressed, good-looking, always spotless, rather intolerant (but not for long) and pretty sophisticated. For a few weeks, we thought nothing good of each other. I could not believe that such an individual could have any real heart for the railway and all the work entailed, or any real understanding or bond with the Plant workmen; but how wrong I was!

Suddenly and for no apparent reason we buried our hatchets and that was the beginning of a friendship, strengthened from the start because our personalities were so different and because, every now and again, a bit of Basil's make-up rubbed off on me and vice-versa. We had some wonderful times together off the job, and particularly on the footplate in our spare time, before he went off to the RAF, in 1943. For very practical reasons, he never returned to the railway, but after the war went into the City for a little while (which didn't suit him) before returning to the RAF.

But he never lost his love for railways and certainly became the only member of the British Embassy in Paris (where he was assistant air attaché in 1959-60) to drive a De Caso 4-6-4 of the SNCF: this he did on several occasions under the guidance of our dear friend André Duteil, Mécanicien de Route, of La Chapelle Depot.

However, at Doncaster Miss Marsh took the dimmest possible view of Basil's off duty activities, comparatively innocent though they were, and he was nearly out on his neck within the first week or two, to my smug and secret delight. But MM relented, which was just as well for me, and before long, she, implacable opponent though she was of the gay life for her younger lodgers, had succumbed to Basil's charm.

Basil will remember, quite apart from MM's Yorkshire puddings, her celebrated sandwiches. These unfortunately were not in the same class. Butter was short, so come to that, were margarine, cheese and jam. After I had been at Doncaster a few weeks, apart from occasional week-ends at home, I used to go out each Saturday afternoon and evening on the footplate, coming home about midnight. Miss Marsh used to fit us up with a stack of sandwiches, thick dry bread and scrape, to keep us going against the strenuous work we were called upon to perform.

However, locomotive cabs are not the best place for sandwiches which started life on the dry side, and sometimes I went home hungry knowing there would be fish and chips in the oven, whilst at other times I was given cheese and onion or some similar delicacy by the people with whom I was working. Anyhow, one night there was a painful scene when I inadvertently opened my little case and MM saw in it a package of her sandwiches, uneaten. After this, neither Basil nor I dared take anything home, but I've wondered from time to time whether those packets I threw over the parapet at Wakefield Westgate ever hit anybody on their way to earth, or what the householders near Netherhall Road Church said next day when they found their little gardens strewn with sandwiches.

With week-ends on the footplate during those few years I covered over 60,000 miles, a wonderful grounding on which to base a knowledge and understanding of the work and personality of that greatest of individualists, the British engineman. They became so much a part of my life that I will say no more about them in this chapter: but I mention this to make it clear how much my life centred round my work and training and around the people with whom I worked. This was ideal, though a complete change of environment now and again was essential.

Through some friends of my mother, I was introduced to a Doncaster family, John and Amy Johnson, he an architect, she a remarkable bridge player. They lived in a lovely old house in Regent Square and they worked hard on me! I loved the meals I had there, and the long evenings afterwards. I always seemed to talk too much and they must have listened to far too much of my excited chatter. But they civilised me quite a bit, broadened my mind, and widened my interests, pulled my leg and introduced me to other families; and I always looked forward to the chance to go there to laugh and be laughed at, to be smartened up and to talk of things other than railways.

Three years completed my works training: I had enjoyed it immensely and made many friends. I was not a polished erector, or anywhere near it, but I had a grounding in human relationships denied to many people whose training in their career was far more sophisticated and far more expensive. Theoretically, I earned my living; in practice I had to be assisted by my mother for a time, but in addition to all that it cost her – and that was enough as things were – there was the lump sum of 50 for the premium. I believe I was the last paying premium apprentice, certainly until after the war, for the lucky folk that followed me had the same training without the premium.

During those years I had dealt with dead locomotives; they arrived for repair on our pit and we did our stuff, saw them lit up in the weigh-house, and then only very rarely had to go to the Running Shed – Carr Loco – to put something right. Naturally, that meant trouble for us, but the joy of freedom from the heat – or indoor cold – of the Crimpsall Shop made up for any recriminations. Edgar Elvidge had his views on Running Sheds and Running Shed staff, whom he regarded as a lot of ignorant bodgers; indeed he used to get out of the place and back to the plant as quickly as he could before he fell out with somebody. But I longed for the rigours – and romance – of the Running Shed. I waited patiently for the day on my programme when this was where I should find myself and somewhere about the end of 1943 it came.

III

Running Shed and Breakdown Gang

LIFE IN a running shed, I felt, would be romantic, exciting and free from the discipline of works life. Nor was I wrong, but there was far more to it than that, for the life of a running shed artisan, working as he had to do in a shed without end doors, facing north-south, could be hard and cold. It goes without saying that the working conditions were filthy, and that there were very few amenities provided for the benefit of the staff. Nevertheless the romance, at any rate to my starry eyes, was there all right and so was the excitement. A works is a factory where discipline and precision and planning must predominate, but a running shed has an apparent freedom from the more obvious form of discipline.

The enemy of the running shed man is time, whether it be the need to carry out a valve and piston examination in time to enable a locomotive to go into traffic an hour or two later, or whether to complete a minor but necessary repair to keep in service a locomotive whose failure might cause a train to be delayed or even cancelled. The running shed artisan has to know the short cuts, the quick, may be, temporary repair, without which another locomotive would have to be found by a harassed foreman in order to save a delay. It is a situation where the works-trained artisan with time on his side could be at a loss – at least for the first few months. Not all running shed work was done at high speed; the more extensive examinations took time and normally were very thoroughly carried out.

So there I was, one November morning in 1943, working on the drop pit. I cannot remember how long I stayed there for I was moved from job to job; working on big ends, injectors, valves and pistons, examinations, valve-setting, regulators, brake adjusting, tank cleaning and so on, and for each I was attached to a fitter or mate. I cannot pretend that I was skilled enough by October 1944 to be regarded as even a potential running shed fitter. The work had interested me, I had learnt a great deal, but, once again, it was perhaps the human side of the job and the personalities with whom I worked that made the greatest impression on me.

In any case, an apprentice really begins to learn when he ceases to be an apprentice, and takes responsibility. Then he stands or falls by his decisions and his mistakes: he makes plenty of the latter, but if he is lucky he has a good mate. I had but three weeks as a running shed fitter and made many mistakes; I would have made many more but for my mate. In the boxing world, his name will never be forgotten: nor in the railway world by those who knew him – Roland Todd. I think Roland had run through a fair amount of money, and eventually came on the railway in the 30's. He was a Londoner and a wonderful man to work with. Sometimes he talked boxing but not a lot: he was universally liked and he loved his job: he kept me on the right lines.

Our foremen were Charlie Walker, another Mexborough man, and, after a little while, the vigorous Cyril Palmer, who, I believe, originated from Wrexham but was largely King's Cross trained. He had had a bad time in London with the bombs, but survived them, and early in 1944 was transferred to Doncaster. He was a fine foreman: he saw to it that our lunch hour cricket and football stopped at the appropriate time, that we came out of the canteen smartly instead of hanging about; you may not love your foreman when he chases you, but you can respect him for his knowledge and his dedication.

Cyril Palmer, who later rose to become Motive Power Officer on the Great Northern section, took a personal interest in me which persisted until his tragic death in his early fifties in 1965. People seem to have enjoyed pulling my leg down the years: Cyril was no exception to the rule, but at the same time he kept a firm hand on my activities. He was a fine breakdown foreman and as I was the apprentice attached to the gang for some months, he saw to it that I had my share of work. This interest in a youngster by a foreman was something new to me: it could never have happened in the main Plant Works, where the whole outfit was larger and less personal, and where there were far more apprentices.

Our breakdown gang under Cyril Palmer was first-rate. We got plenty of variety: we rerailed wagons on colliery branches, sorted out piles of wagons which had derailed when points had been altered under a moving train, and dealt similarly with locomotives and carriages: we put in the new turntable at Lincoln; under strict secrecy at 6.0am one Sunday we loaded a midget submarine at Gainsborough; and one beautiful summer dawn in 1944 set off for Keighley.

This was a grand day out, but I was to go through the hoop before the sunset! I travelled on the crane for part of the way, after we had done our usual stint with sand and ballast to help our old K2 climb away from the Carr Loco up to Balby Bridge and across on to the down slow road. When we got to Keighley (which being outside our normal area, was new to the gang, but not to me), we found a standard 20-ton brakevan standing slap across the road 60ft below the railway, which at that point runs into the passenger station from Bradford Exchange round a sharp curve.

A freight train going towards Bradford had stopped to pick up at Cullingworth. The guard had unhooked his brake, leaving it on the main line, but had forgotten the little matter of a hand-brake, and away the van had gone down the 1 in 50, next stop Keighley. The signalman had done the only possible thing and put the van into the goods yard, whence it whistled through the goods shed, through the end door and the retaining wall, and into the street, landing four-square and remarkably little damaged. Mercifully, the street was clear at the time. Thus the task of lifting from the main line was fairly simply achieved in easy stages.

After lunch, we were ready to go; the K2 had the job of hauling the 45-ton crane and runners and three 8-wheel coaches up the 1 in 50. The driver was Charlie Hook of Bradford. I knew the road well enough, but not that type of engine, and so I took no part in the proceedings. We did very badly for steam but just kept going at walking speed, and a slow jog-trot at that. Our crane-driver had all the time in the world to get off his crane, pick a splendid bunch of wild flowers from the bank side, and get in the riding van as it passed him. As for me, my stock as a fireman fell a long, long way, for nobody would listen to my explanation that I was a non-combatant on this occasion, certainly not Cyril Palmer, who never lost an opportunity over the years to bring up that little story.

Nor did he ever forget my excuse for being late for a breakdown call out one summer morning. I was called 2.0am and found I had no cycle headlight. My cycle was a bit rough anyhow (it had cost me 15s second hand): I seem to remember it had a fixed wheel and no brakes, no back mudguard and no rear light and, of course, no headlight. However, a torch did the necessary. I got to the Gaumont traffic lights, which were red. Nothing happened and there was no traffic so off I went; but seemingly out of a manhole cover in the Great North Road there rose a gigantic policeman who booked me on what subsequently turned out to be six charges. I nearly missed the vans and, of course, I had to live that one down as well. I must say that I thought at the time that a 10s fine was very reasonable under the circumstances.

After a few months at the depot we were out one day on a breakdown and Cyril Palmer called me into his compartment. He wanted to know what I intended to do on the railway. I have asked the same question of many people over the years, and sometimes wondered why they hesitate to reply. And yet, at 20, I had no target. I had ambition; I knew I had to pass my examinations first time, and that I wanted to go into the Running Department; but the higher posts were a long way beyond my reach. Though I certainly started my career with the intention of being Chief Mechanical Engineer, that aim disappeared after a few months, for I realised fully my technical and mathematical limitations, now understanding that such a position must call for great skill and knowledge of management and administration. And yet here was Cyril saying that he felt I should go in for the "administrative side". I was quite hurt, thinking that all my urge to get practical experience had gone for nothing.

About this time the great John G Robinson, long since retired from the position of Chief Mechanical Engineer of the Great Central Railway, wrote a letter to the *Railway Gazette* putting his home address in Bournemouth at the foot of the letter. I had done a great deal of road work on his locomotives, as well as running repairs to the O4 and Q4 types that worked in and out of Doncaster frequently. When you are young, you don't stop to think, and I wrote straight from the heart to the great man telling him that his locomotives were without compare (at any rate some of them), that I had spent many, many happy hours working on them, and so on.

Why did I do that? I don't know. Probably I never stopped to think, but it seemed to make the old man's day, for by return of post I had a lovely letter, telling me amongst other things about his early days on the railways, and how he had gradually worked his way up the many rungs of the ladder. He finished up as follows: "Young man, you will never forget that there is no finality to what you can learn about the railway profession". Thus spoke the octogenarian to the boy. How right he was! One of the many joys about the railway life is that you never need to stop learning; you can never know it all, or be an expert on everything.

In November 1944 I left the running shed for the drawing office where I spent my last nine months at Doncaster. This was a different world; it was not my world, but one which contained some engineers of great skill, who in those days, I felt, were grievously underpaid and underrated. But it was not my life. I knew that the years to come must be spent in running sheds, amongst the sort of men that I had just left, many of whom, fitters and boilermakers, were resourceful craftsmen of the highest order, individualists, capable of doing the impossible to avoid a delay or a cancellation. I was to learn that not everything was golden, not everybody all they seemed to be, but all that as yet, was in the future.

Up to now I have confined my story largely to the works and my own daily existence in the running shed, but during my last 18 months at Doncaster I naturally got to know quite a few of the Doncaster enginemen. My weekend and night activities on the footplate had mostly taken place away from Doncaster. Towards the end of my training I had a few weeks on the footplate during normal working hours and I made the most of this, working 16 hours a day or more. It was during this time that, in addition to working with all my existing friends in the West Riding, Neasden, Grantham, Sheffield and so on, I worked with some Doncaster men.

Some of their names were famous. There was Joe Duddington for example. I had one trip with him to Leeds on a stopping train; he had a Pacific and four little Great Northern orangeboxes. He found it a lot easier to do 126mph down Stoke bank on his world's record run than to stop in the right place at Hampole and Fitzwilliam Halt with his tiny train. There was also Jack Sheriff, who had been a hard runner before the war and who had an excellent reputation as a main line driver. It always amazed me that though he had done much of his firing and early driving on the Great Central Hexthorpe pilots he was so much at home on main line work.

Then there was Harry Moyer, brother of Bill Moyer of Grimsby. Harry was a good engineman, and also a terribly hard taskmaster. You either got on with him as a mate or life was pretty rough. One Saturday afternoon, we set off with "Green Arrow" 2-6-2 No 4802 and 18 coaches from Dancaster. We got to Newark, found the left hand big-end very hot indeed and Harry decided to come off. Newark was far from being a main line depot and they did well to find two engines at a moment's notice. These backed down, with plenty of black smoke and not much steam. The front engine was a GN 0-6-0, Class J4, No 4041; we got on to the second engine, an 0-6-0 superheated J6.

The driver on the leading engine was Harry Moyer's equal in pungency of expression and had strong views on the instruction that the leading driver should handle the brake, but Harry doubted the ability of a mere shunting driver to handle a passenger train at all, never mind 18 coaches. Whilst they were shouting at each other, we got busy raising steam. The J6 engines never did steam very well, and we were not ready to set off. Nevertheless, when our two jockeys had settled their differences, away we went, quite unprepared. There was a lot of poking and darting of fires, black smoke and not much steam. And after three or four miles, we were ready to give up the ghost, and so was the crew of No 4041, but in the end we got to Grantham. There have been faster climbs of the Peascliffe bank, but we never actually stopped even if our maximum speed was no more than 25mph. Luckily Grantham had another "Green Arrow" in readiness, and that was that.

IV

Early Days on the Footplate

EXCEPT FOR a short period in 1943, I had to wait until near the end of my Running Shed training to gain footplate experience in the Company's time. But since the spring of 1941, I had spent evenings, nights and week-ends of my own time on the footplate with an infinite variety of engines, gaining experience that was going to be of immense value to me: value because it enabled me in time to understand completely, perhaps almost by instinct, certainly by experience, the work and personality of enginemen in this country. Over the years, in my time as a Shedmaster, Assistant and then District Officer, I have been able to build on that grounding so generously given me during the four-and-a-half years of my Doncaster training.

And I say generously, for a very good reason. These men had no need to befriend me, nor had they any specific authority for taking me in hand. But they did so, and one of my most treasured possessions is an album of photographs I made up not so long ago, using the negatives taken between 1941 and 1945 by my little old eight-and-sixpenny box camera. Each of the 260 photographs tells a story and so now I have a pictorial history of the men I worked with in that period, and of some of the machines on which I worked. I have never kept a diary, but in the years to come that album will be better than any diary. People used to say to me: "Ah, Dick, when you're a boss you'll forget about us". That I certainly never have done for that period of my life is indelibly clear in my mind; many of the people with whom I worked are now dead, and yet seem so very much alive.

I had better begin at the beginning. In April 1941, I stood with a friend of mine from the Works at Doncaster, Derek King, now at York, on Bradford Exchange Station. We had travelled the West Riding in the afternoon, the GN West Riding of course, and we were on our way back to Leeds to catch the Doncaster slow at 7.24pm. An Ivatt N1 0-6-2 tank, No 4364, stood in the platform at the head of our train and the fireman, one John Albert Walker, looked down at Derek and me. My hair was unkempt and I had plenty of coal dust in my eyes: he took one look at me and said: "Heaven help the poor starving barber"! This made me laugh and we got talking. And so we made our first trip with a Leeds engine and men.

It was not much of a journey, and we had only a little train, but in a month or so we were back again in the West Riding and, this time, at the end of the day we stood on Wakefield Kirkgate as the London mail came in from Leeds. The engine was the beautifully clean ex Great Central B4 No 6100 on which I was to work many, many miles, and as the train was booked at Kirkgate for some minutes we went up to the engine.

Immediately, the fireman asked us who we were. We were "apprentices from Doncaster" and before we could say any more, we were in the cab of a J G Robinson masterpiece. It was a Saturday night and the regular fireman had asked for leave and by the good fortune that has so often come my way, the spare man covering him, a so-called "young hand", was Stan Hodgson. Stan would talk to anybody and subsequently we became great friends and have kept in touch over the years. Had the regular man been there, we might never have been invited up by the driver, Bob Foster, around whom the first part of this part of the story must certainly be sketched, for without this meeting my life in the years ahead might have been quite different.

We had a big train and were soon on the move, purposefully and without a slip, although I remember she had a heavy bang in the right hand trailing axlebox, but we climbed away by Crofton and Hare Park with a wonderful throaty roar. I remember too, that when Bob turned the summit at Nostell, he put her in the first valve and by the time we were by Hemsworth Junction we were going very hard. Before South Elmsall Bob shut off, and we rolled right through to Bentley and nearly into Doncaster without steam. There were at that time five of these engines in the West Riding, Nos 6098 to 6102. 6099 and 6102 had balanced slide-valves and the other three had piston-valves. The former were very free running on the easy gradients with just a breath of steam and would seemingly coast for miles, although the piston-valve engines seemed stronger on the banks and consequently were lighter on coal and water. No 6100 worked the mail pretty regularly for several years until superseded by the K3s and later on by the Pacifics on the through workings to Grantham.

At 11.0pm we stopped in Doncaster station, and it was still daylight. "You can't get off here", we were told, and away we went to the Garden Sidings, turned on the triangle, took water, trimmed the coal, and sat down to talk until the down mail at 2 o'clock on the Sunday morning. We talked about engines, about Bob's early days as a driver at Mexborough, about his move to Wakefield GC and then the closure of Wakefield and his transfer to Ardsley with a job. Some antagonism existed at that time between GN and GC men, arising from the transfer into GN depots of GC men, which deprived the existing younger men of driving or firing positions that they had hoped were theirs. Later on, Bob went to Leeds and but for the war would have been in the Pullman link.

Having met Bob, I would not let him go and saw him at Doncaster time and time again; from time to time, indeed, my youthful devotion must have seemed a burden to him. He used to bring a K3 2-6-0 from Leeds, arriving about 7.0pm, and returning with the 8.8pm from Doncaster; another turn was the train arriving in Doncaster about 8.30pm and leaving at 10.1pm for Leeds. Both the down trains were London trains and very heavy, never less than 16 or 17 bogies. After battling with my homework, I would often be waiting during the 1941/1942 winter on the Hexthorpe bridge at 9 o'clock to meet the old man, to help him take water, trim the coal and build up the fire. Sometimes I was allowed to handle the engine from the Garden Sidings along the western platform and into the spur behind the signalbox.

In time I made several trips across the West Riding to Wakefield on the K3s. They were very fine engines indeed; there were only four of them at Copley Hill, No 91 with a GN tender and right-hand drive and Nos 202, 203 and 231; later on No 135 came along from Doncaster. They were in excellent condition, and with first-class Yorkshire coal, tubes clean and valves well set, the crews were masters of their work, the firemen being mostly highly experienced men who had been on the job over 20 years. Time was rarely dropped: certainly the engines were rough-riding, but the results were there for all to see. Through Bob, I got to know a great many Copley Hill men, with whom I worked at weekends and who taught me a great deal. I was introduced to the Great Northern Atlantics, of which class there were three stationed at Copley Hill, Nos 3280, 3300 and 4433.

Anybody who has fired an Atlantic will never forget the remarkable personality of these wonderful old engines; they were all that people have said about them – and more. They were uncomfortable, and rode badly; they threw you about; in many ways they were crude; but there were few machines built that could take such a beating uphill, or fly downhill, or roar across the level, as a GN Atlantic, and always with steam to spare. Through Bob, I met one of the classic Atlantic drivers – an ex Great Central man from Retford, the deep-thinking Bill Denman. He knew these machines intimately and preferred them, much to my disgust, to the GC "Immingham" 4-6-0s and also to the four-cylinder "Valour" class, which were operating from Leeds at that time. Bill Denman had worked the 7.44pm from Doncaster with 4433, made up to 17 bogies of nearly 600 tons weight, and had started from Wakefield Westgate on a 1 in 100 gradient without assistance. Certainly this could never have been done with a slide-valve Atlantic, but with a piston-valve Atlantic which could be reversed with the regulator wide open and the cylinder cocks shut, Bill Denman got this lot away at the sixth attempt and without a Bradford portion standing detached behind him to use as a catapult. And then he ran to Ardsley summit in full gear nearly all the way and full regulator but never the remotest shortage of steam, nor with excessive consumption of coal and water.

I had a similar experience with the same engine in 1944 when we had come out of Bradford with 12 bogies up the 1 in 45, with an N1 0-6-2 tank in front of us. The rail had been bad and we lost some time slipping, and so we were behind time when we got to Wakefield. To cheer us up, five more coaches were attached. Then, Charlie Hook of Bradford rolled up his sleeves and got busy – and so did I. The downhill start over the arches helped us and once more nearly in full cut-off and with the throttle wide open, we roared up to Nostell at what must have been a steady 45-48mph. Yet the work was simple, neither unduly hard nor nerve-racking, for the water level remained constant, and the steam pressure never moved from 170lb/sq in. The coal was nothing special, but the trap door of the firebox was open throughout and we could have gone on like that for miles and miles and miles. It was when an Atlantic got flying down Beeston bank towards Leeds, hit Beeston Junction at over a mile-a-minute and took the curve through Beeston Station at that speed that you thought your last moments were coming! But they never did, and the riding was accepted by most enginemen as something to be tolerated in a world where nothing, not even a GN Atlantic, can be perfect.

Bob Foster liked Atlantics, but he liked his Great Central engines and even more, when he got his hands on them, an A1 or A3 Pacific. At that time, Copley Hill worked a Sunday turn to Grantham with a Doncaster Pacific, which came down the previous night on a slow train. My first trip on this train was with Bob and stands out in my mind because we had the famous 4472 *Flying Scotsman*. She was a very fair machine at that particular time, although not one of the best and, of course, she still had a low pressure boiler. Bob worked his engine delicately, with small alterations in working constantly made to suit the road and the fluctuations in steam pressure: such was the text-book method of driving; theory and practice were matched to perfection and his work was a delight to watch. The original A1 Pacifics, with their 180lb pressure, were splendid wartime machines because with their three 20in cylinders they could rub along with the huge trains and just about keep time with no more than 140lb of steam, whereas a really shy-steaming A3 or A4 Pacific or a V2 could be the very devil.

It was Bob who had taken me to Mexborough and introduced me to the Hepworth family who lived in a terraced house in Wellington Street high above the centre of the town. Charlie Hepworth had fired for Bob at Wakefield and had come to Mexborough after the grouping: his wife, Emily, was the daughter of a real old timer, John Duckmanton, who ran a Robinson 4-4-0 No 105 from Wakefield to Cleethorpes for many years. She had been reared on locomotives and railways as the old man lived for his work and his engine. Indeed, from the day she and Charlie met, they had the common bond of railways and locomotives and loco men between them. Ralph, their son, was brought up in the same mould and their little house became another home for me. Days beforehand I relished the thought of an evening there knowing I should be mothered by Emily and taught the West Riding dialect by both of them, and instructed on locomotives and rules and railway lore by Charlie Hepworth who was a very deep thinking and knowledgeable man. He wrote for many years in the ASLE&F journal on mechanical matters and it was a great tragedy that he never became a driver and later an inspector.

His eyes were damaged a few months before going to pass for driving by a hard thrown snowball which hit him in the face and the footplate career which would have blossomed to everybody's delight and benefit was ruined. But he kept his interest and turned to other things such as instruction, writing and model work, and these two friends made me see the smoky and dirty old town of Mexborough in its true light, friendly, direct, warm hearted, bound up inextricably with the two great industries of railways and coal mining.

Copley Hill Shed at Leeds was full of GC men in the 1940s as, indeed, were both Bradford and Ardsley depots, although most of them were in the tank engine links that covered the local train work in the West Riding and some of the Doncaster turns. Who will ever forget Alf Cartwright? He was a great big slow-moving man, with a wonderful sense of humour, who came from Staveley to Leeds. Once he stopped in error at Beeston on a Castleford train; the stationmaster came to the engine, let off at him alarmingly and Alf's reply without a suspicion of expression was "Well, we had better go then; we'd only called to see how you were getting on". He always wore the same celluloid collar with the tie knotted over the top, seemed to be years older than he actually was, spent much of his time talking when he might have been preparing his engine, but was a grand engineman and mate to his fireman of the day, Percy Hudson. Percy used to get a lot of driving, as the old gentleman rather fancied himself with the shovel, although he was a bit heavy-handed from time to time. As a driver Alf was very good with the Pacifics and K3s, and excelled with the *Valour* Class B3 4-6-0 locomotive.

Round about 1943 No 6164 *Earl Beatty,* fitted with the Caprotti valve gear, was transferred to Copley Hill. She was still painted green, which was a bad sign, but was followed by No 6165 *Valour* straight out of the shops and painted unlined wartime black. Because of the miles run under wartime conditions No 6164 was in poor fettle, but No 6165 was a fine machine. Over the years, before some of them were fitted with the Caprotti poppet-valve gear, these locomotives had been maligned in the technical journals because of their poor showing on the Pullman trains working out of King's Cross, to which they were transferred soon after the Grouping. They burnt plenty of coal, their design was in many ways traditional, but they rode beautifully, steamed well and ran like greyhounds, whilst the cabs were comfortable, and that is a point because railways do not always run uphill.

I once fired *Valour* for Alf Cartwright from Wakefield to Grantham on a Sunday. She was never converted to the Caprotti gear and retained the Stephenson link motion, with outside admission to the outside cylinders and inside admission to the inside; the booked locomotive was a "Green Arrow" 2-6-2 which had failed and been replaced at short notice. *Valour* was worked with a fully open regulator where there was heavy work to be done, and on the first port downhill, whilst the reversing lever was set as near the mid-gear position as the gradients would allow and constantly altered to suit the road. The load was about 540 tons. The fire was kept right up to the level of the firehole door and above it when possible; all I had to do was to keep straightening my back and slipping the coal just inside the door, whilst the blast and the motion of the engine did the rest.

The injector stayed on for mile after mile, the steam pressure was as steady as a rock and when not engaged in firing we could sit comfortably and relaxed, keeping a sharp lookout, with everything nicely to hand. It was a comfortable, memorable journey! Alfred sat in his corner, with nothing much to say, his movements matched exactly to the requirements of the locomotive and the road. It was a journey certainly more comfortable and less hot than if we had been working at that time on the average "Green Arrow" with no flame scoop and dirty tubes, and intolerable heat thrown back on face, arms and legs. A good "Green Arrow"? Ah, well, that might have been a different story.

The late Chief Inspector Jenkins was a fireman on No 6168 in 1923-24, and once he and his driver had mastered the technique of a locomotive completely different from those to which they were used, they had no trouble and work became a pleasure. The same thing happened with several crews, but one or two of the famous names of the day were incapable of modifying their tactics and were violently prejudiced against the GC engines because of their heavy coal consumption, and because the method of operation they adopted resulted in shortage of steam, black smoke and loss of time. Eventually the locomotives had to go, replaced by 4-4-0 "Directors" and then again by the Atlantics. The name-plates of *Lloyd George,* No 6167, were taken off this engine in 1923, and lay behind a cupboard until the Top Shed at King's Cross was flattened in 1962.

Jimmy Simpson came from Barnsley to Leeds; he had a dark moustache, a fierce reputation, was very gruff, but was a fine engineman; on the Pacifics, he became especially good. He usually greeted me with "What the bloody 'ell's tha' want", but the twinkle in his eye was always there, or nearly always. Not one night on the Mail, however. We had No 91, a K3, and at Hemsworth we had about half-a-glass of water and a fair fire, and with Doncaster in mind, and a turn-around time of some 2/hr there, I shut off the faceplate injector on the driver's side to allow the water level in the boiler to come down another inch. But I should have left it alone. Jimmy immediately told me so in no uncertain terms, so I tried to put it on, and as so often happens at the wrong time, it wouldn't strike the water. This engine had a very difficult "F" class exhaust injector on the left-hand side which would not always work reliably, so for a while neither injector was working.

Practical experience is always a boon, however, and never more so than in times of crisis. This time we got out of trouble by a trick I had never seen before or since. The faceplate injector that will not strike the water has to be cooled down, and quickly at that. You can use buckets of water or the text-book method, but never before or since have I seen the water-cock below the barrel of the injector opened, and a sharp brake application made, followed by an equally sharp release. The jerk apparently sent the water flying up into the barrel of the injector, cooling it down and enabling the expert Jimmy Simpson to get the feed on at once. "Allus do as th'art told, Dick" was his comment, and I was sufficiently humbled. Firing a locomotive was so like other worth-while skills; you could be on top of the world one day, a bit cocky perhaps, but then next time you would do something daft and get into difficulties from which only experience could extricate you. So on this run we flew by South Elmsall and into Doncaster none the worse and on time. I liked Jimmy Simpson; he was a grand mate to those that stood up to him and who did their job.

The standard of work by the Leeds men during the war was uniformly high; the older men retired, the next line stepped up and the job was carried on just as reliably and soundly whilst the standard of firing by men, most of whom were in their forties, was exemplary, whether on tank engines or on the larger designs. And to a man, they went out of their way to help me master the practical side of locomotive handling.

One day however, I thought I would travel with Bradford men. Maybe this was another turning point, for after a while I met the remarkable Ted Hailstone, but the first encounter was fascinating enough. Leeds men had stuck to the N1 and N5 tank engines for the local services. Bradford men got everything, J1, J2, J3 and J6 GN 0-6-0 tender engines; N1, N5 and GE N7 tank engines; an occasional A5 (not for long); and, of course, GC "Immingham" 4-6-0s and occasionally GN Atlantics; all these for passenger work, never mind the goods. But whereas the Leeds tank engines had the best Yorkshire coal, and it was traditional never to shut the firehole trap door unless in dire trouble, many Bradford engines ran with razors permanently across the blastpipes, used very small coal, had very little paint on their smokeboxes, and yet worked the majority of the local expresses as well as all the Bradford, Keighley, Halifax services.

I shall never forget my first day with Bradford men: it started with a blood and thunder journey with an express from Wakefield to Bradford on No 3005, a J1 saturated 0-6-0. The driver, George Stoyles, was a Brontë expert, and the fireman was George Barker, who retired not long ago from Low Moor. We set off full of hope with our five bogies from Wakefield and this old thing. She felt every bump and corner of the road; the fire was disappearing as quickly as it could be kept replenished, the water came down the glass alarmingly fast; and we just made Drighlington summit but only just. All the while, George Stoyles was lecturing me on *Wuthering Heights*, the Brontë sisters, the best way to fix a razor (an art you learnt quickly at Bradford), and how much easier the work would have been with an N1 tank.

When we got to Bradford, I joined George Coulwell on No 6101, an "Immingham". He was working through to Doncaster, and our load was about 320 tons, for which we had no assistant engine on the front, no push at the back, and up a gradient off the platform of 1 in 45. No blowing-off was allowed in Exchange Station, and a tremendous fire had been made up in the back half of the firebox. On starting the fire was pushed all over the firebox (and that took some doing), the door was shut tight, the regulator was opened wide, and until St Dunstans was passed the lever was in the corner. After we had travelled some 200yd the injector was set and firing started, almost continuously to the top of the bank at Dudley Hill, where the cut-off would still be about 60 per cent. Steam had to be maintained against the injector and there could be no question of shutting the feed off. Curves, rock cuttings, a thunderous exhaust, a great cloud of black smoke going vertically upwards, a high degree of skill and application, the steam locomotive hard at work, in all a great struggle – the kind of thing you don't forget in fifty years.

Next time we had the same engine but a different driver, one Ted Hailstone. But there was an interval of several months between the two journeys, during which time I had met Ted, had covered many miles with him on smaller engines, and had been instructed and drilled to the required standard by that remarkable man, who finished his career in the mid-fifties at King's Cross on the A4 Pacific No 60014 *Silver Link*.

V

The Art of Firing

EDWARD CHARLES HAILSTONE, born in 1892 into a railway family, was an outstanding man. He became a regular fireman at Gorton in 1919 and worked with the martinet George Bourne on the Robinson 4-6-0 No 428 *City of Liverpool*. It was a great experience to work for a hard taskmaster who knew his job on a fine machine. Via March, he came to Bradford in 1927 as a young driver, and in later years transferred to King's Cross. At Bradford he established a reputation as a first class engineman, although some people disagreed with his methods. For even as a young driver, he was a disciplinarian and would not tolerate slaphappy methods of firing and driving, or indeed of railway management. In general, he was disliked by those who did not measure up to his standards and worshipped by those who could.

When the GC influx hit Bradford in the middle and late twenties, the drivers brought with them a route knowledge far beyond the bounds of the West Riding, to which most of the Bradford men had been confined throughout their careers. Long-distance special train work was transferred to the depot, to be handled largely by the "foreigners", and this sharpened the feeling that already existed against the GC men. However, Ted and his regular mate, George Howard, went "specialling" all over the place and feared nothing – to Skegness, Cleethorpes, Bridlington, Scarborough, Liverpool, Manchester, Banbury, Marylebone and even King's Cross. This was usually with an "Immingham" or with one of the three ex-GC "B6" 4-6-0s, Nos 5416, 5052 or 5053, but sometimes with a J39 0-6-0, a K2 2-6-0 or even an old Ivatt 0-6-0 goods engine, like No 4094, on a Bridlington or Cleethorpes excursion – a long way for one of these old carts.

The first time I met Ted, he was standing beside an N1, No 4594, in the bay platform at Wakefield Westgate, talking to a man deeply interested in railways, Alastair Kerr. Introduced to me by Alastair, I remember a kind but penetrating eye and a handshake that pulverised. According to my mother, who met him on several occasions, he "had the look of a Bishop". That day I spent with him, the first of many, his mate, Maurice Saunders, did much of the driving whilst he fired with skill and precision. In the evening at Bradford, we found a quiet little pub near the depot and sat talking, and it was then that I heard for the first time of his years at Gorton and the long days before the war on the specials to all parts of the LNER. Thus we began a lasting friendship, for no man did more to help me master the engineman's profession. He taught me the finer arts of locomotive management, the building of a fire with the utmost care, thought and precision in firing, and the correct use of fireirons, cleanliness and tidiness.

If I so much as failed to sweep up after firing, he would stand and shuffle his clogs until I did so.

When I was up to his standard as a fireman, then I went across to the other side and I had to achieve the same high standards when in charge of the locomotive. His methods of handling engines were based on the full open regulator theory, provided of course that the engine concerned was neither rough in the axleboxes, nor off the beat; and that a fire must be in perfect shape *immediately* the journey began. In 1956, not long before his retirement, I made my last journey with him from King's Cross to Leeds and back, his skill had not diminished over the years, and I experienced the same delight in working for him then as I had in the 1940s. We had surprisingly few rough trips; mostly they were carefully planned and executed although I remember vividly three journeys which could have been quite humdrum but turned out very differently.

One in particular I shall never forget. We had arrived at Wakefield with an N1 0-6-2 tank, No 4593, and should have worked a slow train about 6.30pm via Dewsbury to Bradford; but we had somehow got off our diagram and were called upon at very short notice to work the 5.54pm Bradford portion off the 1.10pm from King's Cross. We had no real body of fire, and because of some very poor coal we had not done too well coming from Bradford, but we started off with a boiler full of water, a full head of steam and a heart full of hope. I darted the fire as usual and fired very, very lightly, but very frequently, four small shovelsful, one in each corner. Ted set his injector very fine indeed, and notched up as high as he dared for this particular bit of road.

The fire must have been very thin, but as she warmed up the steam stayed constant at about 160-165 lb/sq in, and Ted opened the setting of his feed, lengthening his cut-off little by little. We went through the junction at Ardsley with the controls unchanged, using the brake against the regulator in order to maintain the same condition of fire, steam and water level, and so we made Morley after climbing 8 miles at 1 in 100 and 1 in 60. We left in the scheduled 13 minutes with flying colours and continued thus over Drighlington summit and down into Bradford. It was a blazing hot day and those few miles had taken it out of me. Why, you might ask? Well, we did the job as it had to be done in the most adverse circumstances over a road where, with small engines, one false step so easily made by a fireman could result in loss of time and frequently with a need to stop altogether for steam and water before turning the top of the bank.

One of the last trips I ever made in the West Riding was on V J day 1945, with Harold Binder. We had No 4567; and from the start nothing that I did went right. We turned the top of the bank, but ran so short of steam that we had to stop to blow up by Tong Cemetery on a falling gradient. This was because we had had to use both injectors, after letting the water go in our efforts to get over the top without stopping; but the second feed knocked the boiler back and the brake went on and stopped us when actually going downhill.

One night in the black-out, on the same stretch of railway, we left Bradford for Halifax with our N5 GC tank, No 5901. The uphill part of the journey was perfectly normal, with plenty of steam, and a chimney full of fire thrown high into the sky. As we approached Queensbury, we got the distant on, the home on and then we were cautioned into the station. Queensbury used to be a triangular junction, and our line was joined at the far end of the station by the line from Keighley. When we ran into the station, the starter was at danger. But it soon came off, and with a green light from the guard off we went, quite gently. On this occasion, I was doing the driving, and Ted was on the other side of the cab. Round the corner was the advance starter, near the entrance to Queensbury tunnel.

Soon I pulled the wheel up and opened the regulator a little more. Ted came across and had a word with me and then something compelled me to look out into the blackness; we were just passing under the advance starter showing a red light! Despite my efforts to stop, we ran four coach-lengths past the signal, a serious irregularity. I had neglected to concentrate on my job, had allowed my mind to wander, had forgotten we had come into Queensbury on a yellow light and had failed to observe *all* signals. This taught me a lesson I shall never forget as long as I have anything to do with railways. It is a lesson I never hesitate to pass on to others: never relax your concentration when on a moving locomotive or train.

Harold Binder was another driver who helped me in many ways that it would be impossible to forget. In the top link at Bradford along with Ted, he was a different type. For years he had fought ulcers and stomach trouble, and yet had taken his turn of firing with his 40-year-old fireman. He was another father to me, always kind and generous. He didn't believe in short cut-off working, and I remember pulling up an N5 a little too tight and getting roundly rebuked for my pains. My Tong Cemetery disaster had been when firing for Harold. On another occasion we were about to work the 4.39pm Bradford to Halifax with the N5 5901. Just before we left, along came one or two other interested folk, and getting out my camera, I very quickly took a snap or two of them, leaving the very quiet faceplate injector on my side working. Coming back to the cab, I shut off the feed just before departure, and starting up the 1 in 45, Harold opened the second valve.

That did it! She caught the water and we roared up the bank at walking speed with clouds of steam and water billowing from the chimney, the safety-valves, and the cylinder cocks. We just managed to get up to St Dunstans, a mile or so away, but with practically no steam, no water, no fire. It seemed an unending drag to the top of the bank. But I had been the careless one, and now I had to work twice as hard for half as much result, and again I learnt a lesson.

One of the last journeys I made with Harold was to Cleethorpes in July, 1945. We started off with a J6 0-6-0 – a clatterbang of the worst order, which made him laugh and swear most of the trip. We got to Wakefield, where we joined up with the Leeds portion, headed by B4 No 6099, with Driver John Smith and Fireman Sid Watson. They backed on to us, and so took the brake, which meant that we got their smoke and coal-dust, and went at their speed most of the time, which was not conducive to comfort. Our fireman was a well-known Bradford man who had some malady that sent him to sleep as soon as he had nothing to do; not only so but he went to sleep with his eyes open showing nothing much besides the whites of his eyes.

I shall never forget our journey, with this little engine crashing and banging and throwing us all over the place, whilst Herbert B - - - - sat on his seat, seemingly in a trance. How he managed to sit there at all was a mystery to me. Occasionally he woke up, looked round to see if we were still there, and went off again into oblivion. Luckily for us we got a hot box and a hot big-end on the "A" engine and were given a GC compound Atlantic, No 5258 of Class C5, to work home. Harold knew these engines well. You left them in full gear from start to finish, making all the necessary alterations with regulator and change valve. Herbert and I shared the work; we got stopped by communication cord at Aldwick Junction for no apparent reason, lost No 6099 and the Leeds portion at Wakefield, and took our six coaches up Batley bank at 1 in 38-42 without trouble or assistant engine. A grand day with dear, lovable, Harold Binder!

But there was another depot in the West Riding, Ardsley, where the work was largely freight and where the passenger work was confined to 20 turns which included three from Wakefield to the munition works at Thorp Arch, and three to a similar factory at Ranskill near Retford. Ardsley depot monopolised the ex GC 4-4-2 tank engines, class C14, most of which were in the West Riding, although three were still at Ipswich at that time. Splendid machines these were, but pressed at only 160lb/sq in so that you had to be on the ball. They worked the Ranskill trains, while the "Imminghams" monopolised the Thorp Arch services. There were also two N1s at Ardsley, and I believe one C12, although I would not be sure about that. Incidentally, Leeds had Nos 4010, 4014 and 4020 of this class, whilst Bradford ran Nos 4018, 4524 and 4536. If you got the fire thin and bright, pulled them up so that the front end of the engine was bouncing up and down, and did not overload them, they were grand little free-steaming engines. How on earth did the Brighton Railway contrive to produce its I1 4-4-2 tank based on the same design, and yet quite impossible in its performance?

The Ardsley passenger men had either had sufficient freight work to make their little pile and were now after eight hours a day, or simply preferred passenger work to freight work, or had other occupations. The gang was voluntary, was kept up to strength by advertisement, and all things being equal the senior applicant got the job. So you had senior men in their 60s and late 50s, whilst, at the other end of the scale, there was one Benny Faux, who had started in 1917, and was aged round about 42-43, while his fireman, who had started in 1919, was much the same age.

Benny Faux has now retired, and I have never seen him since the middle of 1945, when he went back to his seniority position in one of the goods links. He may remember many things we did together, some of which could never go into this book! He was the direct antithesis of Ted Hailstone who used to say "I don't know what you can see in that Faux." His motto was "we never bother, it'll be reight." But whilst he may have had a lighthearted attitude and a certain recklessness, he was nobody's fool and knew what he was doing. Being young and active, Benny did most of the firing himself, whilst Percy Thorpe, his regular fireman, took on the driving.

The first time I met him was one foggy morning around 8 o'clock in Doncaster station, maybe in the autumn of 1943. They were working the early Ranskill turn and this was the return service with the night shift from the factory. They stopped for water at the column, and as I stepped on, Benny's first remark to me was to ask if I had ever had hold of an engine before. The answer being "yes", he took up a position on the top of the tank on the fireman's side of our C14 No 6124, and away we went. These Ranskill trains stopped at the closed station of Bentley to set down the factory workers, and this took a bit of finding in the fog as it was the first time I had ever stopped there. It is one thing time and again to pass a closed station, as I had done for three years, and quite another to find it and stop just so without losing time.

These little engines were very strong; they had a good front end and climbed the banks well with a crisp beat. The screw reverse had to be watched: being slide-valve engines, the gear kicked hard as soon as you took out the catch, and you had to move it bit by bit and slip the catch in quickly where you felt she liked it best. It was better to have a catch rather than the method on the GN tanks, on which the gear was locked by a hand screw and clamp on the shaft. I once caught an awful packet on my leg when the screw flew into full gear on an N1.

Life was never dull with Benny. One night, we set off for Thorp Arch with our ten coaches of munition workers and No 6101. We stopped at Wakefield Kirkgate, Normanton, Castleford and then ran to Thorp Arch via Church Fenton and Tadcaster. Getting to Thorp Arch without incident we went round the factory loop and set down our passengers; then out again, we did our work, had some food and started off again tender first round the loop to pick up the afternoon shift. This would be at about 10.30pm. When we stopped at the River Station, Benny got off the engine and started talking to the folk in the train. Now, the passengers knew their man; they also knew that we had the railway to ourselves in the middle of the night, and they suggested to Benny that if we ran like hell and got home some twenty minutes before time, there might be a collection for him.

This was too much for Benny to resist. He came swaggering back to the engine and told us that neither of us knew how to run a train hard, so that I had better get busy and fill the box. And so I did! We got away from Thorp Arch and as fast as I put coal into the firebox, Benny knocked it out the other end. We went round Tadcaster and Church Fenton curves far, far, too fast; then we dropped our folk at the various stations, never waiting for departure time, and thus we landed at Wakefield with a beautifully red smokebox door and a very hot fireman. In a flash, Benny was down on the platform with his cap in his hand, but, bless his heart, I who had done the work never saw its contents.

But he made up for that one Saturday evening by taking me home to a tremendous meal, and to what I think was a family birthday party, without any warning to his wife and family. We went through the back door, both of us in overalls, and into the living room and my introduction, black as I was, was the simple but complete welcome: "Here's Dick", and to me, "Take us as you find us! Make yourself at home."

Another day we had the N1 No 4581, and were hooked on to the front of a Pacific working the 3.20pm from Leeds to help him over Ardsley summit to Wakefield. When we left Holbeck, I pulled the regulator wide open to climb the heavy gradient past Copley Hill depot. Benny called out "What d'you think you're doing? Put her in the first valve." I said I didn't think we'd do our share like that but he was adamant and said these Pacifics were plenty powerful enough to do two engines' work. When we stopped at Wakefield, we unhooked and got away before the Doncaster man could get at us!

And so on: I couldn't keep away. I had never handled a "B" engine, Class J3, on a passenger train. Benny worked a very early morning workers' train from Carcroft to Wakefield, and they always got an Ardsley "B" on the job. I remember it was a weekday and I got up at some unearthly hour, caught a miners' bus at about a quarter to five from the Palace theatre or thereabouts, and landed myself at Carcroft in time to catch this train and sure enough, I had my little drive and found the old "B" engine so willing and strong for its size. But the fire had to be just so; the grate was very small, the same size as a C12 and a D3, I think: and the fire always seemed to be going up the chimney. But like most firethrowers, they would nearly always steam if the fireman did his stuff. Back I came on the first Doncaster slow and straight to work. Sleep did not matter in those days, and I would willingly work the clock round for there was a lot to learn in a short time.

Our Benny was quite unique. Percy Thorpe was the ideal foil for his dashing temperament and they taught me a great deal. With them there was no rule book, no "round the wheel", but they gave me practical experience I shall never forget, with many an insight into the short cuts and dodges that enginemen almost inevitably use if they are cute enough and know their job.

The West Riding was a great school. They said that if you could fire an engine here, you could go anywhere. This was not quite accurate, perhaps, but the point was that no engine would do its stuff unless the fireman completely mastered the principles of combustion by knowledge or by instinct, unless he was concentrating on his work and unless he realised that one mistake, a little too much coal here or too little there, could result in an inevitable shortage of steam. I can see a superheated N1 standing across Drighlington crossing in the moonlight. We stopped at the platform on the up road not far from their engine, and could see the fireman having a good poke at the fire; the roar of the blower, a cloud of smoke, a little steam at the safety-valves, the smokebox door warped and red, and then suddenly a short whistle, a snap from the snifting valves and they were on their way again. We would laugh and say a bit smugly "old so and so's stopped to blow up." It might happen to us next time.

But there were other lines, too. In the summer of 1941 I was going south to my home on the 6.15pm from Doncaster, which ran in from Leeds with Grantham men and 4-6-2 No 2549 *Persimmon*. The driver was the wizened and skilful Jack Dodd, otherwise known as "Kruschens", and his fireman, Alf Rudkin. I remember little of the journey except that I travelled on the fireman's seat, and some very hard work was done. After a year or so, Jack Dodd went up into the top link and his place was taken by J. A. Thompson, always known as Bill; as I write, he must be 78 and I shall soon be receiving yet another of his marvellous letters from Australia.

As a globe trotter, he takes some beating. Half his life these days is spent in Enfield, and the other half sailing to Australia, living it up down there and coming home again. A fine man was Bill Thompson, and a grand mate. He was a purveyor of apples and pears while you worked for him; he never got ruffled and if things were going wrong, as they often did during the war, he simply sat up in his corner, gave her a bit more, looking at the same time surprised and sympathetic, his eyes twinkling and his moustache adding a touch of distinction to his kindly face. "Are we winning, Richard?" he would ask when we were in dire straits, as though it was quite normal to be working 18 or 19 coaches with no more than 140lb of steam.

One Sunday night Bill, Alf and I waited at Grantham for one of the night trains to come down. It was over half-an-hour late and eventually rolled in with a V2. The Cockney crew got off at once; all the driver said was "She's a barstard, mate, and you've 19 on, ta-ta." We had a look round; all seemed quiet; she had only 130lb sq in on the clock, an inch of water in the glass, a dead swirling fire and an invisible tubeplate. The first thing to do was to get the dirt off the tubeplate with the slacker pipe, which was normally used for watering the coal and keeping things clean, and was quite illegitimately used for this particular purpose. Maybe it did some good; I don't know. We blew up to 200lb and made a start. That night my arms and chest ached, fit though I was, for there was an endless route march into the tender for coal, on to the top of the dust to see if there were any lumps, while all the while the engine roared its way through the night with 140 to 160lb of steam against the 220lb I would have given anything to have had.

Bill fed me with apples, and I fed the firebox with anything that came to hand. Eventually we stopped at York, where the Geordies got on; we got off as quickly as we could, and with a "She's a barstard, mate, ta-ta," in our turn we retired to the other side of the station. After a quick meal we relieved on a really good engine, and before you could say "knife" I was back in Doncaster, washed and in bed by 5 o'clock but away again and at work by 8am.

The last time I had Bill Thompson on a "footplate" was on an electric Kingston "roundabout" of the Southern Region, 1966. We went from Waterloo to Waterloo with the old Queen Mary stock and the Westinghouse brake. We made some smart suburban stops, making Bill's hair stand on end. He enjoyed his evening, but said it was a lot easier to go from Grantham to Newcastle on a steam engine, than to stop and start all those times with an electric train. As a matter of fact, with the proper use of the Westinghouse brake there was a great deal of interest and skill, not to say panache, in suburban electric train work.

I suppose one of the hardest days if not the longest that I ever experienced started at 4.0am on the up "Aberdonian" at Doncaster. I worked my passage to London and slipped across to Marylebone, from which I went down on the 10.0am to Leicester with Jack Kitching of Neasden shed and a beautiful Caprotti 4-6-0 No 6168, just out of the shops. Work on this engine was a delight. She was free running, easy on coal and water, comfortable and good to look at into the bargain. But it was a different story coming home; we had another Caprotti, and if I remember rightly it was No 6167. Certainly she had a horse-shoe tender, which meant that the coal had to be trimmed by hand after the hole had been emptied. She was in very poor form, and had a hungry reputation, while for good measure, five empty coaches were added to the seven which formed our train.

We were "all stations" from Leicester to Harrow-on-the-Hill, and it seemed a long, long, way. The last of the coal from the back of the tender I shovelled into the hole while we stood at Chalfont waiting for the Chesham branch train, and the last of the coal on the tender went into the firebox about Neasden. She had plenty in the firebox to get to Marylebone, and I had had enough for one day; I never bothered to calculate our consumption, but the coal was stacked high when we started, so you can work that out for 103 miles. In London, there was time for a drink and some food and then over to King's Cross.

The down "Aberdonian" was worked by London men; the fireman, Horace Farey, was very musical, and a rare artist to boot. However, on the "Aberdonian" that night, and with me to do his work, he looked forward to a comfortable journey to Grantham. But he hadn't bargained for those 360 miles that were already in my arms, and when we were passing Peterborough I had to call it a day, which made him laugh then and again in later years at the recollection. "Now, I could never say that, could I?" he would comment. "No, Horace, you couldn't, but I did have 430 miles and 17½ hours to my credit, didn't I?" would be my reply.

Last, but not least, before we leave the north, there was Joe Oglesby of Neepsend and Darnall. Joe was a cavalier on the job, a grand mate, a hard runner, and with all the confidence in the world. He and Gladys Oglesby had their own domestic cross to bear during the war and I remember with happiness and pleasure their united family and how after a 24-hour session, finishing up with Joe on the Leicester Mail, I would go to their home in Firth Park in Joe's little Ford 8 for a wash and breakfast before going back home to Doncaster.

In 1942, Dick Lawrence, who was training at Derby and had been at Marlborough with me, came over for a weekend. We decided to go in the train from Doncaster to Sheffield, then to Penistone, down to Barnsley, across to Wakefield by the L&Y, and then see what the West Riding offered. We struck off from Sheffield behind a GN Atlantic, No 3296, a slide-valve engine and not very strong. When we got to Penistone, we walked up to the front, started talking and away Dick started for Manchester with Joe Oglesby, on whom we had never set eyes before that day, while I hopped back into the train. When we came to Hazlehead, we had to stop specially to pick up. Twenty minutes elapsed before we got away again, for the slide-valve GN Atlantics were brutes to start on a gradient, but after Dunford and Woodhead Tunnel that wartime 60mph limit went for a Burton. It was my turn coming home and my friendship with Joe had started.

Later on we used to go to York and back on a Saturday afternoon via Doncaster, or to Manchester and back, or to Leicester on the 11.58pm Mail, or to Retford or Cleethorpes. It made no difference. Most of our trips were good ones. I remember vividly running 12 bogies from Sheffield to Manchester with a B7 4-6-0, No 5461; she started away from Penistone as though she had no more than three on. Or there was a B17 "Sandringham" 4-6-0 on the Leicester Mail, when we sighted a signal at danger at Woodhouse Junction, and had to drop every anchor and a few more to stop on the right side of the red light. Or it might be roaring across the level to York with a "Director" 4-4-0, perhaps No 5434 or No 5438, or stopping three times to blow up with an impossible K3 2-6-0 between Sheffield and Dunford with, believe it or not, only six coaches on our tail. Such were some of my many experiences with Joe Oglesby, until our last journey together two days before I left Doncaster, on that beautiful machine, the *Valour.* What better engine could there have been to finish up with? I had the pleasure of taking her to Retford, not a long journey but one never to be forgotten.

Such was my Doncaster story. A great deal has had to be left out – journeys on every conceivable GN and GC type, on LNER standard locomotives, on a few of the ex-NER types. As to half-an-hour coughing and retching on an up freight train in the Woodhead tunnel, or Grimsby Bill Richardson and his compound Atlantics and Dos, or my own off-duty life, such as it was, there is no room in this story.

In June 1945, Edward Thompson sent for me. I remember his words distinctly: "I'm sending you to Shenfield to see Mr Leslie Parker, who is the Locomotive Running Superintendent of the Eastern Section." The time had come for me to exchange locomotive building for the wider world of railway operating.

VI

Edward Thompson

BEFORE WE PROCEED to the next stage of my career, however, a brief chapter is needed to describe the man who had so strong an influence on its earliest years.

This is no critical appraisal of the rights and wrongs of the locomotive policy dictated by Edward Thompson. Nor is it an attempt to comment on the methods he was said to employ beyond the fact that I believe he was a splendid administrator, and, in the opinion of many, a fine leader of men. I should not have understood all that, for I was an apprentice and I saw him through the eyes of youth. Towards the end of my Doncaster days I got to know his sister and his niece, Betty Stratford-Tuke, who lived during the war in an old rectory on the outskirts of Balby. Through them I had an insight into his character, although I never visited him at his home. But one or two things happened to me during those few years which convinced me that Mr Thompson was an extraordinarily human man. After I left Doncaster, indeed after his retirement, he wrote quite regularly and was a constant encouragement to me, particularly at a stage in my career when everything seemed to be going wrong.

One winter evening early in 1942 I got off a locomotive at Wakefield. There was nobody on the platform that I could see. Suddenly I was confronted by a tall figure in the gloom; it was none other than Edward Thompson. He recognised me instantly and wanted to know what I was doing. I told him I was working on footplates and that I had no authority for doing so. His reply as the Doncaster slow ran in was simple: "I'm delighted to hear you are interested enough to do this in your spare time. Are you going to Doncaster? If so, come with me." Without hesitation, he strode to the nearest coach, which was an open third, called me after him, and sat down at one of the tables. We were an incongruous pair side by side in that stuffy and none too clean coach. He, as always, was elegant and well dressed, while I, wearing overalls and clogs, and with eight hours dirt on my hands and face, as always was untidy.

The journey passed all too quickly. He told me about the locomotives that he intended to build; they were to become the B1 4-6-0s and the L1 2-6-4 tanks; also he intended to simplify some of the Gresley designs and to rebuild some of Robinson's to form prototypes. He made no mention that I can remember of new Pacifics or Pacific rebuilds. He talked of locomotives past and present, and also encouraged me to tell him about my life and how things were working out. He talked also of his own life at Marlborough, where his father had been a housemaster; this fact may help to explain his interest in me. When we got to Doncaster, he walked with me to the Beckett Road bus stop, and his last words were: "Come and see me when you want an engine pass." I went home scarcely able to believe my ears, or indeed my good fortune. But alas, all life is not so easy as that.

One day in the next week I asked permission to leave my work for a few minutes, and went to the new offices near the Carriage end of the Works where Mr Thompson was now installed. I came straight from the bench, clomped through the main door in my clogs on to some rather nice carpets, and said that I wanted to see Mr Thompson about an engine pass. This caused an unbelievable stir amongst the great man's personal bodyguards. People came out of the shops to see the Works Manager, or at a pinch the Mechanical Engineer, but were quite unknown in the offices of the CME, and most certainly not youngsters in clogs and dirty overalls! I got as far as a Mr Gosling; he was a Londoner and insisted that Mr Thompson could never for one moment have said I could have an engine pass, the while looking at me as though I had the palsy. But with a persistence which was far from popular I got that coveted bit of cardboard, and celebrated it with 28 hours continuous duty. A certain Ivy Shingler, who was Edward Thompson's secretary, took compassion on me that day and indeed whenever I used to arrive in person with my request. She always saw to it that I got my pass in the end, and when she retired about three years ago she gave me the photograph of her boss which had been in her office after he retired. That is something I can treasure. The old man looks at me every day in my office.

In 1943 I decided that I must join the RAF. Basil had gone and I had to follow him. We could not get our release for anything other than flying duties; so I went to see Mr Thompson and explained that I wanted to go, and he gave his permission. I had my medicals and examinations, and was told by the RAF that I would be on my way within a month or possibly less. So I began to go round the West Riding, saying goodbye to all and sundry on the strength of a permit for three days footplate riding that I had been given. The month came and went and nothing happened. The Doncaster RAF folk were encouraging; the local Group-Captain would say "Yes, my boy, any day now." And yet it never came: and I have a feeling that it was Edward Thompson who put the stopper on it. Maybe he was right, I don't know. If I had left and joined the services I should have forfeited my premium and might have ended my Railway service. But my mother had paid my premium, and helped me to exist, and this she could ill afford to do. So I stayed.

In June, 1945, there came from my Chief those words: "I'm sending you to see Mr Leslie Parker". He knew that he was sending me to be interviewed by a frightening little man with a bow tie and a soft voice, who worked his young men desperately hard. He knew he was sending me to the Running Department, which was to become my life for the next 17 years, and he knew that he was sending me to work for the man who had a reputation second to none amongst the railway engineers of the country for training young men as managers. And for those few words, quite apart from anything else, I am profoundly grateful to Edward Thompson.

The last time I saw him was after I left Doncaster. He was standing beside his rebuilt A1 Pacific, No 4470 *Great Northern,* talking to the Neasden driver, Bill Andrews, whom I knew very well. The engine was about to be inspected, I think by the Board, and once again I had the opportunity of a long talk with him, this time all the more enjoyable because Bill Andrews was involved, and because I now had my experience at Stratford about which to talk.

I cannot believe the story that so fine a man as Thompson *deliberately* chose Gresley's first Pacific to rebuild so radically as to make the engine unrecognisable. Chief Mechanical Engineers are unlikely to choose one particular engine of a class on which to experiment, and the probability is that No 4470 was due for a general repair, or had even been remitted to the shops by a grateful Running Department as being too long in the tooth and in need of a rest. I am in no position, however, to say anything about the performance of the Thompson Pacifics. I have never had anything to do with them; indeed, I very rarely saw one, let alone worked on them. But I do know that the Thompson *régime* produced some splendid machines, which stood comparison with anything else in the country of similar size and power.

It was at Doncaster that I saw the first Thompson B1 4-6-0, No 8301, and very shortly after that, on Easter Monday, 1943, I joined a Gorton crew at Sheffield and made my way south on this smooth and comfortable machine. I went through to Marylebone, and remember thinking what an easy job the engine had made of its 12-coach train. Later I was to get to know the new Norwich B1s, always so beautifully kept, with one crew to each engine, and later still, as the numbers of the class multiplied, I was to work with these engines through to the end of steam traction. Without a shadow of a doubt, they were the equal of the London Midland Stanier Class 5 4-6-0s in every way, whether on the road or in the shed. They could always rise to the great occasion, and in the interchange trials of 1948 their work was not to be lightly dismissed.

Thompson aimed ultimately to reduce the numbers of the classes by the introduction of standard types and by rebuilding. His Class L1 2-6-4 tanks were powerful machines, but the earlier examples were the very devil to keep going; the later engines, however, with axlebox wedges, gave excellent service. His O1 2-8-0s I got to know well because of their work on the Annesley – Woodford "windcutters", which were among the fastest unfitted freight trains of their time. Even the most dedicated Great Central drivers had an immense respect for the power and speed that these engines could produce, and the first time I rode on one from Rugby to Woodford I was amazed at the speed with which we tore through Braunston after a dead start from Rugby.

Of Thompson's Class B17 rebuilds, however, the less said the better, for he made a bad machine worse. Not so his Class Q1 0-8-0 tanks, which he rebuilt from Great Central 0-8-0 tender engines into quietly effective shunters, or his rebuilt GCR "Pom-Pom" 0-6-0s. And I certainly remember his "Baby Bongos", two-cylinder Class K1 2-6-0s, based on the rebuild of three-cylinder Class K4 2-6-0 No 1997. These engines ran fast freight and passenger trains with great competence. In later years I was to have many trips on K1s with Clacton trains; some of these were bumpy, when we finished up without any footboards in the centre of the cab, but time was rarely lost, and Stratford might have been in an awful mess at summer week-ends but for the willingness of these comparatively small machines, some of which had been illegally borrowed to tide us over.

I thought Thompson was a great man, though when I knew him, as I remarked in the opening paragraph of this chapter, I saw him as the apprentice sees the man at the top. Many of his locomotives, I would contend, made a great contribution to the successful running of the railway, were built relatively cheaply, were of simple and straightforward design, and will be remembered with respect by those who were responsible for their operation. That, when all is said and done, is the acid test.

VII

Progressman at Stratford

UP TO THE TIME I went for my first interview with L. P. Parker, the
Locomotive Running Superintendent of the Eastern Region, I had
been working in the Doncaster drawing office, and I was certainly one of
the more incompetent draughtsmen to work at the desks where so many
fine designs had been worked out. Indeed, I achieved very little, for I was
employed on simple and routine jobs in connection with the rebuilding of
the A1 Pacific No 4470 *Great Northern,* the sort of work where my
mistakes could soon be put right on the shop floor.

The interview took place at Shenfield, and I allowed myself plenty of
time to make sure I was not late in arriving. In actual fact, however, I got
there with only ten minutes to spare. My train from Liverpool Street
consisted of eight coaches headed by an old GE 2-4-2 tank engine, Class
F5, a "Gobbler". The journey was terrible. We stopped for steam at Ilford,
Romford and Brentwood, and even then only just got over the top of the
hill at Ingrave. I sat chewing my fingers and hoping that we should not
pack up altogether, but I was to have many more tottering trips with the old
GE engines in the years ahead, so many of which had to be coaxed and
handled with extreme care to get the results they were capable of
producing if in reasonable order and in good hands.

L. P. Parker sat in a small office in what had been a bank near
Shenfield station. He motioned me to sit down and said nothing at all. He
was very dapper, was wearing a bow tie, puffed at a pipe and looked at me
very straight with searching but twinkling eyes. After what seemed an
eternity, he opened fire quietly but deceptively: "Now, Hardy, tell me all
about it". Such an invitation to talk had ensnared many young men and I
was no exception. When I told him I was working on the sand arrangement
for No 4470, he said "Same old George Stephenson arrangement, I
suppose" and when, a little later in answer to some question, I expressed
an opinion on the handling of locomotives, he looked at me fixedly, still
puffing away, and then said, very softly, "Tell me why".

Such was the first of many pits that I unwittingly dug for myself during
my years under this remarkable man, to whom a statement made would
almost certainly be followed by the simple, and sometimes unanswerable
question "Why?" In this case, my views were diametrically opposed to his,
and so I was ultimately sunk without trace. However, he told me that I
could go to Stratford Motive Power Depot as a Progressman as soon as the
results of my Higher National Certificate in Mechanical Engineering were
known and provided, of course, that they were favourable. He set great
store by academic achievement in engineering, not just for its own sake,

but because he knew that, in the long run, the qualified man stood a better chance of reaching the top in engineering or engineering management if his academic backbone were matched by practical managerial experience, above all in handling the staff who were the background of the then Locomotive Running Department.

In August, 1945, I started in Stratford Depot as a Progressman, which was not a salaried position. My job was to chase material from the main works for locomotives under repairs in the Stratford district, and to make sure that the out-depots were adequately supplied with stores. In 1945, Stratford Depot was an extraordinary place, and with all its ramifications and outstations, it must have been one of the largest depots in the world. Certainly the running of the district as a whole, in those immediate post-war days, with the safety-valves lifting from pre-war repression and the intensity of wartime conditions, presented an extraordinarily difficult management task for the District Locomotive Superintendent, Teddy Ker.

In those days all I saw of E.H.K., and rightly so, were short and painful interviews on the subject of 4per cent waiting material instead of 2per cent and, occasionally, a view of him through a crack in the wall between the general office and his sanctum. A hole had been cut in L. P. Parker's day at Stratford to ensure that if you squinted through it you could just see if the great man was free before summoning up the courage to knock and enter. But I knew nothing of management difficulties. I thought that all the world was marvellous and full of "grand chaps"; so a description of the real Stratford, so good and sometimes so impossible, must wait until its proper place in this story. I was given a place at a high Victorian desk in the general office, next to the sleeping Ted Foreman, opposite Kitty Keegan, daughter of a retired chief clerk, and Arthur Scillitoe, comedian and saw-player: all three were clerks, responsible, if I remember rightly, for engine availability, usually a masterpiece of deception. We laughed a great deal, and my companions helped to ease me into what was my first position of responsibility.

Next door, in a dark and overheated little den, were the four artisan foremen, two mechanical and two boilersmiths. As it was my job to see they had the material they needed to complete repairs, on my first morning I was taken to be introduced to them. The air was thick with the smoke of Nut Brown, the strongest tobacco in all London, and all four sat at their desks wearing bowler hats. They were the old school of foremen with a vengeance, Charlie Greenwood, Harry Bull, Fred Lucas and Archie Harper, whose bowler rested on his ears. They were part of the Stratford scene; they knew all the moves, all the wangles and all the short cuts; they had a profound practical understanding of locomotive maintenance, and they commanded a very real respect both upwards and downwards. Fred Lucas died very young. He was a humorist – indeed he had to be to succeed at all – and the last time I saw him he was in the office shaving in shirt-sleeves, braces, bowler hat on the back of his head and a cigarette in his mouth. So I found myself in an office full of characters, and in a depot of characters also.

What of the outside staff? I had very little to do with the general running of the depot and so my circle of associates was necessarily limited. My immediate boss was the machine-shop chargeman, Jack Welsh; he had directed material chasing over a number of years. It was an innovation, I suppose, for me to be doing what he had done, but he treated me very well. His right-hand man and chief robber was Albert Allison, once an Ipswich fireman, who drove the Lister truck on which material was sent across to the works "over the other side." The three of us worked well together and our liaison with the works people on whom we depended for quick return of material was splendid, although our relations stretched a bit from time to time.

Albert knew where springs were to be found, and we operated on the dishonest principle that material lying about near the spring shop must be "Stratford Loco" property. As for poor old Cambridge, Ipswich and Norwich, we got our machines on the road at their expense, unscrupulous Cockneys that we were. I was beginning to learn that at Stratford one had to fight for survival; it had always been thus, for the countrymen had had the advantages of a monopoly of the labour market and maintained a standard that Stratford had the greatest difficulty in achieving. Ipswich and Norwich had been unanimous over one thing only in the days of E. L. Ahrons: "Stratford was no class whatever!" But they were wrong!

I had to keep the out-stations fed with material; this involved me in footplate work to Parkeston and Colchester on the B1 and B22 4-6-0s, to Southend on almost anything whose wheels would revolve, to Enfield, often on the Enfield Class F6 "Glasshouse Gobbler", No 7007, driven by Alf Holland, to Epping, Bishops Stortford and Wood Street. I'm not sure whether these visits were of much practical use to anybody except myself, but they gave me the chance to get to know the workings of the small depots in the Stratford District which had been used before my day to train generations of young men on the threshold of a railway engineering career.

The larger depots, such as Enfield and Wood Street, had 36 sets of men, with 12 engines each allocated to three sets of men, and two spare engines. All minor repairs were done by the shed chargeman and the fitter, if he was lucky enough to have one. At the smaller places, such as Bishops Stortford and Epping, the shed chargeman and the general factotum did the work and the latter, usually quite untrained except by years of experience in helping young men, knew the job inside out. The chargeman was artisan, drivers' foreman, clerk, storekeeper, boiler washer, the lot – a manager in miniature. As for the general factotum, most steam depots, except the very largest, had such a man, who could turn his hand to anything, including the drains. The same principle obtained in other countries also. In 1968, I was in Bidart in the South-West region of France staying with some great friends, the Pardos. Monsieur Pardo knew his railways, and one morning we slipped away for an hour to meet the depot manager at Bayonne. He proudly introduced me to his general factotum, Monsieur Alfonso, "without whom his depot would never hope to survive."

The Great Eastern engines of the older classes were very different from those on which I had been brought up. With the locomotives of GN, GC and NE origin, and, of course, LNER designs, the firehole was of the trap door type, which would remain open in normal conditions, leaving a semi-circle through which to work. When firing, therefore, there was no increase of cold air through the door. But with the GE engines it was rather different, for the "Gobblers" on which I started operations had the GE door, an antiquated arrangement with a flap and a huge door which swung on to the cab floor, and was controlled by a chain and ratchet. This took a lot of manipulating at first, and with any but the best steaming engines you could lose 10, 15 or even 20lb of steam during the short time you had the door down, if you left the injector on. I had been used to a fairly high or very high firehole door, with a low shovelling plate, and it seemed extraordinarily difficult to work with a high shovelling plate and a firehole door bottom ring on a level with the floor of the cab.

Most of the old GE locomotives used in the Stratford district on passenger trains were worked with regulator full open and a short cut-off wherever possible, the notching up being done very quickly, often too quickly, after only a few revolutions. In these conditions if the fire was at all heavy, it had little chance to burn up immediately, but yet I found that on certain classes of work a heavy fire, indeed a really heavy fire, suited some of these engines remarkably well. I had tried firing the Enfield "Gobblers" like an N1 0-6-2 or a C12 4-4-2 tank or any other comparable small locomotive. But this did not work, although such light firing methods were *essential* with a J15 0-6-0 goods engine.

The most productive way to go from Enfield to Liverpool Street and back with No 7007 was to fill the firebox with lumps straight from the coal stage. Having done this, one filled the boiler, backed on to ten coaches, and hoped that by the departure time steam would be on the red. For the first few miles, steam pressure would go steadily back and the water would come down to an inch or so above the bottom of the glass, but by the time one got to London, 10¾ miles and 13 stops away, the fire would be in tremendous form. Then when you had unhooked in Liverpool Street, taken water like lightning, followed your train out, slipped into the dead end, and backed on to the next train, the boiler was full and everything in apple-pie order. With luck, that fire would take you back to Enfield with no need to open the firehole door, except maybe for a sprinkle after Lower Edmonton.

This was not the textbook method, but there was no smoke, plenty of steam and it was comfortable work with a wonderfully economical and powerful little engine.

At times, nevertheless, I was very glad of the Westinghouse brake. With the vacuum brake one could never keep going with 80lb of steam. As for the Westinghouse, I had my first instruction on the "Jazz" services on No 7007, nerve-wracking because the Westinghouse brake is a great brake if handled with confidence and a menace if not. Over the years, most of the GE men became rare exponents in the art of using this brake, and once I had mastered the real pitfalls that exist for the comparatively unskilled, I revelled in the fast approach to the station, the single last-minute application of the brake and the immediate release on stopping. Why, before you had wound the wheel into full gear, the guard would be blowing his whistle and you were pounding on your way again and that could never have been so with the vacuum. Yes, locomotive handling on the Great Eastern suburban services was indeed a fine art.

VIII
Mechanical Foreman in Fenland

In January, 1946, I was sent to King's Lynn as Mechanical Foreman (learner). I was a learner all right, but perhaps beginning to get to that dangerous stage of thinking that I knew rather more than I really did, and that the world was of the opinion that I was quite a good chap. In the next three-and-a-half years, however, I was to receive several forcible reminders that I was still at the bottom of the managerial ladder. Sometimes these reminders were hard, but this simply stressed my inexperience. Yet while starting to learn the rudiments of the art of management. I was to cover an immense amount of ground and to tackle many interesting jobs.

It was a bitterly cold January morning when I took the 8.20 from Liverpool Street and reported at the Cambridge District Office to T. C. B. Miller, the Assistant District Locomotive Superintendent; he interviewed me at some length, and then took me to see his Chief, Mr Rees. These interviews over, I set off again on the 1.28pm for King's Lynn, a London train which ran in with a "Sandringham" 4-6-0 driven by the dignified and moustachioed Royal Train Driver, A. R. Smith. My interest was in the rear five coaches, which left Ely behind, of all things, an ex-GN D3 4-4-0. The driver was a well-known King's Lynn character, Tom Kenney, who punched his little machine across the Fens without much regard to his steam supply. The worst of a dead level road like this, however, is that you can seldom shut off steam without losing time, and so we stood for ten minutes at Hilgay while the wind whistled and moaned, and the dismal fen countryside lay flat as far as the eye could see. Later I got to love the fens, but as we stood at Hilgay I felt, as I worried about what the next day would bring, that I had come to the last place on earth! Eventually we got going and reached King's Lynn in the gloom of a late afternoon; there I made my way straight to the depot, where I had been told to report to a Mr Shaw, the Shedmaster.

With some misgivings, plus that empty feeling born of the unknown, the fact that I certainly had no lodgings, and it was getting dark, I hesitantly entered the boss's office, full of tobacco smoke and the smell of snuff. Again I was assessed in silence, for I was to be an extra or supernumerary at the depot for training and Ted Shaw was naturally suspicious. Anyhow, I was fixed up with "digs" and told to come back at 8 o'clock the next morning. I found that my new boss was a fine, practical railwayman, but that he was determined to bring me up the hard way, for he had had it that way himself at Gorton as a fitter before coming, quite unknown, to East Anglia just before the war. On his mantelpiece was a photograph of a Great Central 4-cylinder express locomotive with the Caprotti valve gear, taken near Guide Bridge. No 6168 it was, and a tonic to a young man on his first visit to the distant and frozen fens.

I was set to work on the enginemen's daily list, and my first job was to master this tightly-worked roster. It should have been Ted Shaw's job to prepare the list but he passed it on to me and any mistake was harshly dealt with. However, the lessons and methods that I learnt and absorbed from him in this not so simple task stood me in good stead for many years. The day-to-day vacancies in the main line links were covered by shunting link drivers and senior passed firemen, and the resulting vacancies by passed firemen and passed cleaners working on a three-shift basis. Any mistake was immediately noticed by the eagle eyes of the passed firemen or passed cleaners and a claim for a driving or firing turn or both would follow. It was a tight list, with very little give and take, and men would be called in for driving turns at a moment's notice. For those were the days when passed firemen had been from 25 to 28 years firing, and every driving turn was one step nearer to the top rate of pay and the driver's position.

King's Lynn was in many ways a very well run depot. Much of the pre-war thinking and discipline still remained, in so far as every preparation and disposal man was worked to the limit of his time and capacity in the particularly bad working conditions of a typical GE country depot. Again I sat at a high desk in an office looking out on to the station, and I was soon given the additional job of keeping an eye on the fitting and boilermaking staff. This was not an easy task, for there were two perfectly good chargemen far more capable than I of organising things and who had indeed forgotten a great deal more than I had learnt. However, the job was there to be done, although I had still to learn that familiarity and friendliness do not get things done without steadfastness of purpose and a thorough determination to do what is right.

In the months that lay ahead, I was to learn a great deal from Ted Shaw and his staff. He was a fine breakdown foreman and he was supported by two running foremen who ran the depot between 4pm and 8am. One of them, Fred Jackson, was later to become Mayor of King's Lynn and a good one at that. The chargeman fitter and chargeman boilermaker formed the remainder of what would nowadays be called a management team. And there was, of course, a Chief clerk, Reg Herrington, who worked in a tiny office full of files, papers and of locomotive availability records. His availability records were the neatest and most perfectly prepared that I have ever known, and Reg would not tolerate any of the short cuts that were normally used to improve the figures. If a locomotive was not available for work by the appropriate time, then it was not available, and he saw to it that no pleading from anybody else – even from Ted Shaw – would induce him to commit a deception. Reg from time to time would talk to me like a father; indeed they all did, for I was wayward, if enthusiastic and a glutton for work. He loved quotations and proverbs and might say, "You know, Dick, you can't make a silk purse out of a sow's ear" when referring to manners, and then sometimes, much more mysteriously: "Attitude is the art of gunnery".

The first week I was at King's Lynn, I was sent on a J19 0-6-0, No 8250, to Roudham Junction to bring home a J17 which had failed on the Thetford goods. Our driver was one Arthur Harrington, known to everybody as Satan, whom he strongly resembled in appearance. He was a tall tough Bradford man, who had fired many times on specials to all points of the LNER with Ted Hailstone, and as he was amusing and outspoken I heard the other side of the coin – what it had been like to get far afield with Ted and the antics they got up to when lodging away or on book-off jobs. And when we got to Roudham Junction, I met a fireman who was an immigrant from Barnsley, which made two Yorkshire contacts in one Fenland morning.

The locomotives stationed at King's Lynn were an interesting, if not very lively bunch. Most of the "Claud Hamilton" 4-4-0s were very dull steaming engines. We had several of the original D15 4-4-0s, with superheaters, extended smokeboxes, and level grates, still with slide-valves of course and air reversing gear, and some attached to the oil-burning tenders with which they had been built. The two best, excluding the excellent Hunstanton No 8896, were Nos 8893 and 8895, followed by 8881. The two worst, which would neither steam nor run at this period in their career, were Nos 8890 and 8891. I took issue with 8890 and tried every conceivable dodge to make her steam, but with never a hope: the needle would stick at around 130-140 lb/sq in and water at half-a-glass. There must have been something wrong, though the tubes were clean and the front ends were in reasonable order; also there were no joints blowing in the smokebox, while blast-pipes were plumbed until we were blue in the face. At that time the locomotives were common-user, and it is only fair to add that the story was a different one, in some cases, once regular manning, with two sets of men to an engine, had been introduced about 1948.

There were quite a few J15 and J17 0-6-0s, and three or four J19s. The J17 was the maid-of-all-work and well suited to the type of straight up and down freight working on this part of the Eastern section; The J15 was an excellent and reliable stand-in and had a splendid steam brake. I remember that one evening Arthur Harrington and I, and I think, No 7875, were running pretty hard with a heavy freight train when we came round a corner and found some gates across the track. Under the influence of steam brake and reversing gear, and of the guard once he had picked himself up off the floor of his van, we stopped about a foot from the gate when it was about to be opened by the scandalised crossing keeper. Then there were four F3 2-4-2 tanks, Nos 8046, 8048 and 8092 at King's Lynn and the beautifully burnished 8095 at Hunstanton. No 8048 was a poor steamer and frequently lost time on the 5.43pm to Hunstanton, which was a through train from London. You had to start with a very hot fire, and operate on the well-tried Hailstone method to succeed at all, but I did have one satisfying journey on this train with No 8048 when we managed to keep the pressure and water up from start to finish. It could only be done, however, if you concentrated very hard and made no mistakes.

The yard shunting was all J67 0-6-0 tank work, while the Wisbech tramway, with which unfortunately I had very little to do, was operated from the sub-depot near Wisbech GE station. The six-wheeled and four-wheeled tram engines did prodigious feats of haulage during the fruit season, when we had to send extra sets of men to Wisbech to cover the extra engines in use. A day's work for a fireman on a tram engine at the height of the fruit season was no joke. You fired in a little pit, either on your own or with the driver according to the direction of travel, hot, bent double, with a short shovel, knocking your knuckles and scorching your pants, often on a twelve-hour shift when we were tight for men.

In early February, 1946, I made a journey on No 8896, the burnished and polished *Claud Hamilton* 4-4-0 worked on the early and late turns by the three Hunstanton drivers. There was a night cleaner who did his stuff well, and the three crews, who shared 8896 and 8095, were all extremely conscientious, even if they never could get on with each other, as so often happens at a small depot. This particular journey with Frank Futter was a delight, and I used, for the first time, the air reversing gear which was so effective, reinforced by a fine thread handwheel for small adjustments. But in a month or two, the jealousies and feelings that existed between Lynn and the gently nurtured Hunstanton men flared up when the Royal "Claud", No 8783, was sent to Hunstanton for Royal Train working, which had been traditionally handled by the Royal link at King's Lynn. In effect, there was no alternative. A Royal engine could never have been a common-user machine and King's Lynn had no regularly manned engines, indeed no Royal Link since before the war.

To cut a long story short, 8783 was cleaned and polished lovingly at Hunstanton and her steelwork and the interior of the cab were a joy to behold. Her day's work was two journeys to King's Lynn and back. But never once did the Hunstanton men work the Royal train to Wolferton with their engine, for after several months the Royal link was reformed at King's Lynn with two engines and four sets of men. But the Hunstanton episode was good while it lasted, and for those few months I saw every day an engine cleaned to a state of perfection that I had only witnessed before in a museum; this memory was an example to me in later years when engine cleaning became very much my business as a Shedmaster and then as an Assistant District Officer. Jim Defty died tragically one evening by the side of his beloved engine in King's Lynn Depot but the others carried on the good work until she came to Lynn for good.

There was very little train work on Sunday; and on that day the fitters and boilermakers worked to make up arrears in maintenance. One of the great difficulties in a small depot with limited pit room is to get what you want where and when you want it, when the place is jammed solid with engines. Locomotives even stood in the goods yard adjacent to the depot, but once the depot was full up and the engine fires had been drawn, everything stayed put and some degree of planning and co-ordination was necessary to get the locomotives set right for the fitters and for boiler-washing. If this was not done, Sunday work would be a fruitless, frustrating, and unproductive exercise.

On Sunday evenings, there was a turn to Ely and back, the return working connecting with the 7.40pm from Liverpool Street, leaving Ely at 10.15pm. As I returned to King's Lynn on this train after a weekend at home, I used to try out different engines and to get to know men with whom I had not previously ridden. Once or twice I experimented with the wicked old No 8890, quite the worst passenger engine I had yet encountered. She would neither run or steam, and I was determined to beat her to it. We used to run non-stop to Downham, and the next stop was King's Lynn; when we shut off for Downham, we were always "on the floor". One night, with Alf Birdseye driving, we reached Downham down and out, but the guard said we were to go so that was that. I was torn between running the fire down and keeping going with a heavy fire, the only thing she seemed to like, but I compromised and as we approached Lynn we had no more than 90lb of steam. Somebody in his wisdom had fitted this engine with an old solid jet Davies & Metcalfe ejector, very slow in releasing when the pressure was down. The long curve brought speed down to the 20s as we looked out for the three stop signals, one after the other, that we were certain would be "off" on a Sunday night, but then, horror of horrors, the last one, beyond the bridge, was at danger! It was pulled off just before we stopped, but by then our driver had made the fatal application which brought us to a stand with the regulator wide open an engine-length short of the platform! It took me months to live that down!

No 8890 was not the only engine to beat me to it. One evening, I tried out a Great Central D9 4-4-0 No 6018, a good free-steaming engine. We came up from King's Lynn in fine style, but on the return trip, I pushed too much fire forward with the poker when leaving Ely, and killed her stone dead. We had a dreadful trip, with the fire level with the brick arch, and it would neither burn away nor make steam. It took me months to live *that one* down, too!

On a moonlight night, there was a great fascination about that little-known stretch of main line railway, with the fens stretching away into the distance, and only the very occasional light of a farm or cottage. The speed gradually fell away as the train mounted to the river bridge between Hilgay and Denver, and the trees on the left, standing out, helped one to look for

Downham distant. The change to 30-foot rails was the sign to shut off steam for Downham, and then came the staggered platform, ill lit, where the driver looked for the stationmaster's front door on the up side as his marker. Then off again, through the fens until the lights of King's Lynn showed up ahead and you rattled over the rough old Harbour Junction.

Life at King's Lynn may seem to have been a trifle light-hearted for a young supervisor in the making. I had never taken charge and when I did so, I realised that things would never be the same again. I did what most relief men do, and that was simply to keep the seat warm, but during that short period, I had my first violent difference of opinion, quite different from the normal run of arguments and disagreements when one's responsibility is limited.

The Dereham branch engine was Class J15 0-6-0, No 7549, a good little engine which started the day with the 6.50am from King's Lynn and returned in the evening. Ted Shaw was adamant that the running repairs on this turn should be done at night. But as the valve spindles had to come out and required turning to take new packing, the work was not done. Whatever the rights and wrongs of the case, however, the glands really did blow, and on misty mornings the crew were unable to see a thing against the wall of steam ahead of them. A driver called Ernie Dix was on the early turn. He was an excellent engineman, a splendid mate, but at times he could frighten those who did not know him with the force of his invective. One day he came into the office, tossing his head and storming about the dangers of working "blind", and about the stupidity of leaving such young men as myself in charge of anything more than a wheelbarrow. My youthful pride was knocked sideways and we went at it hammer and tongs. He disappeared as abruptly as he had come, leaving me shattered but determined to put the engine right. This we did, quickly but quite unforgivably in the shedmaster's eyes by having the valve buckles removed, the spindles turned and fresh packing fitted *during the day,* which meant the loss of a day's work to the engine. A weak decision? Ted Shaw left me in no doubt about it when he came back from his holidays!

During my King's Lynn stay I had a brief interlude at Cambridge in August, 1946, for the purpose of carrying out some comparative tests between B2 4-6-0 No 1671 *Royal Sovereign* and B17 No 1638. The main line passenger work at Cambridge was covered by a large fleet of B17 "Sandringham" three-cylinder 4-6-0s, nearly all of which were shared by two crews. Most of these engines still had their original boilers, but with the pressure knocked back from 200 to 180lb/sq in. This was a wartime measure, to minimise, if not to cure, the trouble with throat-plate and wrapper-plate fractures which had affected the boilers of the light construction needed to conform to Great Eastern weight restrictions. Even so, they were excellent machines, for in general they had been well maintained.

VIII

The reason for the test just mentioned was that No 1671 *Royal Sovereign* had been converted by Thompson from three-cylinder to two-cylinder propulsion, so becoming dimensionally similar in some ways to his own Br, including the provision of a boiler carrying 225lb pressure; it was now the engine used to work Royal trains between London and King's Lynn. The trials against No 1638 were carried out over a period of a fortnight, with Inspector Houghton in charge. The results revealed a marked superiority in favour of the two cylinder locomotive, with its 225lb boiler, but I think there would have been little difference between the two engines had No 1638 been working at 225lb instead of 180lb pressure.

The driver on these trials, Herbert Spicer, was an ideal man for the job. Nothing dismayed him, he was always cheerful, and he co-operated to the full with his fireman, Chris Pilsworth, to get the best possible results. We had many journeys together. One of the most unusual was coming up the Lea Valley with B17 No 1643, a spare engine, on the 8.20pm from Liverpool Street to Cambridge; she was not fitted with wedges, was due for the shops, and was riding very roughly indeed. We had about 450 tons behind us and it was about as thick a winter's night as you could imagine.

So we clawed our way up the Lea Valley, and after that up the valley of the Start, with a succession of distants dimly showing green, even that at Burnt Mill, which was difficult to see even in the best conditions. Then we passed the Bishop Stortford Distant, with a fogman on duty – the first we had seen – at caution. This meant that the Bishops Stortford Outer Home could be on, and the tell-tale rattle of the bridge over the river warned us that we had nearly reached it. The fog by now was so thick that when we stopped I had to climb the signalpost to see if the arm was on or off. None of us could see the Home signal through the murk, and it was very gingerly that we entered the station.

Before we got to Stansted, however, the fog cleared, and we roared up Elsenham bank under a full moon, with the peaceful Essex countryside unbelievably beautiful. Down to Audley End and beyond No 1643 bucketed, and twice down Great Chesterford bank the reversing screw did one of the typical B17 party tricks, the catch lifting and letting the screw fly into full gear. Fortunately Herbert, knowing his machine, was well out of the way and his legs escaped injury. Eventually we ran into Cambridge with the footboard displaced, a cab door missing, the bucket and brush up in the corner, and the footplate deep in coal. Such was one of my most vivid recollections of Cambridge in those days a main line depot of high standing.

IX

From King's Lynn to South Lynn

NOW AFTER no more than eight months at King's Lynn, there came a transfer to another depot, actually no further away than the other side of the same town, but a depot with very different traditions, equipment, engines and men. This was South Lynn, the old Midland and Great Northern Joint depot, where I had the much more rewarding experience of taking charge for several months during two periods in 1947 and 1948. Indeed, this might well have been a continuous spell of nearly ten months had I not been called away to undertake two special jobs, as described in Chapter 10.

As already mentioned, South Lynn was different in every respect from King's Lynn. Whereas King's Lynn had been cramped, there was plenty of room at the M&GN depot. Hand coaling had been the only method at King's Lynn, but South Lynn had a small coaling plant. Then whereas the King's Lynn rosters were very tight indeed, there was a certain latitude at South Lynn. The work at the latter also was quite different, for the most part with heavy trains, especially in summer time, which had to be worked over a hard and hilly road. Norfolk is by no means the flat country that might be imagined; in the 45 miles between Grimston Road, Melton Constable and just beyond North Walsham were many lengthy 1 in 100 climbs. In addition, over much of the distance single line operation compelled many slacks for tablet-changing at the station loops. Finally, even the men were different; the indifference of the average GE man for the "Joint" was only exceeded by the contempt that South Lynn felt for anything GER or, indeed, LNER.

There was a remarkable old character called Luke Watson on the day shunting turn at the shed. He had been a fine engineman, and had spent some of his time at King's Lynn, returning to South Lynn until a painful recurring illness compelled him to take a regular day job. He and I worked closely together, to my delight, for he was a man of some education and ability who would occasionally give me a serious lecture, starting something like this· (Age 61 to Age 23) "My dear sir, permit me to inform you that the M&GN was one of the finest railways in creation and that the LNER" (here he would bare his teeth as though in acute suffering) "did nothing, nothing at all, my dear sir, but encourage laziness and bloody idleness. And it was a bad day for us, etc, etc." Certainly no love was lost between South Lynn and King's Lynn.

It must be said, however, that the long main line and the infinite variety of locomotives and work on the M&GN presented a challenge to all concerned that was lacking at King's Lynn. No "Claud" could lollop along the level on the M&GN at 45mph and hope to survive, though when I first went to South Lynn, in September, 1946, to learn the job, the main line

64

passenger locomotives were not "Claud Hamilton" 4-4-0s but our old friends the ex-Great Central Class D9 Robinson 4-4-0s, splendidly rugged machines capable of climbing banks with a full head of steam, even if at the expense of a fairly heavy coal consumption. No 6040 worked the first regular Yarmouth-Leicester train after the war, but regrettably ran hot, so that the return working was with a London Midland engine, almost certainly a Class 4F Fowler 0-6-0, about which the "Joint "men, who were really Midland men at heart, spoke so well.

The freight work was in the hands of GE J17 and J19 0-6-0 locomotives, which were well-liked and powerful machines (although "not built for our road"!) but we had a few J17s fitted with the vacuum brake for passenger work. This was all very well, for, these engines were never designed to chase down 1 in 100 gradients at 60 to 70mph, even if they were splendid for climbing the banks. Their axleboxes and bearings were very small, and we suffered broken and overheated boxes galore. It must be said that the routine examination of oil-ways, axleboxes and the like was not maintained at a high level and we were bedevilled with constant difficulty; the drop pit was always occupied, whilst one or two casualties were used on shunting or pottering about at Sutton Bridge until their turn came for lifting. For passenger and freight work of the lightest nature, there were 086 and 092, little GN "Yorkies", in a very poor state of repair and, of course, there was what we collectively and wrongly called the "W" class.

Actually, our representatives of the GN 4-4-0 breed were the Class D3s with the short firebox; Yarmouth and Melton ran some D3s and also a few of the superheated D1 class, of which more anon. I would not hesitate to say that the Joint men thrashed work out of these old things that had not been coaxed out of them for many a year on their own territory. There was the pioneer 1896 D3, No 3400, and she was one of the wildest riders I had travelled on up to that time. We carried out some very Heath Robinsonish work on the bogie axleboxes and, fondly imagining we had solved the problem, put her to work one day on the 8.20am South Lynn to Yarmouth. Now this train was often pretty heavy and I remember thinking, as Jack Thurston (who liked to hear them "talk") opened the regulator wide up the initial 1 in 100, that 280 tons was quite a load. Despite stopping at most of the stations, and once or twice having a job to start, she played with the train, but it was on the down gradients that I realised just how rough these little engines could be. Although a GN Atlantic at speed could seem wild, but would feel safe, No 3400 was downright dangerous. So we did not repeat the experiment, but kept her in cotton wool for the summer, when Spalding and Bourne became her limit on short trains, with which she could not even set fire to the fields.

I learnt the job at South Lynn under the kindly guidance of Peter Glenister, the permanent Shedmaster. In the wintertime there was no foreman on days, but only the Shedmaster and, of course, his ADC, Luke Watson, who would have been insulted if one dared to think of him as a general factotum. The Shedmaster and the clerk, Arthur Bettinson, shared a long office with a circuit telephone that had no bell but a species of foghorn, the noise of which is still with me to this day. The rostering of the enginemen was very different from King's Lynn; certainly there was more of what is often referred to as "give and take", and yet this depot still worked with strict seniority as opposed to shift or 12-hour seniority for passed firemen acting as drivers.

In other words, if there was a turn at 2.0am on a Monday and the senior passed fireman firing was rostered for 8.45pm he would be moved from the latter to the former time. He could remain on this for five days if the vacancy turned out to be caused, for example, by sickness, and then because on Saturday there would be a driver spare at 3.30am off a five-day roster, he would drop back to 8.45pm firing on a Saturday night. This was in 1946. It could be extremely hard on a man's domestic arrangements, and yet it gave the senior man the driving turn at a time when, as at King's Lynn, these folk had been waiting for 20 years or more to get a driver's rate of pay and a driver's check, a situation which is comparatively rare these days. There was no booked Sunday work, although there was quite a lot of special freight from Colwick with empty wagons back to the GN; this was nice work, changing over with Colwick men at Boston, I believe, and bringing to South Lynn all sorts of outlandish machines such as Class J1, J5, and the remarkable "Pom-Pom" Class J11 0-6-0 freight locomotives.

The artisan work was a problem, not because of any marked lack of ability on the part of fitters and boilermakers, but because most of the types of locomotive stationed at South Lynn normally needed a fair amount of attention whilst the staffing was tight, indeed too tight to overcome the backlog. What was really needed was the means to "rebuild" some of the engines running into high mileage to get useful work from them in the last nine to twelve months prior to a general repair. At that time the LNER did not have any scheduled intermediate shopping in the main works, and the running sheds were normally expected to get 60,000 to 70,000 miles out of a J17 or J19 0-6-0 on the M&GN. This took a great deal of achieving. With the exception of the chargehand, Bob Emms, who had served his time with Dodmans, the King's Lynn engineering firm, the fitters were all M&GN men trained at Melton Constable or Peterborough. There were two boilermakers, both with interesting backgrounds, the chargeman, George Fuller Pilch, being Melton-trained, and a descendant of Fuller Pilch, one of the famous names in cricket, whilst his assistant, by the name of Alf Milnes, was a long-headed old emigrant from Tuxford of the Lancashire, Derbyshire and East Coast Railway (never mind the Great Central), where he was trained under the famous Robert A. Thom.

George Pilch and Bob Emms shared a little office under the water-tank at the back of the shed. Bob's politics were true blue, while George was towards the other end of the scale, but apart from this basic difference (which occasionally caused a conflagration more severe than any normal difference of opinion as to who did what first on a locomotive about to be washed out and repaired) they worked wonderfully well together. Whatever else was wrong with the South Lynn engines, they nearly all steamed very freely, because the boilers were kept clean and in excellent condition, and I rarely think of George Pilch without recalling an incident which showed a railwayman in his true colours and yet which involved a risk that no man should voluntarily have taken.

This was in the spring of 1948, just before I left South Lynn for the last time. By then some ex-Great Eastern Bit 4-6-0s had been transferred to Yarmouth, shortly to be followed by some ex Great Northern K2 2-6-0s to South Lynn. Immediately prior to the arrival of the K2s and the alteration to the engines' and men's workings, the two Yarmouth Bits, Nos 1530 and 1545 as they had become, worked regularly into South Lynn. Now on this particular day, No 1530 had arrived about 1.0pm as booked with Yarmouth men, and should have returned with the 2.23pm from South Lynn, a Peterborough train. The only engines in steam at South Lynn were the yard pilots, a J17 0-6-0 with a hot box on the afternoon Sutton Bridge job, and a "Claud" with a hot box on the King's Lynn to South Lynn feeder service. The nearest available engine was a J5 0-6-0 on the other side of Sutton Bridge with a Colwick – South Lynn freight, and we had been wired from Wisbech that the Peterborough engine would be a failure on arrival. This mattered not at all until the relief crew, who had stepped on No 1530 for a few minutes in the depot, reported a lead plug weeping slightly. One thing only could now be done; it was to have the fire out, an examination, and the plug changed.

But George Pilch knew perfectly well that a delay of up to half-an-hour might be caused, because we had nothing in the offing to work the 2.23. I had accepted this situation, but George had the low fire of the B12 pushed up to the far end of the firebox, wrapped his wrists and neck in cloths or sacking, got through the firehole door, and very delicately tapped the lead of the plug so that it was no longer leaking. He was in and out again in a trice, having worked under appalling conditions of heat, and in the firebox of a boiler still carrying 75lb/sq in of steam pressure. Why did I allow it? Well, perhaps because I was weak, perhaps because George was determined to avoid a delay to a very ordinary passenger train and I could see he knew exactly what he was about to do. Such is the real tradition of railway service.

It will be understood that I was still very inexperienced, but full of enthusiasm and interest and devoted to a railway career. I was long enough at South Lynn, young as I was, to impose, if that is the right word, some of my own ideas as to how a depot should be run, how people should be handled, how the good men should be encouraged and the not so good handled firmly and, if necessary, severely. From time to time I made mistakes, but I learned from those mistakes while I began to feel I was doing a good job when, as a matter of fact, I was only on the fringe, and in any case life is always easier for the relief man, even if he is a long term relief. The reason is that he can change and improve the things that are comparatively easy to change and improve, but he can leave the difficult or nearly insoluble problems to be faced by the regular man when he returns. Before I describe how my South Lynn period of office was finally terminated, however, and how I received one of those very sharp lessons that are so necessary in life, there is more to say about the men and work at that splendid little depot with an allocation of some 35 engines.

Here I began to savour the very real pleasure and satisfaction that comes the way of any man responsible for the management of a large body of men, of close co-operation and understanding with the accredited representatives of the staff. The enginemen's representatives at South Lynn when I first went there in 1947 were, strangely enough, all NUR men; at that time amongst the old M&GN footplatemen there was a large proportion of NUR staff, although there was a swing over to the ASLEF a year or two later. The three LDC men in 1947 were all railwaymen to their finger-tips, strong trade unionists, straightforward and extremely direct in opposition when necessary, but who had the interests of the depot at heart. They preferred discipline to facile management, and on many occasions a civil word of warning to me of trouble brewing, with reasons and suggestions for the elimination of that trouble before it developed, enabled me to maintain a stable and fair level of management.

The three just mentioned were Ernie Drew, whose son is now at Enfield on the electrics; he was the chairman and a fine engineman; Les Beales, who died very young in the early 1950s; and Bill Pooley, who became a close friend and whose advice I sought on more than one issue long after I left South Lynn. His personal letters over the years to me, re-read in recent weeks, make a fascinating commentary on such things as the introduction of rest days, link-to-link working, cleanliness of engines, and so on, things which, together with conditions of service, are the real *raison d'être* of the staff representative. To take responsibility and to have authority can give real pleasure and also nerve-racking worry, but it can be a great comfort to have the help and guidance of the practical, thinking railwayman in taking those decisions for which one is nevertheless alone responsible.

Apart from Alf Milnes, the old boilermaker, there were a number of very senior citizens among the enginemen, moustachioed men of experience and dignity of a type rarely seen in the late nineteen-forties. It was one of my private worries that these men, old enough to be my grandfathers, might reasonably have had a contempt for so young a shedmaster, but this was not so. They were so exemplary in their work, so conscientious, so interested in locomotives and so courteous that any request, simply and straightforwardly presented, was always accepted and carried out. Lew Bell could talk engines by the hour, and was widely informed; Tom Stokes, always spotlessly clean, was indeed a grandfather on the job; his son, young Tom, was a shunting link driver; and young Ray was a passed cleaner in whom the fires of youth had not died down, and who, like many worthwhile boys, needed watching as a cat watches a mouse. He must have been one of the first people I ever sent home for being, as the record cards say, "Missing from your post of duty". In fact, he was inside a firebox and when he eventually came out, he found he had been booked off for the last hour.

For years after my apprenticeship, I built on the thousands of miles experience of work on the footplate that had come my way during the war; to this day, indeed, I am still building on it for one's experience of men and machines is never complete. But this interest has served me in good stead on many occasions.

The South Lynn main line men in 1947, with the exception of the four seniors who were in an old man's gang, shared passenger and freight work in one link of about 36 turns. One of these turns involved the working of the Birmingham – Yarmouth train from South Lynn to Melton Constable with a Yarmouth "Claud", and then a return with a vacuum-fitted J17 0-6-0 on a passenger train due at South Lynn about 9.15pm, after this the engine ran round its train and worked into King's Lynn and back again. The Yarmouth "Clauds" were a well-kept lot but one or two of them were extremely shy for steam. At 5 o'clock one evening, one of the members of the LDC came to the office to tell me that there had been complaints about the steaming of No 8833, the engine booked on the job and if I could spare the time, would I go and see for myself.

Now the best way to see for yourself is to do it yourself, so we set off for Melton with about 300 tons of train, and "Mo" Seaman as the driver. No 8833 was a "Super-Claud", with Belpaire firebox, and a single very long gauge glass. Many of these engines were notoriously shy steamers, and one had to take every possible precaution. This meant setting the injector fine, shutting off when firing, playing on the water in the boiler, going through the loops at the stations with the injector off so that the blower could rally the pressure to 180lb/sq in by the time the driver opened up again, and then coasting as much as possible with the minimum of braking for the loops. In these ways I schemed for my steam with every trick I knew, and got all I needed. Nevertheless, the engine could be regarded as a non-steamer, and needed attention and cure; but after this experience I was able to describe certain symptoms to Yarmouth, which endorsed what drivers had said and booked, and this resulted in a marked improvement. The point of this story is that a young man can gain respect by doing a practical job competently and by sharing the hardships and difficulties of the men under his control.

However, it did not always work out like that. Indeed, on many occasions my efforts were not to prove successful. I never minded if the machine beat me to it after I had tried all my tricks, but when the engine was just passable and my performance had been indifferent, it was most unsatisfactory. A few months afterwards, on the same train, there were complaints about the booked engine on this occasion, which was an ex-GN Class D1 superheated piston-valve 4-4-0. I was rather partial to these crude old things, rough, uncomfortable but strong and lively, and yet again very difficult from time to time in steaming.

But whereas I had tackled No 8833 with the greatest care and forethought, I allowed myself to be hoodwinked, by a late arrival of the train from Leicester and my own rather good opinion of my capacity, into a situation in which the fire was not white-hot right at the start of the journey, and this after all that my mentors had taught me. So I had one of those dreadful trips, with no steam and no smoke, where 130lb of steam is a luxury and with a yellow pudding of a fire. Everybody was slightly embarrassed, squinting at the fire, the pressure gauge and the water level, but hoping for the best. This was a badly organised trip, and if the crew had said afterwards that the boss had made a real mess of things they would have been right.

There was an interesting sequel on the return journey, however. The fireman was one George Edge, who was going up to be passed for driving very shortly. Occasionally you could find a man, who, when he came in front of the inspector for his practical driving test after years as a fireman, had never had hold of a locomotive working a passenger train. Such a happening might be rare, but this was the case with George Edge, and so he had his first main line passenger driving experience that evening under the driver's supervision while I did his work for him!

A few days before I left South Lynn, some time in the late spring of 1948, I was to have yet another practical lesson. It was on this same return working, when No 5588, a vacuum-fitted J17 0-6-0, ran into South Lynn on time at 9.15pm with five coaches. The fireman unhooked immediately, and very quickly indeed the engine was attached at the other end. After working late at the depot, and wanting a ride into King's Lynn, I walked up the platform slope and got up on to the engine. Something was wrong and there was the usual crowd of people offering advice and hoping for the best; they included the station inspector, the fireman, the guard, and one or two others. And only 15in of vacuum could be created. I said that I couldn't hang around all night – a decision everybody accepted – and after we had eventually blown about 17in with the large ejector, we found that we could move if we continued to blow hard. Into King's Lynn we had a clear run, though gradually losing speed, due to the fact that the brakes must have been rubbing all along; nevertheless when the time came for a severe reduction in speed, very little happened, not even when the vacuum was completely destroyed. So we entered the platform quite smartly, but mercifully the steam brake, tender hand-brake, reverser and regulator all but stopped us, and the buffer-stops did the rest. One or two passengers who had been shaken up gave us black looks but that was all.

What was the lesson? Why, when there is trouble with "creating a brake", there is an age-old test laid down in the Appendix which tells you exactly what to do. There are no short cuts, and had I said at South Lynn "uncouple the pipes, leave the engine-pipe off the dummy and try to 'create' a brake", I should have found what was wrong. The trouble was a partial blockage in the front elbow of the pipe by a piece of waste, which shifted enough to allow a fair brake to be 'created' at South Lynn and shifted again to prevent a proper reduction throughout the train at King's Lynn when the application handle was placed in the emergency position. And so we were very lucky!

It was at South Lynn that I first learnt the joys and pleasures of working closely with the traffic people. As a Motive Power man, I had always been brought up to keep a wary eye on our uniformed friends in the yards and on the stations. For were they not traditionally the enemy – the "Traffic" – whilst we, needless to say, were the "Bloody Loco"? They knew all the dodges, how time lost through operating causes could be laid at the door of the poor old "loco "department. But then we were just as bad – or good – for we knew exactly how often a Distant signal at caution (sometimes a fixed signal) could save the day, and the two minutes lost rolling up to the Home which by that time was as clear as daylight, could give us 20lb of steam and another inch of water in the glass, and yet be put down quite legitimately to "signals". The station-master at South Lynn, a Mr Shirley, was a charming man, though nevertheless implacably opposed to us engineering types. The power behind the throne, however, was Jimmy Greaves, the Yard Inspector. He was keen, he wanted his engines at just the right time, and if there was any messing about he would be on that dreadful old circuit telephone with a "Where's that so-and-so pilot, Dick?"

When the summer of 1947 came along, and the service really warmed up after the wartime recession, we used to get together as soon as we received the week-end workings from the diagram office and rebuild them so that they met our dual purpose; this was to minimise locomotive and carriage movements at the busiest times, when engines from Leicester and Yarmouth had to be turned round, coaled and sent on their way in next to no time. We used to think these Headquarters folk knew nothing much about the "Joint" and its requirements: in fact, our alterations did not amount to very much, but they enabled us to think of the service as our service where time was utterly important, and where failure to cover a train adequately was a "Joint" failure. Likewise with the Norwich and Cambridge controls, where we had to balance our power resources against the-service and the single line and our little coaling plant on which so much depended, it was a real joy to rise to the occasion and to beat the clock. Over the years, I have had a wonderful time with the operators, for it has been a pleasure not only to watch them at their little games, but to help them to improve train running and standards of punctuality, and to cope with their sometimes all but impossible demands.

And so, lastly, we come to what were to be known as "Hardy's Follies". It must have been early in 1948 that I presented to our Cambridge Headquarters a set of proposed engine workings for South Lynn. These were paired diagrams, included three Yarmouth turns, with regular men on regular engines, and also combined passenger and fast freight workings. On these suggestions I had burnt plenty of midnight oil. Soon after this, I was sent on a special job for three weeks to the Motive Power Headquarters at Liverpool Street, but when I got there something had gone wrong and I had no work. By an extraordinary stroke of good luck, I was given my own diagrams to examine and if "found satisfactory to ask the engine diagram people to introduce them". Needless to say, I found them most satisfactory, for they gave South Lynn the cream of the Yarmouth work! The only problem was that "Clauds" could not operate the mixed workings; so we decided that K2 2-6-0s should do the job.

In due course, some Stratford "Ragtimers" were transferred, and what a collection *they* were! However, when the May diagrams came out, we set up the rosters, pencilled the engine numbers up against the men, and started off full of hope and propaganda. No 1738 was the roughest, Nos 1743 and 1757 were the best, and 1748 and 1766 were half-way between. The enthusiasm that our crews felt for their vastly improved work was largely tempered by the physical discomfort of standing on an old "Raggy" on an all stations trip to Yarmouth and back lasting eight hours, after months on the luxurious "Clauds". On the first Tuesday, I met a driver called Freddie Fisher, who had just come off No 1766 on the 1.17pm arrival from Yarmouth. He looked like a sweep, and his language was violent; he was black and blue, he said, he had ruptured himself trying to reverse at stations where she refused to start, and everything was wrong. This I wouldn't tolerate and said it was just prejudice against the engines and I'd come to see for myself.

Next day I couldn't leave the depot until lunch time, but this enabled me to get as far as Hillingdon, where No 1766 was already standing in the loop waiting for us to arrive from South Lynn. Fred and his mate again were black and fed up, the cab was filthy, steam pressure was about 140sq in – a despairing situation. The rightaway was given and Fred, a little man, stood on his toes to open the regulator somewhere up in the cab roof. Nothing happened. He closed it, turned to me and cursed 1766 and everything and everybody connected with it. "You try it" he ejaculated, "you got the bloody things sent here". Now, I had driven Great Northern Atlantics many miles and knew how the piston-valve engines could be "poled" with the regulator open if they would not start in fore-gear. The "Ragtimer" was a piston-valve engine, and no doubt Fred had messed about opening the cylinder-cocks and closing the regulator, as if he had had a slide-valve engine. So, I thought, "We'll show him how to do it". Opening the regulator, I got hold of the reversing lever, and nipped it into back gear. Immediately she barked up the chimney, kicked back on to the train, and at the psychological moment I tried to whip the lever forward. It was stuck fast and I couldn't move it. Diving for brake and regulator I stopped, reversed, opened the regulator again, and off we went. Two minutes had been lost, and I think Fred Fisher must have enjoyed that little story for many years after. My opinion of the K2s was that they were strong and would steam, but they were never universally liked, and when the B12 4-6-0s came in greater numbers to the M&GN the day of the "Ragtimers" on the "Joint" was virtually at an end.

In May, 1948, I left South Lynn for the last time. Though I had learnt a great deal, I was still very young in many ways and had yet to learn that, although I had worked very hard, although I had managed to improve several aspects of the working at South Lynn Depot, and although I had achieved very satisfactory staff relationships, I was still a bit wild. One or two rather stupid escapades on my part came to the notice of my District Officers at Cambridge, and it was decided that I should be transferred to Headquarters as a Technical Assistant. I realised that I was sailing far too close to the wind, and the move brought me to my senses. Overnight I grew up a year or two, and I am now truly thankful that those responsible for my performance at Cambridge in 1948 took the action they did. Not only did it bring me on to L. P. Parker's staff at Liverpool Street, but it also brought me back home to Amersham. Within a few weeks I had met Gwenda, who became my wife in the following year, and who has shared the domestic burdens and helped me over the hills, in a career which has been hard at times, but which has given us both endless satisfaction and happiness.

Pre-Steaming and Oil-Burning Tests

BEFORE MY TRANSFER to Liverpool Street I was sent to carry out two special assignments in the Cambridge District, both extremely interesting, but neither of which bore any fruit at the time. The first, in March 1947, was to March depot to supervise the introduction and working of the pre-steaming apparatus which was being introduced to reduce the turn-round time of locomotives passing through the depot for wash-out and repair. Pre-steaming had never been tried before on the LNER, and this was a bold attempt to introduce what was once fairly common practice in America. The method was simple. A "Wath Daisy", as the ex-Great Central 0-8-4 tanks were known, was stationed at the side of the main shed, and a steam line was taken from a flange in the cab where the steam brake valve should have been, through the side of the shed and down vertical pipes at the bottom of which was a wheel-operated stop-cock. The vertical pipes served two roads only but that mattered little in the initial stages. A flexible reinforced copper pipe was screwed to the bottom of the downpipe and to the blow-off cock of the locomotive at the base of the firebox, and steam was then blown into the boiler. My job was to supervise all three shifts for a fortnight – a typical Parker assignment – and to report on performance. I was not required to say whether the boilers and fireboxes were affected by the hammering they seemed to be getting.

The night we started operations was unforgettable. The winter of 1947 had been desperately hard and the weather changed over the week-end, from bitter cold to rapid thaw with high winds. As I returned from Buckinghamshire on the Sunday night, because a tree had been blown down at Great Chesterford, there was no service from Liverpool Street and so I caught the 8.20pm from King's Cross to Peterborough. By now it was really blowing and we ran into some trouble at Abbots Ripton, so that I only just caught the London Mail out of Peterborough East of 11.0pm. I remember running over the Nene Bridge as I crossed the town, and wondering at the height of the water. By morning, the Fenland rivers had burst their banks and the water was all over the land, houses had been demolished, railways were submerged, communication had been severed between King's Lynn and Ely *via* Downham and between Ely and Cambridge, and all but severed outside Ely between the station and the North Junction. It was a shattering time for the Fenland people, for animals and wild life, and many hopes and ambitions were swept away in the darkness of the night.

At 11.20pm I arrived at March, hooked on to a driver going on duty, and arm-in-arm we fought the gale as we walked across the flat to the depot. But before midnight, I was helping Driver Fred Stables, and his mate on the "Daisy", to couple up to our first locomotive for steaming. We worked hard until past 6 o'clock in the morning and by that time we had worked out a very effective technique. Up came a dead locomotive; we checked the level of the water and if necessary let some water out through the blow-off cock until the level in the gauge-glass was below the bottom nut; then we coupled up our pipe, opened the cocks on the "Daisy", the stand-pipe and the dead locomotive, while the fireman outside shovelled for all he was worth as the "Daisy's" boiler did its stuff. Meantime the steam passed into the dead locomotive, thumping and banging in the water spaces, whilst Fred Stables and I went down the shed with long shovels to fetch fire from the sand furnace. Already we had coaled the firebox and soon we had a nice fire going; up came the water on the gauge-glass while the needle of the pressure-gauge quickly began to lift; then on would go the blower, and even at low pressure this would have just enough effect to hasten still further the rapidly-evolving process of steam-raising. Smoke would be reduced, although certainly not eliminated, and within an hour, the locomotive could be presented to a grateful Running Foreman.

At about 7.0am I got a fresh crew and another set at 3.0pm, and eventually I left Fred Stables to it at midnight on the Monday, after having worked round the clock. Because of the floods, traffic was disorganised and it was not possible to get a really accurate assessment of improvement in turn-round time against basic diagrammed work, but there is no doubt that some amazing results were obtained; indeed, our record during the second week of the tests would take some beating. At 4.0pm one afternoon Fred Stables and I had no engine connected up, and I went to our Foreman about it. He told me there was no engine for the 5.30pm to Peterborough, and he couldn't see how he was going to find one unless we pre-steamers for once justified our existence. So we were given No 1863, the 2-cylinder rebuild of a Gresley K3, and got to work on it at 4.20pm. The water in the boiler was not cold, but it certainly wasn't blazing hot. We let some of it out, coaled the firebox, got several shovels of fire and then some more and more again. We poked and darted, while the "Daisy's" chimney poured black smoke as she forced steam into the violently-vibrating 1863. The prospective driver was highly suspicious of the whole business but nevertheless he was away from the depot by 5.28pm, and the train was only about six or seven minutes late away. We never beat that record, and it was combined operations with a vengeance. Many doubts were expressed about the wisdom of such quick steam-raising, but we found that, if quick turn-round was to be the main criterion, the availability of locomotives for wash-out and repairs could be improved out of all recognition. But the system was not applied extensively.

In May, 1947, I got another assignment. It was to go to Doncaster to work back on the one and only LNER oil-burner – of this particular generation – for I was to be head cook and bottle-washer for three months on this rough old machine, ex-WD 2-8-0 No 3152. We came down the Joint Line with a freight train, together with our fuelling tender, and I thought the engine did quite well, holding her own against the injector with very little smoke, although the demands on the boiler were not heavy. Two days after, with some trepidation, we lit up without singeing any whiskers, gently raised steam and backed out of the depot to go to Peterborough, with a "Crab" trip. With us we had Horace Emery, a Headquarters Locomotive Inspector of some distinction, to make sure the locomotive was worked in the L.P. Parker tradition with wide open regulator. We made a lot of smoke and not much steam, particularly when working tender first. A long list of experiments then began with the setting of the Weir burner, glazing the bricks in the firebox, reducing the number of air-vents in the floor of the firebox, and fitting a trough in the ashpan to scoop the air.

The results got steadily worse until one day with 60 wagons of coal we very nearly stopped altogether on Great Chesterford bank and had a terrible time in the Audley End tunnels. It was a very hot day with a following breeze, the cab was like an inferno, and we were right down for steam, pouring out greasy black smoke. As we entered Littlebury tunnel the train was moving at no more than 8 or ten miles an hour. The heat became quite intolerable; there was no air, nothing but ghastly black oily smoke. I was doing the firing at the time and was loath to leave the controls, but we were forced to get on to the floor of the cab to try to get some fresh air that way. We had very nearly reached the point of no return when suddenly the engine, left to herself, crawled out into the sunshine once more leaving us gasping for air, soaked to the skin and exhausted. This was a harrowing experience, but it was all over in five minutes.

Next day we returned home from Temple Mills as a light engine and in disgrace. The following day, in desperation, we tried one last change of settings of the burner, striking the brickwork well up towards the firehole. The locomotive was completely transformed. Now we could go to London economically, on the minimum of fuel. We ran with the water solid in the gauge-glass, with never a suspicion of priming, never any exhaust visible at the chimney top, climbing the banks with the barest trace of the regulation brown smoke, showing the white feather at the safety-valves and always ready to lift but kept in check by delicate control of the burner. One day we went over Elsenham Summit with a heavy coal train, shut the fire down, closed the dampers, and ran through Bishop's Stortford and on to Broxbourne and Waltham Cross – without a fire! Here we still had 150lb of steam but very little water, so we lit up on the move and finished our journey. Those white-hot bricks had done the trick, and we were mightily pleased with ourselves. After a while enough firemen had been trained to handle the locomotive and my sphere of usefulness diminished, but it was all most interesting. Nevertheless I longed, in that blazing summer, for the physical work of a coal-burning locomotive instead of the stifling cab through which no air seemed to pass, but there was plenty of interest in going to Temple Mills and back, for we were always the object of attention, making new acquaintances and talking about experiences and lessons learned.

One day we climbed up to Elsenham, on the way to London with a coal train. It used to be the practice on passing Elsenham to apply the tender hand-brake, and the engine brake if necessary, to keep the train in check until well on to the falling gradient to Stansted. After getting through the bridge on the London side of Elsenham station and round the corner, one could see Stansted Distant, and if it was green the brakes were released and the train left to coast down the bank, accelerating all the while; for if the Distant was clear there was nothing to stop for until Bishop's Stortford. It was essential to adopt this drill as otherwise it would be difficult to stop for signals at Stansted with a heavy loose-coupled train. On this occasion, however, our crew took no such precaution, went ahead through Elsenham, and came round the corner to face Stansted Distant at "yellow". All efforts to stop was futile. Brakes, reversing gear, regulator had no lasting effect, and the heavy load pushed us by the Distant and on up to the Home signal at danger. Mercifully our prayers and whistle calls for help was answered, and just in time the Home and Starting signals came off, so that we were able to carry on with reputations untarnished. Another lesson learned – or witnessed – was how easy it is to take chances, and with loose-coupled trains how difficult it is – if not impossible at times – to avoid or extricate oneself from run-away trouble.

One week-end I was asked to take my oil-burner to King's Lynn for the benefit of the Enginemen's Mutual Improvement class. The idea was to set off for Ely on the Sunday morning with a couple of coaches, and let the class members have a look at oil-burning and try their hands. We had the railway to ourselves and enjoyed the freedom immensely. The preliminaries to this trip – and the sequel – were amusing. The tubes on No 3152 were prone to leakage, but only when the fire was shut down, and this never worried us on the road. At March we always took our steam for lighting-up direct from the pre-steaming equipment through the blow-off cock, and rather than rig up a special arrangement to take steam through the oil-burning control manifold after a night in King's Lynn depot, I preferred to shut down altogether and relight during the night as the steam pressure fell. This time, however, No 3152 decided to leak abominably, and I had to go to her every two hours during the night, light up and fill the boiler, and then go back to sleep on the floor of the shedmaster's little office. After the trip to Ely, I made arrangements to be dragged to March early in the morning to refuel, fill the boiler and steam in readiness for the 8.5am from Whitemoor to Temple Mills on the Monday. When I came back to the engine in the early hours to ride on her across to Whitemoor, the level in the boiler was almost where I had left it!

No 3152 was destined to be the only LNER and Eastern Region oil-burner of her generation. Had the oil-burning scheme gone through, we should have achieved some excellent results. It is far easier to achieve the precision of operation and understanding that an oil-burner needs when there is a number of such locomotives allocated to a depot in regular operation. I had an excellent time on her, avoiding the terrifying explosions that I had been told came the way of all oil-burning participants, and I certainly learned the road from Whitemoor to Temple Mills!

XI

Leslie Preston Parker

F<small>ROM</small> J<small>UNE</small>, 1948, until I was appointed as Shedmaster at Woodford Halse, in the autumn of 1949, my working life, and indeed my off duty hours, were dominated by one man, Leslie Preston Parker, Motive Power Superintendent, Eastern Region, British Railways. And as his influence was felt throughout the Region, I was very conscious of his presence right up to the time when I left the Eastern for the Southern Region in August, 1952. In Chapter 7, I have described my first interview with this formidable man, which will help to explain how he dominated me as he has dominated many other men. At times he made life almost intolerable, but one had to remember that he was a great administrator and manager and that he was passionately devoted to the training and advancement of his young men, from whom he expected complete loyalty to the cause. Indeed, one's domestic life in the context of one's training and future advancement was of no consequence whatsoever to him, although no man could be kinder or more considerate in cases of real need or domestic difficulty.

He had been a Stratford works apprentice and Whitworth scholar, and until 1948, when he took over the Western section also, had spent all his working life on the Great Eastern Railway and the Eastern section of the LNER. This was no handicap to him, for he was extremely well read and followed the wise policy of watching "the rest of the world" to see what could be learnt from the activities and mistakes of others. He had been District Locomotive Superintendent at Stratford for many years; his influence on men had been enormous and in that vast Stratford complex he had been all-powerful and had ruled by fear but also by example.

With the aid of a handful of inspectors, by means of improvement classes, by his own lectures, and by the force of his personality, he had impressed on a whole army of enginemen that the "full open regulator" principle of operating steam locomotives was the only correct method in normal circumstances, and to have driven this home to the point of almost unanimous acceptance by his enginemen, by tradition the greatest of individualists, was a remarkable achievement. His influence on his mechanical staff was no less marked. He was for ever experimenting and was a willing collaborator in this respect with Edward Thompson, at that time Mechanical Engineer at Stratford, and A. E. English, the Senior Technical Assistant to the Mechanical Engineer. The latter, a fount of knowledge, experience and ingenuity, was a technical power behind the throne at Stratford Works and greatly respected by Sir Nigel Gresley.

As for Parker's young men, in the Stratford days, their life was fraught with difficulties and dangers. All the experimental jobs on the Motive Power side fell to them, an answer was required within an impossibly short time, and the first results of their labours were often consigned to the waste paper basket. They had to cover relief jobs, or they received appointments at the smaller Stratford sub-depots, where L. P. would visit them, and often reduce them by question and probe to speechlessness, but he would leave them wiser if chastened, with improvements to make. If these were not dealt with quickly the fat would be well and truly in the fire. All this was before my time and they say that he mellowed when he became Motive Power Superintendent. Whether that was so or not I can't say, but I do know that he sometimes drove me desperately hard for the comparatively short time I was directly under him; even so, the treatment that I received was mild compared with that suffered by one or two of the really brilliant men who, in his opinion, were destined for the highest peaks of railway management.

Many things that were good could be traced to his influence. The insistence after the war that engine diagrams must be double or sometimes treble-shifted so that the locomotives could be rostered to their own crews brought a reliability into locomotive performance that would have been missing otherwise, as well as engendering a real pride of craft, economy of working, and achieving maximum mileage on main line work between shopping. Was it not the retention of treble-manning that kept the "Jazz" service going as a reliable entity right up to the last day of steam working in 1960?

Detailed attention was given from 1948 onwards to the elimination of hot axleboxes and bearings by the oil squads and oil inspectorate, to the stores organisation, to detailed and regular examination of late running of both passenger and freight trains, to the closest co-operation with the operating people, with freedom to experiment, to innovate, and above all an insistent interest in men's performance, their future careers and development, not always known to or understood by the person concerned. These and many other things were typical of L. P. Parker's interest in affairs. Certainly he made mistakes and maybe the greatest criticism that could be levelled against him was that he failed to realise that the Stratford district of his day, up to 1940, was vastly different from the wartime and post-war Stratford, particularly the latter, when there were serious labour troubles and labour shortages. He never would allow that the position of District Management in a difficult area had changed from his day, and that it was very hard to regain and maintain pre-war standards throughout the district.

In June, 1948, I was put in Room 128 at Hamilton House, Liverpool Street, along with the Headquarters Locomotive Inspectors. We were a splendidly happy if underpaid crowd. I now received L420 a year as a Class 2 Technical Assistant and we were responsible for train working, punctuality, engine availability, freight loading, locomotive trials, liaison with the train planners and a thousand other things. Within a day or two of my arrival several letters and files on a variety of subjects marked "please speak" had arrived on my desk. I ran my eye over them, asked L.P.P's secretary for an appointment, and eventually got inside. Again, the little man sat at his desk, smoking his pipe and fiddling with a circular slide rule that he alone was capable of understanding.

He completely ignored my presence until, after what again seemed an eternity, he said "Well?", looking at me searchingly. I told him he had sent me some papers and would he please advise me on the line he wished me to take. This provoked the simple reply, "No, you tell me the line I am to take". For nearly two hours, I stood by the side of the desk while I forced myself to think and speak on subjects about which I had failed to brief myself, while he asked, time and time again, the same question: "Why, why, tell me why?". He would not permit me to take a note of any description and I was pretty well done in when I made my way back to the office to get to grips with the resulting correspondence. But this was not the end of my ordeal, for the letters had to be signed and I had missed one or two fairly important and one or two not so important points in the replies.

So the letters came back to me unsigned once, twice, indeed three times in one particular case, without a remark of any sort and without any further guidance, so that I had to think again and again until I got what he wanted. It was a shattering experience, but the next time I did a lot more thinking before I went inside. Nobody, certainly not the youngsters, ever sat down in his office for this sort of interview, but the "please speak" interviews at least had the advantage of preparation. But when that buzzer sounded in Room 128, and you went along post haste through the front door of his office, knocked and without waiting for a reply went straight in, you never knew what was going to be asked. I was responsible for the Eastern Section punctuality records and I might be summoned to explain why the 5.6pm Liverpool Street to Norwich had lost time on Monday and Tuesday, or, as happened at 9.2 one morning, to explain what had caused the 8.51 arrival up from Clacton to arrive at 8.55. He expected an answer, and if one was not sure and tried to get round the question one courted a disaster too overwhelming to contemplate.

One day he summoned me to relate that he had come into Liverpool Street alongside another train, the engine of which had coasted down Bethnal Green bank with the lever well down. The Eastern Section men, over the years, had been trained to coast a piston-valve engine with the lever just outside midgear, for the very good reason that the anti-vacuum valves thus were held in the closed position and that the pressure this built up in the steam-chests kept everything tight and minimised knock and play in axle-boxes, big-ends and valve gear. Running alongside an N7, he could always see what was going on.

L. P. Parker was a great man: I have come out of his office after a roasting which had finished with a kind word or a mild commendation; the reaction brought me near to tears when I got outside. But sometimes, after lunch, one might be called inside for a pleasant fireside chat about this and that. All seemed well though one learnt that, if one dropped one's guard for a second, the cosiness would vanish and one would be on the spot once more. I have been instructed at 9.0am to prepare a report on some difficult subject to be typed and on L.P's desk at 5.0pm that day; no man on earth could have achieved what was an impossible assignment in the time, and so I had to work all night. Next morning the results would be typed and inside by 10.30, but only to be thrown on the floor as useless. There was no respite and by 5.0pm, after 30 hours continuous work, the final result would be on L.P's desk. Again there would be that simple word of mild – even grudging or sarcastic – commendation, but that was enough.

It may be wondered why we stuck it, why we did not answer back. Now and again, we ventured to stand our ground and, if we felt we had a case or even if we did not but were obviously sincere, it did no harm in the long run. Some of the senior people would not tolerate this treatment, whilst others did and suffered accordingly. I stuck it because I had to and because I realised it was being done for my own good, and that I had been pulled back from the brink after juvenile acts of stupidity in the Cambridge District, and that here was a man who was determined either to make or break me. I had to survive.

Another practice of this remarkable man was to send his young men to hold down temporary jobs far above their grade or to represent him, with full powers, at meetings where the other representatives were often senior officers in their own right. I shall never forget being sent to represent him at what must have been the Motive Power/Operating meeting to decide on the allocation of the Eastern Region share of the 1949 Locomotive Building Programme. I was a Class 1 Technical Assistant, whilst some of the other people at the meeting were at the top or nearly at the top of the Operating and Motive Power Departments. But I was expected by L.P. to stand my ground and to see that we got what was our due. We did, though I doubt if I had really much to do with it!

On September 29th, 1948, L.P. sent for me to tell me that I was standing up well and that I was to be regraded to Class 1. It was that night, after Gwenda and I had been to see "The Happiest Days of Your Life", that we decided that we were going to get married. Next day I thanked L.P. for the promotion and told him what had happened on the evening before, and to say that he was genuinely pleased was to put it mildly. Next, in April, 1949, not long before our marriage, I was sent back to Cambridge to replace Bob Gardiner for a fortnight as Assistant District Motive Power Superintendent. The Superintendent, A. H. Rees, also was away, and as his place was taken temporarily by George Weeden, the Shedmaster from King's Cross, we were a comparatively inexperienced pair of officers. Inexperienced or not, however, we were expected to do the job. George also had had the experience of working under L.P. at Stratford, and delighted in telling me that the I P. of 1949 was quite a mellow old gentleman compared with District Officer L.P. in the 1930s!

After my marriage in April, 1949, I returned to Hamilton House, Liverpool Street, while Gwenda and I settled down in half a flat at Westerham in Kent. No more than a week later L.P. sent for me, and delivered this blow: "Hardy, I've a pleasant surprise for your wife. I'm sending you back to Cambridge to relieve Mr Gardiner for several weeks", and, if that were not enough, "Your main difficulty will be to get on the right side of Mr Rees".

The subject of this last gloomy forecast had been in charge at Cambridge when I was removed from South Lynn, and he regarded me as quite inadequate to serve as his Assistant for an indefinite period. L.P. knew this, of course, and it was for further character building that he had given me this assignment. Well, after a few weeks and a good deal of burning the candle at both ends I began to show some improvement, and eventually I achieved the happiest of liaisons with "Dolly" Rees. Gwenda and I had to put up with the fact that only once in a fortnight, and sometimes once in three weeks, could I get home, and it was not surprising that in these circumstances off-duty life in Cambridge seemed dull, mundane and depressing, living as I did in a small hotel near the station. For several months I was working right out of my class, struggling and at times making mistakes, though gaining in confidence and knowledge, taking responsibility and making decisions, and, indeed, succeeding in my most onerous task, which was to prove to "Dolly" Rees that I was no longer just an inexperienced and irresponsible youth.

So matters continued until September, 1949, when the biggest step forward took place with my appointment by L.P. as Shedmaster at Woodford Halse, as dealt with in detail in Chapter 12. Here Gwenda and I settled down, or thought we had, but if so I had reckoned without my mentor. For no more than seven months later I received a message from my Superintendent, Charles Read, that L.P. was sending me to relieve at Ipswich depot. This really was a blow. The day before my wife had returned from hospital with our first child, Anthea, and I asked Mr Read if it were not possible for someone else to go; surely by now I had done enough relieving and could be left to settle down and consolidate my job as Woodford Shedmaster. The next day Charles Read turned up at Woodford and asked me to travel with him as far as Aylesbury. His advice was that transfer to Ipswich was imperative. When he had spoken to L.P. about my domestic situation, the reply simply was, "If Hardy does not go to Ipswich, he will stay at Woodford for a very long time". There was no possible alternative. Not only so, but L.P. sent a great friend of mine, who was on his staff, to see me at Woodford to make certain that I realised how much my future lay in his hands, and how much damage I might do to it by refusal.

L.P's personal letters had a style of their own. You could almost hear him speaking in his reply to a letter of mine in May, 1952, when I was wondering whether to apply for a job on the Southern Region to which I was ultimately appointed. Here was his advice:

"At the same time you must not forget that there are many – too many, I think – steps on the ladder, that you have to climb them and that life is short. I am convinced that in about 20 years' time my young men will be running a large part of British Railways, but they must seize any opportunity that is going".

And then in May, 1953, more lightheartedly after I had written to him from the Southern Region to ask him to propose me for the Institute of Locomotive Engineers, and had told him of the activities and of hot axle-boxes on the B1 class locomotives on loan to Stewarts Lane Depot during a Bulleid Pacific axle crisis:

"It is pleasing to know that the Southern men got on all right with the B1 class. Of course, these engines like all others will get hot axleboxes if they are lubricated with water or not lubricated at all".

L.P. did not have what is known as the common touch. The maintenance and footplate staff respected him for his immense knowledge, his influence on their affairs, his fairness when they were in real trouble, but they feared him because of his power, his sarcasm, his biting tongue. As long as they are alive, however, they will never forget him. A few months ago I was talking to an ex Cheshire Lines driver. He had been made redundant in 1931 and sent to Hertford East for no more than six months. The only name he recalled was that of L. P. Parker, but he remembered him clearly and described him with a wealth of picturesque accuracy Not many men got the better of L.P. in verbal conflict but once, in his Stratford days, he was walking across the front of the Jubilee shed when he saw an engine taking water, with the tank overflowing, and water cascading down the side into the axleboxes and along the framing. Suddenly the fireman, who had been in the messroom making the tea, saw what was happening and dashed across to screw down the water-column stop-cock. At this moment L.P. stopped beside him, and, poking a lump of coal lying on the ground with his umbrella, said ominously: "Tell me, fireman, do you let your bath at home overflow like that", to which the fireman replied "Barf guv, barf, ain't got a bleedin' barf".' Without a word, L.P. turned and continued on his way.

Those who took the chances he offered never regretted it. He lived in an era when, on some lines, young motive power men were given little responsibility, and had poor prospects of advancement. Not so under L. P. however. He was a fine engineer, and, where some engineers were just engineers, he was *par excellence* a manager and administrator as well. He retired in 1953 and died a year later. Many of the people who had served, and suffered, under him went to his funeral and were deeply moved. They had lost a champion.

XII

My First Command

IN SEPTEMBER, 1949, something happened which was to mean for me a complete change of life and scene. I was summoned to Colwick to be given a very thorough going over to decide my suitability to become Shedmaster at Woodford Halse, on the Great Central main line of the Eastern Region. When I heard that I was to be given this appointment, naturally I was delighted. Not only would this be my first command, but it would mean that my wife and I would be able to find a house and settle down. Up till then we had both been fed up with enforced separations in the first year of our married life. So now I found myself working on the very line on which, as related in Chapter 1, my earliest youthful love for locomotives and railways had developed in Amersham days.

The Woodford job was an excellent one for a young man. It was a training ground, but not necessarily an easy one – a place where one could and did make mistakes which in some cases had a marked reaction. I was responsible to Charles Read, the District Motive Power Superintendent at Colwick (Nottingham); he had come from the London Midland Region and was forceful, go-ahead, and full of ideas, many of which were new to the LNER personnel in the district. The shed was typically Great Central, comparatively modern, and with a fair amount of room inside and outside; there were separate ashpits, a triangle for turning, a modern coaling plant, a staff of some 270 to 280 men and approximately 45 engines. My office was palatial compared with those to which I had been accustomed, and I was protected by the Chief Clerk, through whose office anybody who wanted to see me had to pass. There was a front door, but this was forbidden territory; it remained so traditionally all the time I was at Woodford and probably for the rest of the time that the depot survived.

The running foremen had once been in the same block but now worked in a new brick building next door to the messroom, about which there was a splendid story. It was built without hot water facilities, or, more accurately, the boiler was installed but the Engineer's people ran out of money before the building to house the boiler was completed. At Woodford there was a driver by the name of George Needle, who must have been well over 60 by this time. Some years before, he had been injured whilst on duty, and he suggested that as he had been treated fairly by the railway during the time of his recovery, he would be delighted to build the boiler-house if we could find the bricks.

This was no problem, and so the Needle boilerhouse gradually took shape, though not without one or two sharp questions being asked me from time to time by my superiors as to what the devil was going on, and why a moustachioed and dignified old gentleman in driver's uniform was working as a bricklayer without authority!

The work of Woodford depot was largely freight, for which we were allocated WD "Austerity" 2-8-0 locomotives, and Robinson J11 0-6-0s, some rebuilt by Thompson with piston valves; there were also two Robinson 2-6-4 tank engines, Class L3, Nos 9050 and 9069, for the High Wycombe freight; N5 0-6-2 tanks for shunting and later on N2 No 9560 for the Banbury "motor"; and for the passenger work B17 "Sandringham" 4-6-0s of the later "Football" series. Their work comprised slow trains to and from London and Nottingham, and the cross-country trains from Banbury to Nottingham and Sheffield.

The fast freight trains were worked to the south and to Banbury by the WD 2-8-0s and this service was a part of what must without doubt have been one of the fastest and most reliable loose-coupled freight services in the country. The northern workings centred round Annesley depot which, rightly, had the cream of the assignments. The Annesley men came south to Woodford with the "Windcutters", on Class O1 or O4 2-8-0s which were carefully controlled by the District Office at Nottingham: they stepped off one engine on arrival, snatched a bite to eat, got on to another and roared away back to the north, whilst the Woodford men got all the fire-cleaning and coaling. This caused some ill-feeling but the working was correctly planned and the results were there for all to see. Furthermore, late starts for the southbound workings from Woodford were not tolerated, and all the staff were on their toes.

Our B17s were a rough-and-ready collection, universally detested. Much of my time at Woodford was spent in trying to get rid of them in exchange for the well-liked Thompson B1 4-6-0s which were stationed at Neasden, Leicester and Sheffield; but I only succeeded in getting two more B17s transferred instead! Their numbers were 61650, 61651, 61664 and 61669. The last two had high-pressure boilers and were both given general repairs during the winter of 1949-50. This, of course, transformed them, but when in February 1949 I was told that two more were coming from Lincoln, my heart sank. My misgivings were confirmed, for Nos 1647 and 1667 were two of the hardest and most uncomfortable old cabrankers it was ever my misfortune to come across.

And thereon lies a tale. A month or two after the arrival of the pair, I received a letter telling me to transfer two B17 locomotives to the Norwich District. My instinct was to forward the two latest arrivals, but I was foiled because the agreement that the mileage before transfer of passenger engines must average 25,000 had been imposed and these two each had run up something like 45,000 miles. No 1647 had to go, and No 1669, quite new from the shops, went along with her. But no sooner had the numbers been agreed, than I was told I was to move from Woodford to

Ipswich, as we shall see later. One of my last acts at Woodford was to see the back of these two, and one of my first at Ipswich was to welcome them as just transferred to my new depot. So I was lumbered with the unspeakable No 1647 for another 20,000 miles, during which time she was confined to the 5.25am Yarmouth goods and the 7.23pm stopping train from Yarmouth to Ipswich. She was a low pressure engine, and in her original condition, would not steam, nor run, nor ride well; yet after shopping, she returned with a high-pressure boiler and ran for thousands of miles after that without serious complaint.

The water at Woodford was extremely bad. It came unsoftened from a number of bore-holes, and the boilers got dirty very quickly indeed. It was necessary to work to a regular programme for sifting, renewing tubes and cleaning boiler barrels. Tube leakage was a continual problem, particularly with the WD locomotives, some of which were confined to the Banbury workings, where we could keep an eye on their troubles. Woodford Halse itself was the junction for the Stratford-on-Avon & Midland Junction Railway, the "Nibble" as it was called, and, of course, for the Western Region and beyond, although the actual junction for Banbury was a mile and a half to the south at Culworth Junction in the fields opposite the village of Eydon. The extensive yards and the depot were built in the glorious rolling Northamptonshire countryside, because it had been necessary to find a site near the junction with the Great Western Railway at which to re-marshal and re-form the freight trains that rolled southwards from the coalfields and industrial areas.

The railway village of Woodford was built alongside the much older and much more attractive Hinton. At the end of last century a nucleus of north countrymen had come south to man the depot, and the Great Central terraced houses for their accommodation were built in parallel rows, looking as if they had been picked up out of Gorton and dropped down in the fields. There were several shops, and I think the White Hart Hotel belonged to the same generation of architecture. Long before my time the real Woodford railwaymen had spread to the surrounding villages, Byfield, Eydon, Culworth, Badby and Moreton Pinkney, but wherever they lived, they were railwaymen through and through, and also by now countrymen, even if they had come years before from the industrial areas.

In Woodford village, the railway was the cornerstone of the lives of the great majority of the inhabitants, and a remarkable loyalty and feeling for the work and for the railway was still maintained and had become a tradition. Many men, particularly enginemen, would rarely be seen out of their uniforms, whether in their gardens, in the pub, in the fields, on their allotments or even at the meets of the Grafton or the Bicester Hunts in Byfield or Eydon. I remember very vividly the sudden death of Ted Preece, a well-respected breakdown and maintenance fitter, and the funeral procession through the village to the cemetery. Never have I experienced anything else like it for there must have been well over 200 people slowly filing along, deeply touched and sorrowing, to pay their last respects to an artisan craftsman who knew his job thoroughly.

In such a close community, families married and intermarried, and he who condemned his fellow men did so at his peril, for words spoken behind the back almost invariably came back to roost. The Stationmaster and Shedmaster both held positions of some importance in the community, and a young man such as myself was expected to behave off the job as well as on it. I believed then – and still believe – that to gain the respect of the men you work with you must know their abilities and their personalities *intimately* while on the job, but there, generally speaking, it should finish. I never went into a Woodford "pub" all the time that I was working in the village, and I feel that I was expected to keep away, so that this was correct procedure.

I was lucky enough to follow my predecessor into some splendid "digs" with a family called Bobby. Fred Bobby was a clerk in the Permanent Way Department; Hilda, his wife, had spent a lifetime in the neighbourhood and so they knew everybody and everything, everywhere. Fred was steeped in the railway tradition, and in the evenings he would talk about the personalities of the past and the present until I began to feel part of the Woodford scene myself. On Friday evenings we used to walk to Eydon, three miles away, the village where Fred had been born, and we would spend the evening in the Royal Oak. It was very old and very friendly, and in time, through the kindness of the Bricknells who owned it, we achieved what had seemed the impossible; we were able to rent a little Tudor thatched cottage along the road. So it was that a few days before Christmas, 1949, my wife and I moved into our first real home, with a hand-pump for the water and a right-of-way through the middle of the garden!

As the weeks turned into months, and I began to get a grip on my job and to know the great majority of the by no means inconsiderable staff, I developed an intense pride in my depot. I would cycle down the hill from Eydon in the March and April mornings and in the distance I could see my depot. I was only 26 years old and I felt great, like a biggish pebble in a smallish pond. But that feeling of pride and contentment had not been achieved without effort, without mistakes, and sometimes not without violent disagreements.

Life at Woodford depot started with a bang with an error on my part that set heads wagging. In Sidney Road, one of the Great Central streets, two houses had been knocked together to form a large lodging house. In the late 40s, young men in the country preferred either the fields or the industry of Northampton and Rugby, and our intake of youths for footplate work was at a standstill. Despite the loss of work to Annesley depot at the time of the introduction of the "Windcutter" freight trains, and redundancy for a number of drivers who went back firing, there was a critical shortage of firemen. The vacancies were filled on a temporary basis by youngsters from Sheffield, Leeds, Gorton, and elsewhere, who were unable to get regular firing at their home depots.

All of them came from the cities, most of them lodged in the dreadful old dormitory in Sidney Road, and only the more resourceful escaped the utter boredom of a life which was to them devoid of everything to which they had been accustomed. It must be remembered that buses only went to Northampton and Banbury two or three times a week and that cars were rare amongst the railwaymen; thus a young town lad was virtually imprisoned. Anyhow, on my second day there was a fight in the ill-supervised dormitory between the night attendant and a youth from Leeds. I accepted the word of the night attendant against that of the youngster – wrongly I felt afterwards – and sent him back to Leeds with a disciplinary charge sheet, which I subsequently had to withdraw. It was a hasty decision and a wrong one, and for a while it did me little good; but by the end of my time at Woodford I felt that in several ways I had been able to improve the lot of the young men in this lodging.

I had hoped to let the world pass me for a while until I had found my way about, but events crowded in at such a rate that this was quite impossible. One evening I was confronted at 5 o'clock by the point blank refusal on the part of a main line fireman to work the 5.30pm from Woodford to Nottingham with No 61651, one of the B17s, on the grounds of excessively bad riding and because the footboard was sloping downwards whilst the flap plate sloped upwards to the tender. I did the only thing possible under the circumstances. The driver, Tommy Faulkner, had not refused the engine, and eventually I prevailed on the fireman to go if I went with him to see what could be done to improve matters. So I shared the agony by doing the firing as far as Leicester. It was no fun to work in an uncomfortable hole, with one's right foot on the wildly lurching flap plate, axleboxes knocking, cab-side vibrating, cab doors flying open and reversing gear kicking – a brute of a machine.

After a lot of argument we got her to the parent depot at Colwick a few days later for the trailing axleboxes to be re-fitted, the cab re-stayed, new springs, new bearer bars for the footboard – a fortnight's work. All this made very little difference in the end, except to show that I did *care*, and was prepared to share and try to improve the intolerable working conditions that some steam locomotives imposed.

A day or two later I was walking down the yard, and one of the Woodford WD locomotives, No 3040, came in as light engine off a Banbury freight. The big ends were knocking to glory, and she seemed in pretty bad order. The driver, one Leslie Bernard White, a member of a well-known Woodford railway family, wanted to talk about the shortcomings of this class of engine on the Neasden services, both as to steaming and riding. I was quite prepared to believe anything about the riding, but I had always felt that the WD locomotives steamed freely enough for the heaviest work.

Again I decided to see for myself and one evening set off with White for Neasden on the 6.45pm from Woodford. Again I did the hard work and there was no doubt that 3188, our engine on this trip, would not steam in the best of conditions. The oscillation between engine and tender at speed brought a constant stream of coal on to the footplate, making it impossible to keep things clean, and I was tired out when eventually I got back home at about four in the morning. I was even more so when I set out for the depot after breakfast to sort things out with Joe Goode and Bill Jeynes, respectively chargeman fitter and boilermaker, who came to see me each morning at nine o'clock and stood in front of my desk like Tweedledum and Tweedledee answering questions that I hoped sounded more confident and incisive than I felt they were. And thus began the battle of the "Jimmies".

Some of the WD engines still had their original numbers, such as No 78514, while those that had been through the Faverdale shops had received new numbers – 63186, 63188, and 90046. Generally speaking those which had passed through the shops had a large blastpipe top, of either 5⅛in or 5¼in diameter. Whatever the size, however, the steaming was very bad. On the other hand, some of the older engines with the WD numbers were fitted with a form of "jimmy", three studs fixed to a ring which in turn was fitted to the blastpipe top. The studs restricted the blastpipe orifice and sharpened the blast so that the boiler steamed freely. Now it was expressly forbidden to fit any such device, and so I was faced with a difficult situation. Time was being lost day after day on the Neasden service through shortage of steam.

I knew perfectly that to obtain authority to alter the design would take months, and I decided that the only quick solution – for we were there to run trains to time – was to make and fit a razor across the blastpipe of the engines with the large tops, bolting it securely in place. As a result, timekeeping became exemplary, but our boilermaker chargeman went about with a long face, feeling that the boilers were being forced and tube leakage accentuated. On the contrary, I felt very strongly that if the blastpipe tops were too large, the "jimmy" simply had the effect of making the boiler do its stuff.

However, our firemen never liked the WD flame-scoop, and the firehole mouthpiece that fell into the fire at the slightest touch. Bill Jeynes had plenty to say, too, about cold air on his tubeplates, so we devised a method of fixing the mouthpieces and altering the set of the flame-scoop. I could stand in my office and look across the front of the shed and if I saw a WD 2-8-0 about to go out without a flame-scoop in place, I would dive out of the front door and up on to the footplate to remonstrate (to put it mildly) with the crew. Maybe they cursed me, but we got those old "carts" to steam, to do their work, and to run to time, and we eased the terrible leaking of the tubes that was a constant worry to all concerned. And provided we took the "jimmies" out when the engines went to the works for overhaul, everybody was happy. Certainly we reduced the number of discarded flame-scoops dumped at the far end of the triangle and in the station yard, and *that* in itself was a major triumph.

There was no doubt that we had several very poor engines. The "Green Arrow" 2-6-2s were mostly at their worst, badly off the beat, with regulators blowing through so badly as to make them exceedingly difficult to control within the depot. They would start a medium weight train quite easily with the regulator closed; while they would lose two out of the six beats when pulled up to 50% cut-off after leaving the station. However, we set to work on the 2:1 valve gear and eventually got some semblance of regularity of exhaust. Many of our drivers were full regulator men, but by no means all. I remember having a go at one of them who had emptied his tender on a V2 and lost time into the bargain. A turn-and-a-half up and the first valve of the regulator was fine for an old GC Class O4 2-8-0, but was nothing short of murder for the fireman on a "Green Arrow".

No description of any GC depot would be complete without a word on the "Pom-Pom" goods locomotives, the Class J11 0-6-0s. We had several, some with slide-valves, some with piston-valves. They could work anything – a stopping passenger train to Marylebone, the Banbury "motor", and those seemingly endless trains of empties from Quainton Road to Woodford. All bar No 64438. When I came to Woodford, she stood dead, rusty and filthy at the back of the shed near the sheer-legs. On the availability return, she was shown as "N" – that is, available as a "spare". After any amount of procrastination by Joe Goode, who knew what was coming, I had her lit up and oiled for gentle work around the yard. Soon I learnt my lesson, for everything blew that could blow, and, in addition, she had a broken axlebox, a hole in the smokebox and goodness knows what else. She had also about 8,000 miles still to do before going to Gorton Works for an overhaul. To repair her would involve breaking into the targets Joe Goode and I had set ourselves with the more important engines, but to leave her "N" indefinitely was unthinkable, and would get us nowhere.

So I had to resort to one of the dodges which certainly used to help us in steam days to improve a situation by bending the locomotive availability statement which had to be sent to our betters each week. Seven miles an hour, twenty-four hours a day, and seven days a week on the shed driver's ticket, as the shed "mike", we soon got the figures going the right way on the accountant's mileage records and our little old engine was eventually dragged to Gorton and emerged ready for anything. I think that one might call this a "management decision". It was, of course, an outrageous fiddle, but it served everybody's purpose and we got a useful engine reasonably quickly without breaking into our self-imposed repair tasks on the V2s, Bus, WDs, and so on. In those days, figures could be "bent" in a way which would be quite impossible – or unnecessary – with the closely-controlled diesel or electric locomotives of today.

Woodford had many outstanding characters, although there were few more so than "Pom" French, our Running Foreman, a great practical railwayman and a local sportsman. Dedicated to the job, he had a rugged, kind face, a stalwart frame, a tremendous voice and at times a ferocious temper. He could not have been mean or petty even if he had tried. Personally he was a great encouragement. "Ah, you'll be all right, Sir, old head on young shoulders", he would remark at times when things seemed to be going constantly wrong. It was said that he had got his nickname as a boy when he said to a friend, "You be a Tiny, I'll be a Pom-Pom".

His methods in dealing with people who were late for work were robust to say the least. On one occasion, when he was at a football match at Northampton one Saturday afternoon, he saw a Woodford fireman in the crowd. Looking at his watch, he realised that the young man should be at work very shortly. So he elbowed his way through the press and with a "Now, my lad, you're on at 5.20 tonight, get off out of it", sent the erring fireman packing forthwith and no argument about it! Dear old "Pom"; he was Woodford through and through. When he reached 65, he retired from the grade of Running Foreman, and started again as a passenger guard at Woodford station. In 1961, my father-in-law, who lived at Amersham, was seriously ill and I used to slip down once a week on the 5.0pm or 6.10pm from Marylebone, both trains worked by Woodford men; sometimes our guard was "Pom", and often I used to do the driving. He stood no nonsense from passengers, station staff or anybody else, and one night he reprimanded a rather dignified passenger who cut it too fine with "Now, come along, Sir, this won't do at all. You'll have to do better than that tomorrow. My driver can't wait and time is time".

Autumn was nearly over and the leaves were so much pulp on the rails as one evening, with the 5.0pm down and driver Harry Holton, the chicken farmer from Culworth, I started away from Chorley Wood with a Standard Class 5 4-6-0. We got into the cutting three-quarters-of-a-mile north of the station, and I had the reversing wheel in my hand, moving it slightly towards midgear as she gradually accelerated up the 1 in 105 bank. Suddenly the wheel was wrenched out of my hand, and the engine slipped like a mad thing. Nothing would stop it, certainly not a closed regulator, and as she slipped she caught her water; so wild had she become that I felt that something terrible was about to happen. It was all over in a few seconds, but they seemed like an eternity. The speed had dropped by 10mph or so, and after that we picked our way very gently through the woods to Chalfont lest it should happen again. With relish "Pom" booked three minutes against me, and followed this with a severe lecture when we got to Amersham.

It was a great advantage to me to have lived for so long in earlier years on the Great Central for when I went to Woodford I knew the road intimately, signals, gradients, and most of the landmarks which you needed with left-hand drive engines when stopping at island platforms in the dark. This knowledge was also to my advantage in dealing with the representatives of the footplatemen. Once more it was a delight to work closely with men who had the interests of the depot at heart, who could see both sides of a problem, but who were determined to fight for what they felt was right and just. Such were George Wootton and Charlie Sanders, who had had years of trade union work and experience; they helped me, advised me over problems which looked like developing and causing trouble, and never let me down. And in return I did everything in my power to improve the lot of the footplatemen at the depot. Our joint fight to overcome the tool shortage by means of allocating equipment to each locomotive, numbered and stamped, could never have been successful without the drive of the toolman and the stores folk, and the co-operation and drive of the accredited representatives of the staff and the men they represented.

The peace and tranquillity of Albert West and Dick Porter, the two drivers who shared the polished-up "Pom-Pom" No 64375 on the Banbury motor, was rudely shattered by the arrival of an N2 0-6-2 tank engine, No 69560. This machine, one of a standard type designed for Great Northern suburban services, had started life after the grouping, and I understand had spent some of her life at Stratford on the GE Section. She was universally detested wherever she went. Thick fire, thin fire, blower on, door shut tight, lumpy coal, small coal, nothing would induce her to steam, not even a razor. Everybody tried something, but as she was unbelievably light on water, she could run to Culworth Junction without the injector, and then on to Chalcombe and Banbury with the feed on when the regulator was closed. So everybody muddled through with the maximum of discomfort. If the fireman had the chance to sit down, it was on a piece of board from which position he could neither see out nor move without getting tangled up with the hand-brake.

Some men are born breakdown foremen. I was not in that category, and although I used to go out with the crane, I lent heavily on the breakdown fitter, particularly in my early days. One evening a freight train was being put across the road at Helmdon, and the signalman altered the points when the train was still moving across. The result, of course, was a pile of wagons and debris in all directions. I was called out and on arriving at the shed found that Ted Preece was ill and could not come. My heart was in my mouth, and to this day I shall never forget my nervousness and indecision as to how that crane should be handled. The breakdown men carried me through, as they always will with a greenhorn, but I felt woefully incompetent as we crawled home, dead tired, for breakfast. As my first job of the day this had to be followed by the task of explaining to my District Officer why we had taken so long to clear the main line.

Our driver on the breakdown train that night was another well-known Woodford character, Norman Hines. Norman was a man of some precision of speech, and he was also a bad timekeeper when booked to work on Saturday nights. From time to time he had to be dealt with over this, and until at last he realised that I was in earnest, his excuse for his absence was always "various anomalies arising"! We had several unusual incidents with which to deal. Maybe we were lucky one night when some wagons got away from Quainton Road on the falling gradient towards Aylesbury, and an Aylesbury driver with a Class N5 0-6-2 tank and some empty coaches ran into them and scattered them all over the place. It was dark and foggy, and the down "South Yorkshireman" was not far away, but very prompt action by the train crew in protecting the down line avoided what would have been a very serious accident. I was lucky too one night when we had been trying out one of our V2s on the heavy fitted freight which left London about 10.30pm. We came into the goods road at Woodford slowly and steadily, and could see no red lights ahead. Then suddenly out of the darkness ahead we saw the back of an unlit brake van, and – bang! Luckily our driver had seen the van just before I did and in the nick of time, so our brakes were on, and we were prepared for the awful lurch which I had first experienced in my apprenticeship days when we ran into the back of a Great Northern Atlantic standing in the western platform in the blackout at Doncaster.

I have said that my wife and I settled in the village of Eydon and in no time we were part of the village life. It must have been in the following January – we had only been married ten months – when a member of the depot staff who lived in a neighbouring village came to see us to ask our advice on a confidential problem that was affecting his domestic life. We did our best to help him, for was I not the Shedmaster? But we have often laughed since at the thought of us two children looking wise and helpful, and yet wondering whether or not we were doing the right thing.

Every so often the Bicester Foxhounds used to meet at Eydon Hall or Culworth, or somewhere round about. One day just before Christmas, hounds got on to the railway somewhere between Eydon Road and Chalcombe Road on the Banbury branch just as Frank Brough and his mate came up with a heavy coal train. Frank saw them in time, and "dropping his anchors" knocked the poor old guard from one end of the brake to the other, but succeeded in stopping just short of the pack! The Secretary of the Bicester, Mrs Johnston, who lived at Culworth, hunted until she was over 75 and knew many of our men; so she sent this crew a turkey apiece for Christmas and a letter of appreciation.

And it was Frank Brough who was on duty as Acting Running Foreman when I received word that I was to transfer to Ipswich as a relief Shedmaster. This deeply upset me, for my pride in Woodford Depot dominated my ambition for the time being, whilst at home we now had our daughter, Anthea, to look after as well. But there it was; as described in Chapter 11, I had to go. In that last week, we tried out B17 No 61664, fresh from the shops, on the 6.20am to Marylebone, with full regulator and short cut-off working throughout. This was much to the interest and pleasure of the fireman, Ron Alder, who was an active local trade unionist and, although he did not know it then, was destined to become an inspector and later a depot manager. I had tried to introduce the L. P. Parker doctrine of engine handling wherever possible and practicable, and here was yet another convert. There was no doubt that the short cut-off method of working, sensibly and wisely executed, saved water, coal, and the fireman's back, whilst the near mid-gear position for coasting with a piston-valve engine improved the riding out of all recognition. The firemen of the day were to become the drivers and inspectors of the morrow.

This same 6.20 from Woodford was booked to reach Amersham at 7.57. If you got moving quickly from Great Missenden, accelerating down the bank to Mantles Wood intermediate, and then letting the engine out up the gradient through the beech woods to Amersham, you could arrive there a bit earlier still. A blast on the whistle short of the platform and a very gentle stop, and you could then see the City gents, some of whom I knew, come running down Hill Avenue for all they were worth. Still, it kept them fit!

And now the villages of Woodford and Hinton have no railway nearer than Banbury, 11 miles away. Old railwaymen may be seen to this day working in gardens, allotments and farms, still in their blue overalls or reefer jackets. They still think and talk of the profession to which they were so devoted. The Great Central main line had to go, for it had no place in the scheme of things today, but I am thankful that I spent those hard but happy months grappling with the day-to-day problems of a fascinating Great Central job. A week before the end I had a last journey from Marylebone to Woodford and back "in the chair", with my hand on the regulator, over a road that now had so many associations for me. During the few hours that I had at Woodford that day, too, I was able to see "Pom" French, by now retired, and my old sparring partners of the LDC, George Wootton and Charlie Sanders, also retired, who had come up to the station specially to see me and to talk over both old times and the latest news.

XIII

Ipswich and the Great Eastern

MY ARRIVAL at Ipswich in March 1950 was a fortnight after a serious accident at Witham, when the up Peterborough Mail, headed by B1 4-6-0 No 61057, had collided at 60mph in thick fog with the rear of a Class A freight train from Whitemoor, just as the latter had slowed down to enter a loop and get clear. The 4-6-0 was so severely damaged that it never ran again, but while the young fireman was killed, the driver made a remarkable recovery and was back at work again a year later.

I had been told that I was moving to Ipswich merely as a relief man, but no more than three days after my arrival I was delighted to hear officially that this was to be a permanent appointment. My District Officer, Colin Morris, was off sick at the time, and when he returned was astonished to hear what had happened. But right from the start he was a source of encouragement to me, and we gradually developed a friendship that was well and truly cemented when years later we worked together at King's Cross.

So here I was, back on the Great Eastern Section, and at the age of 26 in charge of a sizeable and complex depot with over 90 locomotives and some 450 men under me. I had heard that Ipswich men were inclined to be difficult, and what I saw and heard at the start made me feel that I might have done better to stay at Woodford, where there was a discipline that seemed to be lacking at Ipswich. Indeed, my first impression was that foremen and chargemen on the running side had little authority and that there was a great deal of slackness. Soon I discovered that this was not the case. I think that I was influenced by the fact that my office was certainly not the inner sanctum that it had been at Woodford; in fact, the front door opened on to a forecourt facing the main line, and seemed to be used just by anyone who wanted to air a grievance. I may have been very young, but when someone came in, perhaps quite a lot older than myself, and sat down – even when I was not there – lit up a cigarette and generally made himself comfortable, I had to take a very firm stand. Not surprisingly, I imagined at first that the same attitude persisted throughout the depot, but I soon found that it was confined to a few who before very long changed their ways. So eventually I settled down to enjoy my job among some of the finest and most conscientious railwaymen with whom it had been my privilege to work.

Right from the start I realised that we had some remarkably able artisans, and that many of the locomotives were maintained in fine fettle from one visit to the shops to the next. When one was lucky enough to have a large staff of apprentice-trained craftsmen, some of whom had experience with the well-known Ipswich engineering firm of Ransomes & Rapier, and all of them taking a real pride in their work, then one indeed could get things done. So, over the next two-and-a-quarter years, I was able to add my contribution to the splendid standards established by my predecessors.

The actual number of locomotives allocated to Ipswich depot in April, 1950, was 91, which included those at the various out-stations, Stowmarket, Felixstowe, Aldeburgh, Framlingham and Laxfield. There were also the yard pilots and the "tram" engines which worked over the dock lines. Luckily these did not all come to the home shed during the week, or we might have been badly overcrowded.

For though the work at the old Ipswich depot was of a uniformly high standard, it was carried out under appalling conditions. The original buildings comprised a running shed, capable of holding four tender locomotives if they were squeezed together, a boiler shop holding one engine only, and a workshop also with room for four engines, with one standing on a drop pit. No repairs of any magnitude could be carried out in the running shed, which was needed continuously for washing out and water-changing. The conditions in which the boiler washers and tulle cleaners worked in winter beggared description, with boilers being emptied and blown down, others being washed out, and steam and water everywhere. One night I stepped through a cloud of steam into the mist outside, and escaped a moving locomotive by a few inches. Never again!

The pits were shallow, and at both ends of the running shed, out in the open, there were some others barely worthy of the name, not to mention some others near Croft Street, which seemed miles away. There was also, of course, the tar road pit, where valve and piston examinations might have to be made in the teeth of the easterly gales that blew straight in from the North Sea. Daily examinations were often carried out on the level – that is, without the protection of being under the locomotive safely in a pit – by the three dedicated chargemen fitters, Russell Gooderham, Wilf Brown and Syd Lincoln, taking the odd chance born out of years of experience. By the same token, drivers of inside-cylinder engines sometimes had to do their preparation on all fours, squeezing underneath, or again over the top of the valve gear.

With good lumpy coal, the coaling was largely done by hand; but the defect of this method was that at rush periods tenders might be coaled with nothing but big lumps, which would make life very rough for the fireman on the next trip, for he would have to spend much of his time breaking the lumps up. But if there was time at the depot to do the job properly, by "cracking and stacking", the hand-coaled tender became a *beau ideal* for a fireman. To help the coalmen we had two diggers, one a Ruston Bucyrus and the other a Fordson, both of which had seen much better days, though their assistance was much appreciated when it was available. But their success depended on reliability and on the off-loading by hand of mountains of coal on to the ground in a part of the yard known as "Down the Digger"; if the depot was short of coal in wagons, the digger was of little value.

As a result of constant supervision and attention to detail, and the adoption of practical suggestions put up by the men who had to do the work, the cleanliness of the yard was gradually improved. Because of congestion no ash wagon could stay put in the same position even for as much as an hour, and as fires often had to be dropped and ashpans cleared all over the yard instead of over the one ashpit which could take no more than two locomotives at a time, the ash-gang and their engines were very much part of the Ipswich scene. As for the locomotives, they really were *clean.* In the wintertime we sometimes had 22 or 23 cleaners on a shift, so all the passenger engines could be properly cleaned every day, the goods engines every two or three days, and some of the pilots once a week.

The lot of the cleaners was anything but a happy one. They had no mess-room worth the name and their working conditions were as bad as they could be. Many an engine had to be cleaned out in the open in all weathers, perhaps in the teeth of the east wind, and to be cleaned properly – paint work, framings, springs, motion, brasswork, the lot. Some of these cleaners were very young; others, in their twenties, were waiting to climb the long ladder of promotion. It was a hard life, and no sinecure either for their chargemen, who not only had these young men to deal with, but also to put up with any complaint that some cleaning had been neglected. Nevertheless many of these cleaners will remember with pride their contribution to Ipswich efficiency, for our locomotives could stand comparison with any others in the country, and some of them right up to pre-war Ipswich standards.

Because washing out was carried on continuously, 24 hours a day, and because ashes were regularly getting into the drains, we had our moments when the pits got flooded. When this happened, work would come to a standstill under the locomotives in the running shed, and Jack Finch, *factotum par excellence*, would be sent for. Wearing a determined expression, *plus* an old officer's tunic, overall trousers and bicycle clips, he would get his rods, disappear into the sumps and work his magic – a closely guarded secret. Whereupon the water would go down and work would be resumed.

The depot offices were not too bad. Ipswich had been a separate District up to 1937, and the Chief Clerk sat in an impressive little sanctum amidst a sea of papers and a cloud of cigarette smoke; his helpers enjoyed some degree of comfort in a sizeable adjacent office, as also his List Clerk and an assistant, with ready access to me. And next door to me were the Mechanical Foreman, the Leading Fitter and the Chargeman Boilermaker Charlie Winney. The last-named was the son of a former driver, and was conscientious to the last degree over his boilers. "Bloody blazing hot" he would ejaculate, throwing his pencil on my desk in desperation after the shed driver had dumped his next load of engines into the shed amid clouds of steam and with hot water pouring in all directions. Over against the side of the Running Shed was a dreadful enginemen's messroom, and also the Running Foreman's office, the centre of the web.

Many of our engines remain clearly in my mind, and will continue to do so for a long time to come. When engines are associated with particular men, both become as one. It is a delight to foster this kind of union, and I went to great lengths to see that crews got their own engines, and that, if humanly possible, these engines had their shed days and examinations on their crews' rest days. In April, 1950, our B1 4-6-0s were Nos 61053-59, 61201, and 61252-54, 11 in all; 61054 and 61056 were the "spares", and each of the others was shared by two crews. Our B17 4-6-0s were all spare engines, and covered for the B1s, B12s and any special workings; Nos 61600, 61668 and 61669 had the higher pressure Type 100A 225lb boilers, while some of the others were similarly fitted during the next two years.

The ex-Great Eastern B12 4-6-0s were marvellous machines provided that they were well looked after, and here was one of the great advantages of regular manning. Their valve rings were sensitive to wear, and when wear was increasing their performance would fall off quickly, but invariably their drivers would give the mechanical people the tip when such symptoms began to show. For a certain reason mentioned later No 61535 was our pride and joy, but any such claim would call for ribald comment from the crews of Nos 61561, 61562, 61564, 61566, 61569 and 61570; No 61577, another spare, made up the total of eight of this class. We had two "Claud Hamilton" 4-4-0s, 62526 and 62590, the latter a "Super-Claud" with the large Belpaire boiler. For freight we had J39 0-6-0s (the best of which were kept for the 7.45pm to Whitemoor, a fast freight service), several J15 0-6-0s fitted with the Westinghouse brake and both J15 and J17 0-6-0s with steam brakes, a few N7 ex-Great Eastern 0-6-2 tanks and three C14 4-4-2 tanks from the Great Central, and finally the six-wheeled "tram" engines for the docks. with their completely enclosed motions, remarkable little engines which hauled tremendous loads round the sharp curves of the dock lines to the detriment of their main steam-pipe joints.

Some of the B12s had been shedded at Ipswich for many years, so that they had been handed over from one generation to another, and from their original condition to their rebuilt state after attention by Thompson at Stratford Works. In their early days the "1500s", as they were always known, had not been easy engines to maintain, but even then, when they were bearing the brunt of the express passenger work between London and Ipswich, Norwich and Yarmouth, often with very heavy trains, they had been capable of great work. From Ipswich they worked the lodging-away turn with the "North Country Continental" through to Manchester and back. This turn required craftsmen in the true sense of the word, and men who worshipped their engines; their firemen were enjoined not to waste an ounce of coal, and to avoid all blowing off from the safety-valves and smoke from the chimneys.

This generation of firemen had become the B1 and B12 drivers in my day, and they had been well trained.

One day in 1927 No 8561 had worked through from Ipswich to Liverpool with 13 heavily-laden Great Eastern coaches conveying 340 Canadians and their baggage. The driver, Jack Packe, had not yet learned the road to Manchester, and so took a pilotman from Lincoln, but remained in control of the regulator throughout. His fireman was Frank Cocksedge, who, in later years tended No 61569 with loving care and later B1 No 61059, and he made some notes on the run; he had never before been beyond Lincoln, so the road was strange to him. The quotation that follows is a remarkable illustration of the interest that a fireman in steam days might take in his work:

"Belpaire firebox, no drop grate, 180lb/sq in boiler pressure. Engine went through on one fire, pricker only being used at Lincoln over the firebars. Big-ends filled up there also. Engine completed trip quietly with no blowing off, but time was kept correctly. Engine took 13 bogies up from Sheffield to Dunford unassisted, but I must say she was beginning to feel the effects of miles covered, and I had to ease the injector a time or two to give the driver the maximum pressure on the bank. At no stage was I in any difficulty, but strict attention was paid to the amount in the firebox at any one time. With a bright minimum fire, one could make a stop with the engine perfectly quiet or continue with the speedy application of a dozen or so shovelsful of coal. My opinion of 8561 and the driver – superb!"

It only remains to add that whereas the run from Ipswich as far as Lincoln and Retford for the most part was over dead level track, from Sheffield up to Dunford, at the mouth of Woodhead Tunnel, the engine had to face a climb utterly unlike anything on the Great Eastern line, with 18½ miles right off at between 1 in 120 and 1 in 132, and on this particular trip nearly at the end of a continuous journey of 242½ miles.

The unrebuilt B12s were good, including those which had been fitted with Lentz poppet-valves, which ran with a lovely rhythm, but I am convinced that mechanically the Gresley rebuilds, classed as B12/3, were better still. For the enginemen they were every bit as good, clean, comfortable, economical and fast. It was always found advisable to attend to the valves after the engine had run 15,000 to 18,000 miles, in order to avoid any blowing through, and it was essential to take the big-ends down after 10,000 miles to deal with any knock or wear that might be developing. One had to watch for any knock or wear at the small ends too, but apart from this, and from frequent scrutiny of the oil pipe-lines to the axleboxes, maintenance of the B12s was quite straightforward. When well set up, the riding of these engines was exemplary, and we were able to keep the wedges carefully adjusted from one shopping to the next.

In Chapter 7 I wrote about the B17s which plagued us at Woodford, and particularly about the degenerate No 61647. But these engines were not without their good points. After we had succeeded in wresting the turn from Stratford in 1952, No 61668 headed our first "Easterling" from Liverpool Street to Yarmouth, non-stop over the 109 miles to Beccles; perfectly groomed for the job, she was worked by Alf Alderton (whose own B1, No 61055, was stopped at the time), and arrived on time after regaining 'min lost by signal checks.

And a little while after I left Ipswich, No 61669 was entrusted for a few months to Drivers Cocksedge and Calver, and thanks to the most painstaking efforts by themselves and the repair staff, a locomotive that had been a bit difficult became one that, at 65,000 miles, rode like a coach and ran like a deer with a beautifully smooth rhythm. How was it done? By the most careful adjustment of front end lubrication, highly skilled manipulation of the wedges by Leading Fitter Jack Percy, the use of the driest possible steam, fairly low water levels, and the greatest care in coasting with the throttle kept very slightly open until the locomotive had all but come to a stand. It could be done – but it took some doing!

Up to the arrival of the "Britannia" Pacifics we shared the cream of the Colchester main line work with Norwich and Stratford depots, but lost a proportion of it when the through London-Norwich workings were introduced. However, we did not do too badly. In addition to main line passenger turns we had the Felixstowe branch service, by now worked by some Thompson L1 2-6-4 tanks that had just reached the depot; some of the Cambridge turns, and a great deal of main line and local freight, including two bonus workings, one to Aldeburgh and the other to Snape. The Aldeburgh was a J17 0-6-0 job, while the Snape working was for a J15 which was thrashed day after day, week after week, until it could be thrashed no more and had to be replaced by another. The sooner the job was done, the more those involved got paid and the sooner they got home, which may not have suited the engines though it did suit everybody else other than the artisans.

Though in course of time I found that discipline was not all that it might be, I found also that many of the staff preferred that discipline should be firmly and fairly administered, particularly in the matter of timekeeping. Most of the enginemen needed little encouragement in this direction, though there must always be the type who knows exactly how to run a freight train in such a way as to lose enough time to lose the train's path also, but not the pay packet on Friday! This sort of thing had to be dealt with, but it took some pinning down, for the time lost was very carefully organised.

My earlier efforts tended to be rather more drastic than prudence might have dictated, and, as so often happens, those who normally give no trouble got caught first, much to their wrath and the local manager's embarrassment. But we were there to run trains to time, and to operate everything in the way required by our lords and masters, the operators, even at a moment's notice. This could be done with an adequate and well-kept pool of locomotives, and with our large establishment of train crews. In my two-and-a-quarter years at Ipswich we never had to turn down a single request from the Norwich Control; we ran everything we could. It was good for business, we felt, in more ways than one.

So much for the general scene. We learned how to get a day's work for a day's pay, to inculcate pride in the job, to engender enthusiasm and interest, and at the same time to maintain a just standard of discipline, even in the very poor working conditions of the depot. By spending a great deal of time on good and painstaking maintenance we sought to eliminate failures and delays; we went to great lengths to keep the regular men's engines working on the fastest trains right up to the day when they went to the main works for repairs; but we had no more idea than the man in the moon what it all cost, and whether some of the decisions which we so blithely took were justifiable financially, for we had no budget and no targets.

In two-and-a-quarter years, at a depot the moderate size of Ipswich, one gets to know the men fairly intimately. In the shortest possible time I sought to know as much as I could about every member of the staff, down to the labourers and cleaners. I was vitally concerned with their ability to do their work, with their behaviour, with their good and not so good points, whether they could be relied on in hard and difficult circumstances – indeed, everything that influenced the performance of the depot as a whole. One must never forget that *every* man has something to contribute, and that the old boy pushing a barrow and sweeping up the shed has as much right to attention as the most skilled man on the premises. One's job revolved as much round the men as round the engines – more so, indeed – and one ignored that fact at one's peril.

At one time we had been having trouble with the steaming of 4-6-0 No 61600 *Sandringham*, and I had arranged for the normal working to be broken down for half-a-day to enable this engine to work a heavy train between Ipswich and Norwich to see if we had solved the problem. The previous evening I had been going through the engine list for the day in question, with one of the Running Foremen, Bill Thurlow, who had done his firing on the "1500s" while they were still in their blue Great Eastern livery. He and his driver had had their cab painted white inside, and his remark that they didn't do that sort of thing these days had set me thinking.

Having seen all I needed of *Sandringham's* performance in about half-an-hour I got down at Diss, and waited for the 1.0pm "slow" from Norwich to get me back to Ipswich; soon it arrived with B22 No 61535 at the head of four coaches. Now Jim Calver was the driver on this engine and with his opposite number, Charlie Parr, and their firemen they had got their cab into a beautiful state of polished steel, copper and brass, without a speck of dirt anywhere; the floor was scrubbed clean, the fireman's wooden seat was polished and bright and covered with linoleum, and there was even what looked like a gold chain for the whistle. Perfect! The only exception to this attractive scene was the black cab roof, though even that was tallowed and clean.

So we set off gently from Diss with our four coaches, the old engine riding smoothly, and as we gathered speed I asked Jim how he would like to have a white or cream cab interior for his 61535. Need I say what his answer was? And so next day she was in the shed and our painter got to work, making a very nice job of it. Thus it was that old '35 became the first engine in recent years to have a cream cab interior, with chocolate sides. Each day the hood was cleaned, polished and tallowed, and she looked simply lovely. A small thing, you may say? The answer is that the more you can give a man to cherish, the more reward you will obtain.

Not surprisingly, No 61535 became famous, and started a fashion. The crews of No 61566 asked for a cream cab, but Charlie Joywheel on No 61253 would have none of that "cissy stuff"; his cab was painted royal blue, and he added a touch of distinction by scouring the brass beadings above and below the cab side windows. Even the Mid-Suffolk branch engine, J15 0-6-0 No 65447, had a white cab and every stay-head on the firebox front burnished. It was just like the old days.

Charlie Parr was a little white-haired man with a stumping determined walk; he smoked a very small pipe, and looked very grey in the face if he was given anything but No 61535 to drive. One of the toughest jobs for the "1500" gang was the down Mail from Liverpool Street on a Sunday night, which could have a load up to 12 bogies with a tight timing. Charlie lived in Croft Street, just outside the depot gate, and when he was booked up to London with the 5.25pm and down with the Mail he was always in the Foreman's office by lunch time. "Have I got old '35 tonight, Foreman?" he would ask.

"Yes, Charlie, she's in the shed and she's been washed out", would be the reply, "so you can go and have your dinner". And off Charlie would stump, contented in mind, but heaven help us if he had a spare engine, or, worse still, a "foreigner". This did happen once when he had to work a Stratford K3 2-6-0 up on the 5.5pm to London; that evening he was *very* grey.

One day in the autumn of 1951 E. S. Cox, who was Assistant Chief Mechanical Engineer (Design) to the Railway Executive, travelled down to Ipswich on 4-6-2 No 70000 *Britannia,* then brand new, with my old friend Chief Locomotive Inspector Len Theobald. I was instructed that they would

104

return on the 1.55pm from Ipswich, and that this was to be worked by a "1500". Our luck was in, for the 1.55 that day was booked to be worked by old '35, cream hood and all. The previous day we had her in the shed, scoured her up and cleaned her to perfection, while our Leading Fitter Jack Percy saw to it that her wedges were set to perfection also. When Cox and Theobald climbed into the cab at the station, the veteran was a picture indeed, for as if by arrangement the sun was shining brilliantly.

I stayed at the shed, for I wanted to see Jim Calver and his mate go by, and I was joined by a big crowd of the shed staff, who knew who was aboard, and wanted to see their engine "do her stuff". Well, dead on time she went by, steadily gathering speed, not a knock, or a wisp of steam from anywhere but the chimney, just the merest trace of smoke – the steam locomotive at its best. So she ran up to London with a smooth noiseless rhythm, finishing quietly up to the buffers at Liverpool Street with the reverser near the middle, not a tinkle from the firmly closed snifting valve and not a knock down below, dead on time. Our Governor, L. P. Parker, was on the platform to meet the distinguished guest, and he was pleased and said so. That really *did* mean something! No 61535 was not the only "1500" of ours to hit the headlines; they all did consistently good work and now and again turned out something special.

Occasionally, however, it was the very opposite, as on a dreadful trip that I had in January 1952 on the down Mail one bitterly cold Sunday night. The driver was Bob Riches, and the load about 400 tons. Soon after starting it became obvious that all was not well; the engine had a defective superheater element and by the time we passed Romford the pressure was barely 130lb. Bob had the reasonable idea that the harder you climb a hill the sooner you get the top, and as always put it into practice up Brentwood Bank, which meant that by Gidea Park I had to get to work on the shovel myself.

To cut a long story short, we managed to get the engine and train to Ipswich on time, with one stop at Colchester, on its sharp timing, and with never more water showing in the glass than one-third and never more than 150lb of steam. Long before we finished the run the right-hand injector had frozen, the water handle could not be moved, and I was soaked to the skin and just about all in. To cap it all, my car at Ipswich had frozen to the road, and could only be moved after a struggle; and although my shirt was far from frozen, it also would only move after a struggle when I got home! But, rough trip though it had been, the job had been done and we had arrived on time.

No 61561 went through a bad patch late in 1951; she would not steam, and even in the hands of the regular men began to lose time on the London turns. We tried all the conventional remedies but without success; the only conclusion to which we could come was that the blastpipe was not in line with the chimney. But we were wrong. One morning I tried my own hand as far as Colchester on the 7.0am up, but without success. Actually there was one last resort which we had not tried, and it was Jack Percy who suggested "back ports made up", a rare occurrence and a rotten job to tackle. But a few days later No 61561 was again on the 7.0am, bounding up the bank to Belstead with a glorious sharp beat, blowing off against the feed with the firehole door open, ready for anything. Which shows that the voice of experience should always be listened to.

No 62526 was one of our two "Claud Hamilton" 4-4-0s, a slide-valve rebuild which, like many of her breed, needed very careful handling. She, too, went through a bad patch; even when we ringed the blastpipe there was no marked improvement. One day Frank Burrows took her out on the fast and light 8.28am flier to Norwich and showed that she would steam if thrashed, but if she were handled normally the pressure would drop back to 130lb and stay there. A little later on we thought that we had found the answer, and I had her put on the 1.46pm from Ipswich to Liverpool Street, a Saturdays only train taking about 88min with a Colchester stop and with a load of ten bogies, a fair assignment. I decided to see what she was made of, and joined the train some ten minutes before the start.

Now at that week-end we were very short of coal at the depot; there was nothing in wagons other than stuff too small for the digger and the stock on the ground was very low. In his enthusiasm to pick up the last ounce of coal the digger driver had dug into the ground, and loaded us with a lot of smokebox ash and chalk; then when he realised what he had done he had covered this layer with coal and kept his mouth shut. We started off in great style, and I really felt that we had reached the bottom of 62526's trouble when I got down to the smokebox ash and chalk. After Colchester we had a dreadful time. We rolled through Chelmsford at about 30mph, and started to tackle the long rise to Ingrave.

Reg Crisp was the driver, and he knew – and I learned for the first time – that there are two short stretches on this climb which are actually downhill. In each of these dips he shut off, in order to make a little steam, and with expert handling we were able to creep over Ingrave summit without stopping, but with no more than 80lb of steam and the water playing "Bobby Bingo". Down Brentwood bank we rolled with the regulator shut, and by Romford had made sufficient steam to get us into Liverpool Street. After that, I decided that No 62526 could stay on four-coach trains to Cambridge – no more experimenting for me!

Long after I had left Ipswich for the south I made my last trip on a "1500", in the dark on a fast train from Ipswich to London. The engine, No 61566, had once been the pride of Fred Thorpe and Maurice Hood, when it sported a cream cab like 61535. Now she had a week or two before being withdrawn; she was rough and ate coal at an appalling rate; and her big and small ends thumped an endless war dance underneath her dirty old boiler. The driver, Archie Rowe, was an old countryman from Chelmondiston, for which reason he had acquired the nickname "Speed the Plough". It was a cold night; we were late; and Archie sat on his seat wearing his old cloth cap and fawn mackintosh, the tails flapping in the wind. Down Brentwood bank we flew Londonwards, with steam on, and we finished up before time. Great engines, these were; they became a way of life to the men who worked on them, and indeed all of us at the depot remember them with affection and respect.

But if the "1500s" were good, so also were the B1 4-6-0s. For most of the year round we were able to clean them every day, and the crews went to endless lengths to look after their charges. No 61059 will always be remembered both at Ipswich and at the London end of the line. In 1951 and 1952 she was manned jointly by Frank Cocksedge and by the same Jim Calver that had been in charge of 61535. They scoured and polished No 61059, scrubbed the footboards in the cab, and handled her with the greatest care. We found that the driving and trailing axleboxes of the B1s generally had to be refitted between shoppings, though some of them would go through from one shopping to the next; but even if they got rough their crews would work them on the fastest turns rather than take out a spare engine.

Frank Cocksedge had once fired for the famous Ipswich driver Arthur Cage, when that great man was finishing his days on the Aldeburgh branch. Cocksedge was a wonderful engineman, light-handed, economical, knowledgeable – a good mate who insisted on the job being done absolutely correctly by his fireman, but also a partner who educated and amused, for he had the vital gift of humour without which work can never be perfect. He had read Cecil J. Allen's "British Locomotive Practice and Performance" articles for many years, and I treasure the memory of leaving the two of them together in my office while they recalled the various epic feats that had been achieved over the years by the men of Ipswich. And not least C. J.'s first footplate ride ever, in 1908, with Driver Cage himself on "Claud Hamilton" 4-4-0 No 1871, non-stop from Liverpool Street over the 130 miles to North Walsham with 13 bogies, and on time!

As Secretary of the LDC I was lucky to have Ernie Payne. In the First World War he had been an officer in France; then in the 1930s he had been a fireman on the Ipswich-Manchester run; in the Second World War he was a Major in the Home Guard. He was an educated man who understood both sides of an argument, and who rightly believed that despite all that was being done, more could be done to improve the conditions, both of depot working and on the locomotives. This was before the days of joint consultation, but we consulted all the same. I was young, and he was experienced, and our partnership, for that it certainly was, warmed me and warms me still when I remember that between us we were able to achieve a fair amount for the good and the efficiency of the depot.

Running Foremen have no small part to play in the running of a steam depot. Their job is no sinecure; they are in the front line; they get the brickbats, but if they are good they in their turn can hand out quite a few. They must know where to find a set of men at a moment's notice, how to set up a shed, how long it takes to steam or wash out an engine in any given circumstances, and, with the artisan staff, how to maintain liaison and understanding as well as law and order, indeed, all the tricks of the trade.

We had six Running Foremen at the depot, with one at the passenger station. It was Bill Thurlow who had spoken to me about the white engine cabs. Youngest of the three top Foremen was a character named Syd Gayfer, a crafty compiler of engine lists for the next day. One morning he greeted me on my arrival at 8.50 with "We've slipped up with remanning the Yarmouth up. The crew are here but not at the station, and the train is at the platform. Please take them in your car, will you, Sir?". So I did, and with a scramble the 9.0 up left at 9 o'clock.

"Stump", as he was known, was a chain smoker. We would be working on the engine list; the 'phone would ring – "You're wanted on the 'phone, Norwich Control, Syd". Down would go the cigarette while he went to the telephone, listened carefully, got out another cigarette, lit it and took a puff, and then put it down also while he made a note on a piece of paper. The conversation over, he would go outside into the lobby, leaving his second cigarette behind, and set to work on the duty list to make some alterations for the morning involving advice to early turn enginemen. This needed the services of yet another cigarette, and so Syd had three cigarettes going at the same time in the same office, and he would manage to smoke them all. His No 2 at this time was Bob Fenning, who had been a fireman during the last war. One night, instead of a J39 0-6-0, he and his driver were given a Stratford J15, No 7544, to work 15 coaches of troops from Liverpool Street to Diss. They *did* get there in the end, and though they never stopped for shortage of steam in any section, they did stop at every water column to fill up the tank and finished up 184 minutes to the bad!

Arthur Rumbelow, another of the Foremen, while still a driver was involved in a remarkable bit of slapstick, the story of which was handed down with relish ever since. Years before, he had been told at the shed to take a J15 0-6-0, to get hooked on to the breakdown vans, which were standing at the outlet at Halifax Junction, and to work them to the passenger station. Action was the order of the day. The engine had just come in and was standing on the ashpit; his mate looked at the pressure gauge, which showed a fair amount of steam; and so they moved out of the depot and hooked up. They were moving nicely towards the tunnel when "Rummy" looked at the pressure gauge again, and with his pipe stuck in the corner of his mouth remarked to his fireman, "I'd have a look at that old fire, mate, if I were you". The firehole door, which had been tightly shut, was immediately opened to reveal – an empty firebox! By this time they were in the tunnel, but they just managed to claw their way into the station, come off their train, and dump themselves in the station yard. Needless to say, some very pertinent questions were asked!

A List Clerk's life is one of battles fought and won or lost. He has the almost impossible job – when it comes to keeping the peace – of rostering enginemen, not only for the basic weekly work, but also for the daily changes brought about by leave, sickness and extra assignments. He has to work to a set of local agreements which are supposed to guide his activities through the perilous channels of the national agreements. In next to no time he can start a conflagration as to who should be where and why and how. It follows that he must be a man of personality, capability and with the hide of a rhinoceros.

When it comes to Sunday work, he can have a particularly difficult time. It was on Sundays that our List Clerk, Ernie Rivers, came into his own. He had been a soldier in his day, and acted as a lunch time waiter at the hotel outside the station, but none of these things had caused him to rate clerical neatness and tidiness as a major accomplishment. On summer Sundays he was hard put to it to find enough men, and an examination of his list after it was posted would often show the mythical name "Smith" or the name of a fireman in the Forces to cover a particular job. This list would go through with so many alterations that it would finish up tied together with stamp edging. But despite these idiosyncrasies people would do anything for old Rivers – even the foreman who had to find the men to cover the non-existent "Smiths". The Sunday list was posted up at noon on the Saturday, and Rivers used to hand it to the foreman ten minutes before midday with the remark "Give me ten minutes to get out of the gate, old mate, before you stick it up". So Ernie made his escape before the bloodhounds got to work and the postmortem on his birthright began.

There was little time for relaxation. I worked seven days a week and on every holiday, even including Christmas Day. Indeed, one was expected to do this, and personally I found Sunday work necessary because a lot was going on at the Shed and in summer on the road as well. It was always my aim to see the depot spotlessly clean by 1 o'clock on Sunday afternoon, and I would not go home nor let the staff finish until this work was finished and all the locomotives in for cleaning were scoured and shining. Then one could leave and forget the place for a few hours. There were times when we had differences of opinion; one Sunday a member of the yard staff got so angry as to let fly with his fist and land me one! But the yard labourers took a pride in keeping their patch up to the mark, which was not always easy, especially when at the busiest times fires might have to be dropped all over the yard.

The Norwich District kept us busy. Life is always good under a fine officer; so it was with Colin Morris, but when he went he was followed by a remarkable little Irishman, Rupert L. Vereker. We were still busy and up to a point life was still good, but when the little man's eyes began to blaze, as they did with alarming frequency, one felt that anything could happen. He had been brought up at Doncaster and spent much of his time as the Boss – quite undisputed – of the Crimpsall. He was a most able engineer, and once you realised that his bark was worse than his bite he was a first-class leader of men who never bore any malice.

On one occasion he was swearing on the telephone about a job that had not been done, and I felt that I had had enough. So I put the telephone back on the receiver. Immediately I realised that my own temper had got the better of me, and the reaction set in. The 'phone rang again, and the voice ordering me to Norwich on the next train was charged with such passion as almost to turn me to pulp. Well, I got to Norwich and found Vereker busy at a meeting, in which I was invited to take part. An hour later I was listening to him telling me that in 1909 he had done just the same thing that I had, but that he was the only man allowed to lose his temper all the same! For his ability and straightness he will not be forgotten, but not everybody realised that he was something more than a firebrand.

The bane of his life was Stratford drivers. One day he travelled in the train from Norwich to Ipswich behind "Britannia" No 70005 and Driver A. E. Jones, a newcomer on the Pacifics but one of the best. Due to shortage of steam they lost a few minutes, and on arrival at Ipswich Rupert marched up to the engine, opened the cab door, mounted the footplate, and looked at the fire. "Your fire's wrong", he said to the driver. "And who d'you think you are?" he was asked. "I'm Vereker", replied the owner of that name, to evoke a reply that became a classic, "Never heard of you. Get off my engine!" Here were the perfect ingredients for a first class explosion, in which the little man, now as red as a turkeycock, played a leading part. I don't know the end of the story, but as likely as not, when order was restored, they shook hands and went their various ways as though nothing had happened.

Until now I have said little about the out-depots attached to Ipswich. By far the most primitive, and that is saying something, was at Laxfield in deepest Suffolk at the far end of the Mid-Suffolk Light Railway, which branched from the Ipswich-Norwich main line at Haughley. This was a real "light railway", with very few signals, level crossing gates opened and closed by the train crews, mixed trains, ancient coaches – the Mid-Victorian atmosphere indeed. Yet when I made my first trip to Laxfield, on the footplate of a J15, it was with George Rowse of Barnsley in Yorkshire, who once had been a driver of the great six-cylinder 2-8-8-2 Garratt that assisted heavy coal trains up the 1 in 40 of Worsborough Bank, and who was now pottering along at 25mph through the fields while I listened to his broad West Riding Yorkshire dialect. The railway is indeed a small world!

When we got to Kenton, the only train on the branch that was in steam, the home signal was at danger and we came to a stand. We whistled at some length, and eventually were called into the station by hand. From a muttered conversation on the platform which I was not supposed to hear, I gathered that this particular signal had been passed at danger for months past, and what the hell did George think he was doing to stop at it!

The Mid-Suffolk line closed down in July, 1952, but other branches, such as those to Framlingham and Aldeburgh, stayed open for some years longer. That to Aldeburgh was worked by an F6 2-4-2 tank engine, either No 67239, the regular, or No 67230. The latter was a poor old thing, and never steamed very freely. Her boiler tubes and Aldeburgh water never got on well together, and at the first sign of leakage old Jack Runnacles, driver-in-charge, would be on the telephone to Ipswich to demand a fresh engine "next time round at Saxmundham". Driver Runnacles had spent much of his life at Aldeburgh, and believed in taking every precaution. It was no mean feat to coax No 67230 over the eight-and-a-quarter miles of the branch even with no more than two coaches, including a stop at Leiston and avoiding an involuntary stop for steam; the fire would be made up at Saxmundham as though they were off to the ends of the earth! I used to think that the difficulties were exaggerated, though I must confess that I had once got a very wet shirt trying to get steam out of this dreadful old engine.

The Felixstowe branch was very different, and in high summer extremely busy. There was no shed at Felixstowe, but an L1 2-6-4 tank, No 67719, was worked from that end on opposite turns by two distinguished old gentlemen who never spoke to one another. L1 tanks were the mainstay of this branch, and a very noisy mainstay too, especially when we first received them from Stratford in 1950 and their rattling and banging could be heard for miles. On Whit-Monday, 1952, we found that we could not cover all the branch workings with L1s, and one diagram was left over – a day's work on the afternoon turn. It was a fine day, so we booked J15 No 65467 for the job.

I always went to Felixstowe about 3 o'clock on a Bank Holiday afternoon to see that everything was ready for the evening rush, and to see Jimmy Meehan, the Stationmaster, with whom we had to work very closely. Normally we kept tender engines off the branch, so I decided to ride on 65467 with Driver E. Heifer. Going down we had six coaches. With only about 120lb of steam, it was a poor sort of journey; I had my fire too thick for this class of engine. After washing my hands I called on the stationmaster, and on looking through the workings I saw to my alarm that the return trip of 65467 would be on a London train with ten corridor coaches; at the shed it had simply been arranged that the one single trip diagram, including this train, would be worked by the tender engine, and this was it!

There was nothing more that could be done than for me to tell the crew that fate had dealt them a severe blow and that they had better "get her warm". I decided that I must go back with them. So we backed on to our train with a white hot fire, full boiler, and blowing off – everything in apple-pie order. Steam work in difficult circumstances often used to bring out the best in men and machines, and so it was on this evening. We started off with the injector on and the firehole door shut; with firing little and often, the injector was shut off at the start of the single line station loops – we were nonstop – and the blower was kept hard on until the regulator was opened again, every trick we could think of to keep the pot boiling. Up to Westerfield she came as though her life depended on keeping time, and then we dropped down to East Suffolk Junction and drew silently and smoothly along the length of the up platform at Ipswich. In my mind's eye I can still see the crowd of London passengers waiting to board the train, and looking incredulously at our little engine with its long stalk above the smokebox, and wondering, no doubt, how and when such a Victorian relic was going to get them back to London! They need not have worried, however; old 65467 and her crew had acquitted themselves honourably and could now clear off to the shed.

The little 37-ton J15s were very handy and very popular. We used the best of them on the Snape bonus trip, the worst for shed turning and shunting and for the "wagon wheels" job. On the last-named it was simply a matter of shunting the carriage shop, but the "shed turner" sometimes had to haul eight or nine dead locomotives down the yard – a seemingly endless procedure. Occasionally on the back roads they had the job of pulling a locomotive that was low in steam after wash-out up and down with the gear set in the opposite direction of travel, in a frantic attempt to get the pressure up for a rostered duty. Strictly this was illegal, and I used to walk away when such a ritual was being performed!

And so, after two-and-a-quarter years at Ipswich, I was on my way again, older and wiser. I had worked very hard, and given all I could to my chosen profession. There had been some difficult times and many differences of opinion; it had been necessary at times to take unpleasant decisions and plenty of mistakes had been made. But I had worked with some of the finest of railwaymen, whose only aim was to do their best; and although it is now nearly twenty years since I left Ipswich – before the depot was completely rebuilt – I still remember the place with great affection and gratitude.

XIV

From Eastern to Southern

F OR MORE THAN two years Ipswich was very much my life, but by May, 1952, I began to think that I ought to be on the way again. Gwenda and I had grown very fond of Ipswich and the East Anglians, for it is a fallacy that Suffolkers are anything but the most friendly of people. But as the years slip by, so do the chances of promotion. I let the King's Cross vacancy pass without applying – a silly thing to do, for the only reason against was that the Cross at that time was supposed to be in bad shape, and the job a very rough one. Then the London Midland Willesden came up, and after that Stewarts Lane of the Southern Region. This place was new to me, but as it was in London I applied for the vacancy. I was then 28 years old, but I looked quite a lot more and, after an interview at Waterloo I found that, much to my surprise, apprehension and delight, I had been appointed.

So in August, 1952, I came to London to see the then District Motive Power Superintendent of the Stewarts Lane District to be given a briefing. I was about to start on one of the most fascinating periods of my life, with just over two-and-a-quarter years of never-ending interest, hard work, long hours, worrying times, arguments and verbal battles, all the delights and miseries of working with some 600 independent and outspoken South Londoners, not to mention men of Kent, a few Devonians, one or two Cornishmen, three Poles, three Singhalese, several Anglo-Indians, not a few Irishmen, Welshmen and Scotsmen, West Indians and at least one Nigerian, Mr O. J. Ojo.

At the outset I had been told that there was a good deal wrong with the internal working of the depot. If one has to set about changing practices that are entrenched deeply, trouble usually follows, and it was of vital importance to have the support of a strong and courageous District Motive Power Superintendent. For such a man it was my privilege to work. G. L. Nicholson was under 40 and had the priceless gift of leadership. Right at the start he told me what he expected from me, and then gave me the chance to get on with it. He never interfered with the detail of my work or the detailed working of the depot, but he was a constant source of encouragement and he stood behind me on the many occasions that I had to take disciplinary action with some erring member of the staff.

He acknowledged a job well done and we soon heard about it if the depot failed to measure up to his standards. He could always see the comic side of things and above all, he was respected by the staff throughout the District, which included not only Stewarts Lane, but also Bricklayers Arms, Hither Green, Gillingham and Faversham depots. He took the closest interest in the workings of the many motormen's signing-on points, which were attached for promotional purposes to the steam depots and which came under the control of the Shedmasters. Thus we had the two Victorias (Eastern and Central), Streatham Hill, Crystal Palace High Level, Beckenham Junction, Holborn Viaduct and Loughborough Junction.

Stewarts Lane Depot was situated behind what was then the Hampton Depository, within a triangle of lines. The southern end of the depot was beneath the South London and West London viaducts, and the north end under the viaduct carrying the South Eastern & Chatham main line towards Wandsworth Road.

In the main shed there were 16 roads. It was dark, ill lit and low in the roof, and it was here that all the heavy repairs, washing out and servicing were carried out: but we also retained one of the large workshops of what had once been the LCDR Longhedge Works in which many of the old "Hooley" engines had been built. The small engines, such as the H Class tanks, the C Class goods and the D1 and E1 passenger engines, could be uncoupled from their tenders and taken on the traverser to one of the bays on the south side, where they could be "rebuilt" slowly but surely by the Chargehand, Vic Smith, and his old fitter, Bert Snowball. The lathes and the various forges were situated on the north side, whilst the Foreman's office looked down the centre of the shop.

Among the big engines we had three "Merchant Navy" Pacifics, Nos 35026, 35027 and 35028, some of the most powerful, the most exciting and the fastest locomotives in the country. Our sixteen "West Country" 4-6-2s, most of which had large tenders, were by no means enough for summer Saturdays and Sundays, though more than enough for winter working. Our two "Britannia" Pacifics, No 70004 *William Shakespeare* and 70014 *Iron Duke,* were cleaned to perfection and worked the "Golden Arrow" and the heavy "Second Arrow". We had a good few "King Arthur" 4-6-0s, including No 30768 *Sir Balin*, which also was kept in beautiful condition; several Class N and Class U1 2-6-0s; one "Schools" 4-4-0; several Class H 0-4-4 tanks; most of the Brighton Class E2 0-6-0 tanks; three strong and useful Class P 0-6-0 tanks; and some London Midland 2-6-4 and Ivatt 2-6-2 tanks. The former SE&CR was represented by half-a-dozen E1 and D1 superheater 4-4-0s, true thoroughbreds, and several Class C 0-6-0s, simple and straightforward, free-steaming and free-running, better by far than any Drummond "700" or Brighton "Vulcan". The total allocation was 126 engines, or thereabouts, and in the summer the entire stud got no rest. We worked trains all over Kent and over much of the Brighton Line, passenger and freight, with empty coaches, shunting work, specials, milk tanks, anything and everything, including Lingfield race specials and in particular the Royal Train to Epsom Downs, always a great occasion.

Stewarts Lane was responsible for the "Golden Arrow" both ways, as well as most of the boat trains and Ramsgate workings, the Hoo Junction Goods (a hard job for a C class 0-6-0) and, of course, the "Terrible 6.10". This last was the 6.10pm from Victoria to Uckfield, a ten-bogie train which a London Midland 2-6-4 tank, not in the best of condition, had to work through the middle of the peak hour electrics to East Croydon, and then up the long 1 in 100 to Woldingham. To get past "Balham in nine", as the Brighton men used to say, wanted some doing. We tended to call the Brighton all sorts of names, mostly unkind, such as the "three-and-a-van" railway, but such jobs as these were hard and needed constant attention. Occasionally, Brighton sent an Atlantic up for the 6.10, and the 4-4-2 had one great advantage over the tank engines in that the driver could run hard into Oxted and save a precious minute or two, whereas the tank engines had to draw up slowly to the water column to snatch a couple of feet of water.

My first week at "The Lane" was muddled and fraught with difficulties, and yet remains crystal clear in my mind. I remember a large shed, ashes everywhere, engines belching smoke, and a never-ending battle to find enough engines to go round. There were one or two put-up jobs to test the stranger, pointed remarks about his youth and the fact that he was "off the Eastern", and a bit of feeling here and there but everybody was too busy to worry too much about the new boss.

The steam locomotive bred a race of men apart and Stewarts Lane – and the Bricklayers Arms in the Old Kent Road – had more than their share of personalities reared in the hard school of pre-grouping and Southern train working. Our Chief Running Foreman, Fred Pankhurst, started as a cleaner at Orpington, became a fireman at Margate, and then a driver and later a foreman at "The Brick" before crossing to Stewarts Lane as the Chief Foreman. Fred could write his own name but nobody could read it. Though he could not communicate on paper, he knew but one code of discipline, and he had strong prejudices against people, particularly if they came off the Brighton or from some of the more out-of-the-way SECR depots. But he knew the engine workings without reference to the diagrams, and he knew the inner workings of other depots, together with the spare engines they had up their sleeves, as if he were connected to them by closed circuit television. He knew the route knowledge of every man at the depot without looking at the route sheet, and he saved countless delays which less resourceful men would have accepted. To hear Fred on the telephone to one of his old colleagues at "The Brick", in search of spare engines, was an education, not only in the art of extreme pungency of expression conducted through a smoke-screen of his favourite weed, but in the marvel of his memory and in his grasp of the practicalities of steam railway work. Under his rough and ready exterior, he was the most loyal and kind of men. We had nearly two-and-a-half years together, with a few cross words but a great and lasting understanding.

If you spoke of "Old Sam", the odds were that people thought you were talking about Sammy Gingell. I would not hesitate to say that Sam was unique, not necessarily as an engine-driver but as a person. To-day he is a porter at Victoria and not so long ago, at the ripe old age of 76, he fired a French Pacific from Abbeville to Boulogne though still suffering from the effects of shingles, the first illness of any consequence of his life. He lives for his family and for railways. Sam's record tells one very little. He did not start on the SE&CR until he was nearly 21, and he was appointed driver within what was then the relatively short period of 22 years. When he retired, his record sheet was unblemished except for a pencil note about making smoke at Holborn Viaduct in 1919. But it told very little about the real Sam, who had once worked down the Abergorky pit in South Wales, who was immensely strong and who in his youth thought nothing of cycling to London from Abergorky after work for a long week-end to save the train fare.

In the whole course of his railway career Sam has never been known to get upset, nor to complain about any instruction given him, so long as it involved work. His love for the job was so passionate that one of his greatest pleasures in life was just to be at work, on the road, going fast, the faster the better, with his head over the side, rain or snow – with any engine. However old and rough they all came the same to Sam, and they were all "marvels". A modest man, shy in many ways, Sam revelled in the great variety of people he got to know over the years. He never appeared to exert his authority on the footplate to any marked degree, yet he always had things going his way. For every lump of coal he sent sky high up the chimney, he was always prepared to repay his fireman, and at Chatham or Faversham, on a heavy Kent Coast train, he would slip over the top of the tender and get coal forward for his fireman while water was being taken, or clean the fire at the end of the day, or even do some firing himself.

In 1959, when I was at Liverpool Street as District Motive Power Superintendent, I had a letter from Sam asking me if I would go with him to the House of Commons. By this time he had retired from the footplate, and as a porter at Victoria he had carried the luggage of a certain Conservative Member of Parliament and his wife. Whilst waiting for the empty coaches to arrive, Sam's career as a driver had come up for discussion and the upshot was an invitation to lunch at the House, to which Sam was invited to bring a friend. Many men would never have taken up the offer, but the old man never hesitated and one day we found ourselves in the Members' Dining Room. During lunch, and without any ceremony of any sort, Sam produced a beautiful little cardboard model, which he rather treasured, of the "King Arthur" that Stewarts Lane made famous, No

30768 *Sir Balin*, and handing it to our host, insisted that he should accept it as a gift. And it was typical of Sammy that he was always giving.

But there were many other people in those early days who stood out, and everything was a new experience to me. I spent the summer evenings learning the various roads, the odd Sunday going to Dover or Ramsgate getting to know a little about the soft Kent coal. On the first occasion I found that I had been "had" like everybody else; the engine was a "King Arthur", No 30766, and up the 1 in 100 of Sole Street Bank I found that I had a box full of fire and a falling steam pressure, while the box was still full at Victoria.

During my first week there was a threatened stoppage of the breakdown gang, and I had my first taste of the Workshop Committee. Whereas at Ipswich matters jogged along fairly well, I was to cross swords many times with the Workshop Committee at Stewarts Lane; yet we had some remarkably good times, and acquired a healthy respect for each other born out of argument and negotiation. Bill Brooks and Gerry MacTague had their job to do, and I knew I was in for trouble when the pair of them were waiting quietly outside the door or equally quietly tracking me down in the shed like a couple of detectives.

At the end of the first week came my first experience of a Southern Summer Saturday of those days. There was only one engine failure throughout the day, but it nearly paralysed Victoria. The engine was No 30765, and the train, the 8.46am to Ramsgate, was right in the middle of the sequence of crowded trains. The "King Arthurs" always tended to prime readily on our sometimes overdosed water and on Grosvenor Road Bank this one all but came to a stand in a cloud of steam, water and soap-suds. After this, neither injector would work and Driver Clasby had to come off his train at Factory Junction, near the depot, and take on the first machine that came to hand.

Later on the same day, I went across to Victoria and watched in the background. Two things stand out in my memory. One was Driver Jack May coming up No 7 platform with an "Arthur" a lot faster than Liverpool Street would have thought proper. We did not meet that day, but he was a fine man, the Chairman of the Local Departmental Committee. His Secretary was Cyril Montague Higgins, and they knew what was right for the depot and its men. They knew all the answers, and were both good enginemen and good Trade Unionists, with whom it was a delight to work. Sometimes we disagreed violently, but yet agreed on the things that really mattered, our desire being to see Stewarts Lane as the No 1 depot in the District, and *never* to let Hither Green or Bricklayers Arms work a boat train, however hard we were pressed and short of men or engines.

The second memory that same afternoon at Victoria was of an H Class tank, No 31005, at the back of a boat train in No 2 platform nearly due to depart. The driver was a young man of 24; he must have seen me about the shed, and I could see him shooting bird-like glances in my direction. He and his mate found me so interesting that the fireman forgot about his injector and overfilled the boiler; on the right-away being given, she caught the water, primed like mad and made everybody take cover. The driver was R. J. White, who did some excellent main line work in his middle twenties; later he became a deputy foreman, and soon made great strides up the ladder. But we often laughed about that boiler!

The weeks went by and I found that, although there was a great deal of fine work being done, it was at a price. Not only was the staff timekeeping generally bad, but absenteeism was serious and was not being dealt with. I knew that this absenteeism could be put right but it was but a "sighter" to the real trouble. One day at the end of September, 1952, I stood in the timekeeper's office talking to Ted Betts, who was the telephone attendant and responsible for booking the shed staff on and off duty. At about 4.20pm up went the shutter and in came 13 brass "checks" and a piece of paper, signed A. Smith, saying "All off 8pm". Woodford and Ipswich allowed five minutes grace, sometimes ten, when men were working on a 12hr shift, but three-and-three quarter *hours* was unheard of. So Betts was told by me "You book them off at 4.20". "Very good, Sir" said the bland Ted. "I'm glad I'm on afternoons and shan't be here when they find out in the morning they are 3hr 40min short". They did find out and so began a series of timekeeping battles that I shall never forget as long as I live.

In this case, it was the boiler-washers and tube-cleaners. They had worked on contract since the "black-out" they said, and they got stuck in and always went home when the job was done. Next summer they would see to it that there would be no overtime, and that every Battersea engine would prime itself to a stand inside a month. I won my point, as indeed I had to do, and, strangely enough, when things calmed down and the work was being done properly and thoroughly, a better spirit prevailed. But this battle of wits persisted between many of the staff and myself for the rest of my time at the depot.

I like to think that I won more than my share of the rounds, and that we achieved a high level of work amongst the rougher, harder, dirtier shed jobs, that the engines were well cleaned, and that the depot was tidy and smart. And the extraordinary thing was that although few days passed without my involvement in some "tear-up" with somebody, about timekeeping, or loss of time on the road, or a refusal to carry out an instruction, yet I cannot remember many cases in which people bore malice. Although I may not consciously have realised it, attack, sometimes tactful, sometimes forceful, sometimes violent, was respected whilst weakness was despised. And as I have so often realised, the good men at a depot (and they are always in the majority), have no room for those who do not shape up or who are missing on Saturday afternoons, leaving their work to be covered by the reliable men. If the boss does not tackle this sort of thing with courage and vigour, he will lose the respect of the man who puts service and responsibility first, and that really does matter.

So much happened in the two-and-a-quarter years that I was on the Southern, involving me personally and emotionally, that I cannot hope to unfold my story in a logical sequence. But I became passionately fond of the Southern life, although I had had enough of the intense pressure at Stewarts Lane when I returned to the Eastern Region. The Southern zest for punctuality encouraged us to rise to the occasion, to plan our engine lists with care, and to follow up even the smallest of delays. I worked most Sundays until lunch-time and when it came to high summer, my Saturdays were often 12 to 14 hours long. The order in which Fred Pankhurst and I used to plan our Saturday engine list was first of all engines to jobs, then engines to men, for it was no use presenting an N class with flats on the wheels on an early morning Ramsgate to a more temperamental member of the driving fraternity, and a wordy argument, switch of engines and consequent delay could easily upset the working at Victoria for half-an-hour or more.

I used to arrive soon after 6.0am, work closely with the Running Foreman until 9.0am or thereabouts, go into the office for a while, and then be out and around until lunch-time. Invariably I saw the "Golden Arrow" engine away, and usually travelled to Victoria on it. An hour with the Victoria Running Foreman then would be a prelude to a trip down to the coast on an ailing engine with an inexperienced fireman, to help him out if he looked likely to come unstuck, and then back to London, a wash, another hour at the depot, and home about 8.0pm if I was lucky. A successful summer Saturday, the result of the most careful Operating and Motive Power planning, and also of improvisation if necessary, was an experience not easily forgotten.

Sometimes we used to come unhooked with an H Class tank with 450-tons of boat train empty coaches for Eardley coming to a stand on the 1 in 90 of Tulse Hill bank. This was not, perhaps, a serious matter, but the Southern Motive Power Department operated a punctuality league, with a poster, published each month, giving the time lost per 1,000 miles run by each depot. A few 30-minute delays on short journeys made sure that we were rarely at the top of the league, but this idea was a great incentive to smart working. So also were the meetings I had to attend periodically, chaired by my district officer, at which the Shedmasters had to account in detail for their train running, the actions they had taken. Woe betide me if the "Terrible 6.10 Uckfield" had been playing up, or the boat trains had been losing the odd few minutes here and there.

Our Motive Power Superintendent, T. E. Chrimes, and the Superintendent of Operation, S. W. Smart, were men of high standing known by name throughout the region. Sometime in June, 1954, the last boat train of the day from Folkestone was worked up by Driver Bert Hutton, with "West Country" Pacific No 34071. Next morning there was on my desk a "lost time ticket" for 3min debited to this train, also a note to say that S. W. Smart was to travel by the same train that evening and that we were to give special attention to the running. In order that he might not be delayed by any locomotive shortcoming, I took myself off to Folkestone in the late afternoon to come up with Hutton, who I knew to be an excellent but very light engine-man who had overdone it the previous evening in the interests of economy. We had the same engine, with a load of about 450 tons, and our journey was as good as the evening. We maintained exact time until coming up the Weald and later up Polhill, when we got the Distants against us. In those days, these signals were semaphores, and so we whistled but kept steam hard on. After we had gone by each Distant, we looked backwards, and in each case it dropped off just in time, so nobody was the wiser. When we got to London, S. W. Smart came up to the engine, congratulated us, and asked about our journey. I told him what we had done about his Distants, and his reply was "Splendid, boy! Well done". This made us feel that it had been well worth the time and trouble taken to see that no time was lost and that exact running was the order of the day.

Fairly early in 1953, the Bulleid Pacifics had a spate of axle trouble and were mostly taken out of service; luckily we were not very busy at the time, though I seem to remember a Bank Holiday had complicated things a bit. Anyhow, we got some Eastern, North Eastern and Scottish Region Class B1 4-6-0s, and the opposition establishment along the Wandsworth Road was sent some LNER V2 2-6-2s. Our engines did very well considering that the only person on the premises who knew anything about them was myself. The first journey out of London was on the Stratford engine No 61329 with Driver Charlie Stewart; he and his mate got there after a bit of a scramble, but they did much better coming back. And then in the afternoon Bill Murray and I set off with No 61050 on the 3.35pm Ramsgate. We had a delightful trip and, of course, she was in perfect condition, like all David William Harvey's Norwich locomotives. In two or three weeks the Bis were gone again, but it was rather fun while it lasted.

"Britannia" Pacific No 70004 *William Shakespeare* was the Stewarts Lane showpiece, beautifully kept and a sight to gladden the heart of any locomotive lover. The best 6.0am cleaners were allocated to this engine and after it went off the shed at 12.30 for the down "Golden Arrow", they were usually given an H class tank to finish their day. We went to a great deal of trouble to improve the exhibition finish that No 70004 had been given for the Festival of Britain.

Tom Banton was one of the No 1 Link drivers and worked No 4 duty – the "Golden Arrow" down and up – one week in every twelve. It was my practice to see No 70004, or whatever engine was to work the train, go off the shed every day – cleaned to perfection – and nothing upset me more than to see coal dust on the sloping coping of the tender or on the cab roof once they had been cleaned. I insisted that the fire should be made up and the coal then pulled forward, but some enginemen, if possible, slipped round to the coal hopper to top up when and if my back was turned. Tom and his mate did this one day, but the safety-valves lifted under the coal chute, and the boiler top and cab then had to be cleaned again. I was furious and we came to high words. I accused Tom of being scared of running short of coal. That did it, for he was a man of confidence and was simply aiming to save the fireman's back. For weeks after he would not speak when we passed in the shed and so, to put a stop to this nonsense, I got on his engine one evening at Cannon Street to ride with him to Faversham. We had a "West Country" Pacific and set off without a word said. After London Bridge I picked up the shovel and began to fire, and this had the remarkable effect of bringing Tom to life. Without mercy he thrashed the engine uphill, on the level, and, if he could, downhill as well. I had a wet shirt, he felt better, but he never went round for extra coal again. Honour had been satisfied!

For several months at one period we had been in a terrible state for engine tools; spanners, torch lamps, flares, gauge-glasses, hammers, all of them disappeared we knew not where, but mostly, we suspected, down the road to keep the country depots in clover. Matters came to a head one day when I witnessed an all-too-typical example of our own managerial inefficiency.

Driver Fred Skelly and his mate were getting No 30766 ready for the 12.35pm from Victoria to Ramsgate. She was the regular engine for this train and had been on the job all the week. Arriving at the engine, they looked in the tool-box, which they found to contain a few detonators and a red flag floating in half-an-inch of oil, but no tools. The fireman came to Algy Harman, the Running Foreman, who, disclaiming responsibility, gave him a chit "Full set of tools". The stores were not pleased, but after a lot of searching the storesman banged down on the counter a couple of ring spanners, an oil feeder, two oil bottles and an old gauge lamp. During the preparation, the crew "borrowed" a gauge lamp off another engine and then, this preliminary work done, the driver walked down to the Foreman's office to find out if there were any special restrictions on the way over to Victoria (for there was some engineering work in progress on the Grosvenor Bridge), while the fireman went to make the tea. Five minutes later, they were back – and not a tool on their engine! When this had happened three days running, on the same turn and the same engine, we decided that something had to be done if we were ever to get engines off the shed to time the following summer.

Accordingly we went into battle, and by January, 1953, we were ready. A total of 126 sets of tools was assembled in the stores, whilst propaganda and notices galore were posted for the benefit of all concerned: every tool-box, even that of the P class tanks on the coal road shunt, was to have its own lock and its key, the latter to be booked out to the preparing engineman. Keyboards were installed in the Timekeeper's Office. A toolman, Bill Allen, whose future depended on his success, worked up a lot of enthusiasm and we gradually got into our stride. No new scheme worth its salt gets away easily, but with hours of telephoning, exhortation, and straightforward confrontation, we eventually got on top and after three months we could say that, at Stewarts Lane, a driver could come on duty, pick up the key, unlock the box on his engine and find all the tools and oil he needed, a good prelude to a decent day's work. In a motive power depot of this size and character, needless to say, nothing of this kind could be left to run itself, and constant vigilance was required. For example, during very cold weather we found that gauge-lamps began to run short. We never discovered why, for when the next thaw set in, the shortage became a surplus!

Sometimes a driver would lock his tool-box, get talking in the lobby, and forget the key until he was back home. One day, this happened to a driver called Tom Simmons. Next morning I had to tell him straight that this wasn't good enough; yet a day or two later he did it again. He had come on duty at 3.0am, so we bided our time until we thought he was home and asleep, safely tucked up in bed, and then we sent round to his house for the keys. That cured Thomas! Unless one has actually had to deal with such a problem, it is perhaps difficult to understand one's pride in such a modest achievement, but there is no doubt that its success saved us late starts galore in the busy months ahead. It was not only the challenge and the economies that we made in oil and equipment, but the surprise and delight of our staff in its success, that gave me the greatest pleasure of all.

If I have seemed a bit disparaging at the expense of the Brighton line, it was because the majority of the men and the cream of the work belonged to the Chatham side and because the great strain and challenge of the summer months consequently brought out the best in the Chatham men. But some of the Brighton work was hard and the trains were heavy. The Marsh I3 4-4-2 tanks had all gone, so had all the Brighton express locomotives except the Atlantics. The "Terrible 6.10" had a variety of locomotives, but it was generally worked by an ex-London Midland Class 4 2-6-4 tank engine. In good order, with injectors that were reliable, and in good hands, these were masters of the toughest jobs; nevertheless, timekeeping on this turn was never exemplary in my time, and I spent many evenings battling it out to Oxted with some old machine, trying to beat the timetable. Nor were the Lingfield race trains an easy proposition. Usually, the crews changed footplates at Lingfield with the up specials from Brighton, and after the races were over one was faced with the seven-and-a-half mile climb at 1 in 100-132 through Oxted, and up through the one-and-a-quarter mile tunnel to Woldingham, which seemed to go on for ever if you were not doing very well. And was it not on the Brighton line, at Hurst Green Halt, that I was responsible for stopping short of steam with a load of no more than four – 100 tons – after a tussle with Tilmanstone coal – a battle which I lost well and truly!

The Newhaven boat trains were often Atlantic-hauled, usually with Newhaven men in charge, some of whom incidentally were from the SECR; also we had an irregular turn at about nine o'clock in the morning, worked by a "Schools" 4-4-0, an Atlantic or a U1 2-6-0, and often we did rather badly. The 4-4-0 was No 30922 *Marlborough* until she left us, and after that No 30915 *Brighton*. And mention of Brighton brings to mind the Royal trains that we used to work to Epsom Downs on Derby Day. It took us a week to get *Brighton* ready for a "Royal", when she had been repainted, scoured, cleaned and brought to a state of mechanical and sartorial perfection. Certainly we did not whitewash the coal, but the cab roof was painted white and the locomotive and its Pullman train were a sight for the gods. The last engine we prepared for a Royal train in my time at the Lane was "Battle of Britain" 4-6-2 No 34088. She went light to Portsmouth to bring up to London next day the Emperor of Abyssinia, Haile Selassie. The driver was G. H. King of Stewarts Lane, but the light engine had been worked down by the Nine Elms driver A. E. Hooker, a young man of great dedication with whom I made many journeys over the years, mostly in the evenings after the day's work was done.

I used to go with him to Basingstoke on the 7.54pm from Waterloo, a light and fast train usually hauled by the next candidate for the scrap yard. And so not only did I sample the Urie "City" Class 4-6-0s (the nickname applied to the Drummond rebuilds, Nos 30330 to 30335), which had to be thrashed to reach 70mph, but also the last of the "Remembrance" 4-6-0s, rebuilds of the former Brighton 4-6-4 tanks. These were extraordinary machines. With the first port of the regulator, they would not keep time even with a light train, but with plenty of lever and a touch of the second port, they would leap forward, roaring and throwing fire as if nothing would stop them. One amusing journey was with 32331, when we came to a stand at Woking with time well on our side, despite a speed restriction to 15mph at Surbiton, but ankle-deep in coal, covered with dust, bruised and battered.

For a large London depot, our maintenance was very fair and here an organisation for which I take no credit played its part. A depot cannot achieve success without good artisans and in this respect we were lucky. Nor can one succeed without good supervision, and here again we were lucky.

When examinations drop behind it is a struggle to survive, but with careful planning, and a bit of good fortune and, above all, tactful and firm handling of staff and resources, we got by. Bert Wood, our Foreman Fitter, looked after the job calmly and thoughtfully, and much of the planning was done by the Shops Officeman, Syd Norman, another general factotum *par excellence,* once a fitter's mate, bombed-out like so many during the war, who was also guide to visitors round the depot and flag washer-in-chief for the "Golden Arrow". But despite these stalwarts, my own personal bodyguard on the clerical side, and a list clerk, Stan Reynolds, whose hide could stand up to the invective and threats of those unfortunates he had moved a couple of hours backwards on a Saturday afternoon, Stewarts Lane was always a difficult, demanding depot. There was no let up, and I saw comparatively little of our young family. But the hard work was immensely rewarding, interesting and at times even very funny, not always at the time but in retrospect, for you had to be smart to beat the South London cockney.

One evening, after a rough day, I had settled down to work in the office over some problem that was worrying me. There was a knock on the door and in came a driver by the name of Alf Edgley. He was an old Chatham man, and not a silent type by any means; I was busy, but he was determined to see me and started to lecture me on management inefficiency. I had had a long and tiring day, and when he began to bang the table, with my blood rushing upwards I hopped out of my chair and told him if he thought he could do my job better than I could, he could get on with it *pronto.* Once I had read that the redoubtable Dugald Drummond had done this with great success, but times had changed and I was no Dugald Drummond. Anyway, Alfred nipped round the table and sat down in my chair! I banged the papers down in front of him as unpleasantly as possible, saying "You try your hand at this lot, you won't be so critical in future!" He took a long look at the papers, and then remarked, "Guv, you want to do so-and-so". And he was right! We had a good laugh about it, and he must have had the chance to tell the story thousands of times afterwards. In such ways as these I learnt never to underestimate my fellow human beings.

During a lull one Saturday afternoon in the summer of 1954, Fred Pankhurst and I were talking in the Timekeeper's Office. I had decided to stay at the depot until 4.0pm, and then go home, as we had friends coming in for dinner. Suddenly, the Redhill Control 'phone rang, and as I was nearest I took down a message to the effect that a bridge at Shalford was impassable and that four trains from the Western Region, which normally would have gone across from Reading via Redhill to Tonbridge, were to be diverted via Kensington. We were to re-engine and re-man, and the first train was well on its way. That day we had neither men nor engines to spare, but a little problem like that had to be tackled.

I enquired the Western engine numbers and was told they were "63s". We agreed that two trains should be worked down the Brighton line as far as Redhill with Stewarts Lane men and Western Region engines, for did not the Western 63XX 2-6-0s work across to Redhill over the ex-SECR and was not the Brighton loading gauge wider and higher in any case? I thought so and I said that this would find us two of the engines we needed. The other two trains we agreed to work as far as our men would go via the Eastern side. We managed to get four crews who volunteered for the job, all of whom had already done a day's work, and we fixed up the Eastern men with N Class 2-6-0s. So far so good. My memory is taxed for the detail but this certainly is how they set out from Stewarts Lane less than half-an-hour later.

Maidstone Driver Burrows with an N Class 2-6-0 was to work a Ramsgate train as far as Tonbridge. With him he took Passed Fireman A. Pink of Stewarts Lane to pilot the first Western Region engine to Redhill. Stewarts Lane Driver Gus Wembridge was in charge of another N Class for Ramsgate via Tonbridge, and Driver Fred Holden, a very fine Brighton man, was to pilot the second Western engine.

The first train into Kensington was the Ramsgate via Redhill and Pinky got aboard to pilot the Western men who had thought they were coming off. However, after a bit of argument, they remained on the engine and duly reached Redhill, where they unhooked, left the engine in Redhill Depot, and went their various ways home as passengers. The only trouble here was that the locomotive was not a "63" but a 69XX "Hall", forbidden for clearance reasons to put its nose by Kensington. But we at Stewarts Lane were blissfully unaware of what was taking place, and as no platforms were knocked about by the 4-6-0s cylinders everybody else was happy until, later on, somebody began to make enquiries.

The second train was a Ramsgate via Tonbridge, and Gus Wembridge set off and reached his destination without trouble. But now at Kensington the fun started. Burrows of Maidstone had never been on the Brighton line in his life. He told everybody at Kensington that he was working the next Ramsgate via Tonbridge train, but unknown to him he was put on the next Redhill via Clapham Junction. The guard came up and they exchanged the usual information, except the vital word "Tonbridge". Although the engine was carrying the Tonbridge headcode, on arrival at Latchmere Junction Burrows found the signals off for Clapham Junction, and so stopped. The signalman insisted rightly that he could not allow a Redhill train to go any other way than to Redhill. It was Burrows' lucky day, for at that moment who should see him from the road below but one of the Stewarts Lane Running Foremen going off duty, Algy Harman, who had in fact arranged these locomotive workings an hour or so earlier. He got up on to the railway and, showing great presence of mind, stopped a Willesden-bound train worked by a Southern engine and Redhill men, made them change footplates with Burrows, and away the train went on its way south with poor old Maidstone Driver Burrows heading back to Willesden! Fortunately, there was comparatively little delay, and Burrows got home after he had brought his engine back to Stewarts Lane in the evening.

At the depot, Fred Pankhurst and I were blissfully unaware of this comedy being played out across the roof-tops of South-West London, and were congratulating ourselves on a good job well done, when there was a frantic telephone call from Kensington to say that they had a train bound for Dover via Tonbridge standing in the platform without an engine, and furthermore it had been there a very long while. Our Foreman had an N Class 2-6-0 No 31810, on the disposal pits, facing the wrong way, with a half-cleaned fire. Well, she had to do; so we took coal, finished the fire, and the driver, Fred Morley, went down one side of the engine with an oil-feeder whilst I did the other. Then we turned, got away and dashed over the river bridge to Kensington. There we hooked up, set off again with the train as hard as we dare, stopped at Factory Junction for relief by Dover men, and walked back to the depot, arriving 42 minutes after the receipt of the 'phone call. That might well be a record, but it did not help the infuriated passengers, nor for that matter did it help me when I pieced the story together on a Sunday spent compiling a report which certainly did not please my boss. The responsibility for the delays was placed fairly and squarely on my shoulders, not to mention a little matter of a strange locomotive on forbidden territory. The fourth engine was also a Great Western "Hall", with a Brighton side pilotman, Fred Holden, who had never been over the Eastern Section in his life. Had Burrows been put on the right train, all would have been well, but instead I had had to learn yet another lesson. Had I sent somebody in authority over to Kensington or gone myself, it would never have happened.

There were times in those two-and-a-quarter years when one got ragged, bad-tempered, far too outspoken, and when one fell out with the people who were trying to be helpful. But fresh air and exercise are good cures for liver, bad temper, colds and most other ailments, real or imaginary. And so I would slip down to Dover on the "Golden Arrow" every so often, on the 3.35pm to Ramsgate with No 768, or perhaps on a Saturday morning to Faversham on an E1 4-4-0. If I went on the "Arrow", with Joe Brewer, I used to do the driving on the down journey and the firing on the way back and by the end of the day I was a new man. Going down, Joe would say, after we passed Tonbridge, "Put her on 40, Guv!", and *William Shakespeare* would roar up to the 80s by the time she was by Paddock Wood. The working of these very fast trains presented a hard day's work for a fireman. The start out of London was extremely heavy and went on seemingly for miles and miles until Knockholt was passed.

And then, at the end of Polhill tunnel and again after Sevenoaks tunnel, there were two of the worst Distant signals I have ever known. If either tunnel was clear of smoke, you could see the signal silhouetted against the daylight, but if an up train had just gone through, you could not see a thing. You burst out of the tunnel, with one hand on the brake and the other on the regulator, and did not sight the signal until you passed under it. If it was at yellow, there was only one thing to do! And, yet strangely enough, I do not recall any of our drivers passing the Polhill or the Weald Home signals at danger. Pluckley Distant on the up journey was equally hard to see, indeed, and the problem here was drifting steam and smoke, particularly on a Bulleid Pacific. So, whether at Polhill, the Weald, Pluckley or elsewhere, one had to be 100 per cent on the alert, with just room enough to stop at the Home if any of these hard-to-see Distants was at caution.

In the summer time, we needed every man we could lay our hands on for train working, and the young men had a chance to shine. There were a few older men who had not had the opportunity to learn many roads and the few candidates for Link 3B (Battersea, Brixton and "B – all"), who never wanted main line work. Sometimes there was a spot of resentment, but the summer provided a great opportunity for courageous and confident young men in their middle twenties to take hold of express trains. A young man by the name of Roy Cann had been firing to Joe Brewer most ably, and when he passed for driving took his position as a shed link passed fireman. Soon after, I saw him in the depot and he told me he had signed for Dover, Ramsgate and all round Kent. A fortnight afterwards, however, he told me he had been working nothing except empty carriage trains "round the chimney pots". So I arranged for him to work a "Rounder" the following Sunday.

These turns could be very rough. A "King Arthur" 4-6-0 was the best that could be expected, and one started from Victoria and got relieved at Waterloo Junction after going right round Kent.

Well, Roy Cann had a bad trip and lost time, and though I told him it wasn't good enough I said he should have another go the next weekend.

This time he took a "West Country" Pacific down to Dover on a boat train, and for the return, with about 450 tons, was given one of the "King Arthurs" with a small six-wheel tender. The result was that he had to stop at Ashford for water – because of the small tender – and the guard booked 12 minutes against him and his mate, Jim Williams. In the circumstances no men could have done better and yet their total age was under 50 years. However, these two journeys added greatly to Cann's stature. There were many others like him; being young myself, it did me good to see this sort of thing.

In a large depot, you must have good clerical staff, for without good administration mistakes will be made, letters will go unanswered and discontent will spread amongst the staff very quickly indeed. Besides, you need a personal bodyguard! Charlie Bayliss, the chief clerk, originated from Nine Elms and had the South Westerner's slight contempt for the Chatham man. Johnny Greenfield, his understudy, was one of the few young depot clerks who could and did interview drivers on equal terms on all manner of subjects from lost time to conditions of service. He could hold his own and was accordingly respected because he took great pains to understand the practicalities of the work.

Theirs was a busy and sometimes frustrating life; however, they were not only dealing with casualties, lost time and availability, but also with people, and that was always interesting and often amusing. They had me nicely weighed up. I had a bad habit of chewing pins when I wanted to concentrate. One day I went into the clerks' office, put my hat on Charlie Bayliss' desk and sat on the edge of one of the other desks. We had a problem to discuss and picking up a pin from the container on the desk I started to chew it. Having bent it double, I absent-mindedly dropped it into the lock of one of the filing cabinets and then I went off to my office, leaving my hat on the desk. Soon after, somebody went to the cabinet, tried to pull it open and jammed the lock against my pin. Ten minutes later I went out to the shed in a hurry, dashed into the office for my hat, picked it up and left the brim neatly stapled to the desk. It was an old faithful, but I had to throw the remains in a firebox and buy a new one!

One evening, I sent for a fireman who had come from the West Country on promotion, and who showed signs of being a bad timekeeper. I felt that he needed a warning. When he came into the office, he had his cap on and I told him to take it off, but quite civilly he refused. Pretty sharply I told him to do as he was bidden but again, still civilly, he refused. Had he been coming to see me about some matter of his own I should have told him to go away until he'd learned some manners; but on this occasion I wanted to do the talking myself. So I rang for Charlie Bayliss and told him what I thought of the young man, who stood quietly by the desk. Charlie drew himself up to his full height, looked down his not inconsiderable nose, and advised the young man to do as he was told, and referred him to the dire penalties that would have been incurred in his young day at Nine Elms. But all had not the slightest effect, and I felt I was getting nowhere. So I cut short the interview, having well and truly lost round one.

Round two was on the next evening. I had worked out all manner of threats and disciplinary measures for insubordination. So I sent for the young man, but he proceeded to demolish me for the second time by coming in *without* a cap, and accepting what I had to say in complete silence! This was all the more unsatisfactory, because I was even denied the pleasure of working myself up into a state of righteous indignation! A small incident this, but nearly every day I was faced with a confrontation with some erring member of my staff. This was wearing, of course, and one wondered from time to time whether it was worth all the time and effort, but one was dealing mostly with men of character, sometimes awkward or difficult, sometimes amusing, haphazard, lazy, sharp-tongued. The encounters sharpened one's own tongue, and one's wit. Many times there were heated exchanges and strong language, but never lasting bitterness on either side. I wouldn't have missed it for worlds, nor will I forget the Stewarts Lane men, whether they were my regular adversaries or the conscientious men who rarely if ever gave any cause for complaint, even though the latter were often – and sometimes rightly – our severest critics.

On Christmas Eve, 1954, I was working late, for we were very busy. There was a raging gale in the Channel and the boat had been delayed. The up "Golden Arrow" was over an hour late, and Bert Hutton, the driver, had already threatened to come home "passenger" from Dover. This was the same Bert Hutton who had once broken his arm on the water-column when passing through Tonbridge at speed, but had pluckily continued to Dover. However, this Christmas Eve he wanted to get home, and threatened to leave his engine on arrival at Victoria unless relieved by a spare crew. But we had no relief men available, indeed, there was not a hope at half-past eight on Christmas Eve, so eventually he brought his engine to the shed. Bert had what might be called a pugilistic face and made for my office, his features contorted with rage. I was waiting just outside in a similar frame of mind, perhaps worse. We had a fearful "up-and-downer", both saying unforgivable things, but eventually we were calmed down by Fred Pankhurst, who in fact, could have out-shouted either of us, but instead fussed round like an old hen trying to smooth down her ruffled chickens. Next day I was at work but Hutton was on leave. But we were both there on Boxing Day, and I sent for him to apologise for my behaviour. His answer was: "That's all right, Guv. We were as bad as one another. They tell me you're leaving us and I'm sorry about that, too". So we were back to normal and our anger of two days before was forgotten. He was a good railwayman.

Let the last anecdote be about Sammy Gingell. On all classes of train between August, 1952, and the time that I left the depot in January, 1955 he lost no more than fifteen minutes in all. Three of these were on the "Night Ferry", on a spring night when he dropped the time between Victoria and Herne Hill due to slipping on the climb to Grosvenor Bridge with this very heavy train. On the previous Friday, I had been approached by a driver nicknamed "Count" Harding whose roster for the Sunday was 4.0pm P&D (preparation and disposal), for a day's leave. There was a very firm and fair agreement with the staff that any man who was rostered to work on a Sunday would not be granted leave, but if he obtained a substitute to cover his turn, by agreement with the list clerk, leave of absence could be granted; but the "Count" said he could not find a substitute. So I could not grant him the desired leave, and the position for the moment was deadlock with the "Count" liable to be in trouble. However, I said we would go and look at the rosters together, and thumbing from the top we came to the name 'S. Gingell'. He was not on duty on the Sunday and his Monday turn gave him sufficient rest after covering the 4.0pm P&D. I asked "Count" if he had approached Sammy, and his reply was simply to laugh at the suggestion that a man in No 2 Link, aged 62, would come on at 4.0pm to prepare engines on a Sunday afternoon. However, when Sammy was asked he agreed no doubt knowing that if he was lucky he might get a sprint.

To cut a long story short, Sam was collared by the foreman to work the down "Night Ferry", because the up service was running late and the Dover men would not arrive in time for the return working. He had already dealt with three or four engines in the shed, but in any case, the chance of a main line job was what he had come to work for. I have no idea at what time he and his mate got back on the Monday morning, but I do know that he would not admit he had lost time to Herne Hill until a fortnight later, when at last he was prevailed upon to produce a very soiled lost time ticket. Apologetically he showed it to me, with the remark that he had won the time back by Orpington, but could not quite remember how he had lost it in the first place. So it was to slipping that we put it down!

The steam work and the depot demanded much, both physically and mentally, but this is no reason for failing to mention the electric train work for which I was responsible. At first sight, a motorman's life seemed easy by comparison, more ordered and less demanding, but there were some fast and long-distance duties, and many suburban workings, which with constant stopping and starting and high speed running in between the stops constituted a heavy day's work, particularly if the weather was bad or foggy. In those days, too, there were a good many of the older electric trains still in service. These were hell for the motorman in the summer time; the electrical equipment warmed up in the cab and even with the doors open, the conditions were bad. Also old South-Western type vehicles had pointed noses and a gentle nautical roll, especially when travelling around 30 to 40mph, and could be the very devil after lunch on a hot day.

As to steam, from the outset I liked many of the Southern engines. Collectively the SECR locomotives without doubt were as sound and as simple and as effective as any in the country. Indeed, they were ideal, for not only did they do their work well, but they were easy on repairs also. In particular, the D1 and E1 rebuilt 4-4-0s were great engines; every journey with one was an adventure, and once or twice I had the pleasure and excitement of sharing in the work of No 31019 and No 31165 with Sammy Gingell in achieving a performance that was theoretically quite impossible. Then there were no engines anywhere like the Bulleid Pacifics. At times we cursed the day when they were conceived, with their chain-drive, steam-pipe joints, elements, drifting smoke, shortage of steam, everything about them. But this was a passing phase, reserved for their inevitable bad days, for they were the unique product of a bold man's great brain, and in my judgment there was *nothing* in the country to touch them on their good days. The solid backbone of our work was the "King Arthurs", built like battleships, ponderous but nevertheless fast and economical, and light on maintenance too. Finally, the "Schools", which did so well at Bricklayers Arms, and the "Nelsons", which had handled the pre-war boat trains, made their own unforgettable contribution.

In January, 1955, I returned to the Eastern Region. By now I had learned a great deal about railways, men, management and locomotives. Why fate should have decided to send me to the Southern, I do not know, but for ever I shall be thankful that I went, for those years were packed with interest. Above all, there was the challenge of a job that made me rise to the occasion in such a way that so many of the lessons learned and the results achieved will always remain with me.

XV

Assistant DMPS at Stratford

A N ASSOCIATION with the Stratford District in general, and Stratford Depot in particular, could involve one of many things – strikes, go-slows, critical shortages of locomotives and manpower, nervous breakdown, long hours, and bitter controversy over the condition of locomotives with one or two famous characters who never took an engine off the shed unless they had failed the first two booked to them. All these things could come the way of the man in charge and many more too, but not least the need to laugh with the humour born of adversity and to savour the *camaraderie* amongst so many of the men who worked in this remarkable depot, or who ran its locomotives – men who were extremely able and, at the same time, had a built-in resistance to the frustrations and turmoil of the place.

With my days as a Shedmaster behind me, I walked across the front of the New Shed in a blizzard to join my District Officer, T. C. B. Miller, on 3rd January 1955. Not without some qualms, for I knew the Stratford reputation, and also because I was entering a period in my career where I felt I might not show up to the best advantage. I had been appointed Assistant District Motive Power Superintendent and the District and Depot Offices were combined, as was often the practice in those days on the Eastern Region where the Assistant took some share in the running of the depot as well as playing his part in helping to run the District.

I was to hold this position until the end of 1957, when the Motive Power Department ceased to exist as a separate entity. Diesel main line locomotives then were beginning to arrive, and I entered on a period of my career which started inauspiciously, but developed in 1959 into a tremendously interesting and enjoyable task that was not completely fulfilled until the end of 1962 when I left the area.

In 1955, the Stratford District had come down to some 500 locomotives, and apart from the large central depot, the DMPS was responsible for the sizeable depots of Colchester, Parkeston and Southend, sub-depots at Clacton, Walton, Braintree, Maldon and Southminster, small depots at Wood Street, Enfield, Bishop's Stortford, Hertford and Epping, and any number of places housing one or two engines only, such as Chelmsford, Buntingford, Devonshire Street, Goodmayes and so on. On top of this there were the electric motormen's signing on points at Ilford, Gidea Park and Shenfield, and the terminal point. At Liverpool Street, the scene of many desperate last-minute changes of locomotive. There were diesel 350hp shunters at Temple Mills, and two petrol locomotives, one at Brentwood which worked on one shift only, and the other at Ware, which was often in the news on cold Monday mornings when it didn't work at all.

The running of the District depended very largely on the performance of Stratford, for even Colchester and Parkeston had to send their main line engines up for major work although they avoided doing so if possible. But the Southend locomotives, although they were largely confined to the London services, had to go into Stratford for hot axleboxes and lifting, so that the staffing there and at all the other points might be kept to a minimum. For example, Wood Street, which was a hard job – likewise Enfield – had a chargeman and a fitter working on opposite but overlapping turns, whilst their mate did the boilerwashing with the boilermaker, and the office work was covered by a foreman's assistant working with the chargeman. There were 36 sets of enginemen, 12 booked engines and two spares to cover washout and repairs; this meant precious little spare time for anybody with an intensive suburban service to run. As the N7 0-6-2s and L1 2-6-4s got near to the end of their days, and Stratford took on more and more diesels, the conditions deteriorated at these small places, and their working became a very real problem.

The old Stratford is now a thing of the past; some would say "a good job, too". Indeed, there is no doubt at all that the coming of dieselisation, with a new depot and immensely improved conditions of employment, particularly for the artisan staff, has saved the day. For we were going downhill fast, and providing no real inducements to recruitment in the hard conditions that existed in the late 1950s.

At Stratford there were two sheds, the New Shed, built in the 1870s, and the Jubilee Shed, which was newer, bigger, draughtier, dirtier and on Sunday nights in the summer, after lighting up on a large scale, quite intolerable. The New Shed had no ventilation, and any attempt to light up there, would have led to a threat of stoppage of work. The diesel shunters, heavier repairs, and re-tubing were dealt with in the New Shed, together with the non-steamers, and odd jobs down No 6 Road, whilst the running work was covered by the Jubilee Shed. The offices lay between the two sheds, in line with the extraordinary belt-operated coaling plant which had never been shut down since the day it was built, the hot-water wash-out boilers and pumps (nobody would wash out hot), the machine shop, the Timekeepers' office (the scene of even deeper mischief than at Stewarts Lane), and Oatmeal Cottage, which was a combined store and Chargeman Fitters' office.

Beyond the New Shed was the "Ark", where engines were lifted, and beyond the Jubilee shed there was an extraordinary collection of home-made shacks and buildings housing the oil squad, welding, trimmers and some of the messrooms. This was known as Acacia Avenue, and next door was a new brick building, known as Stanton House, for the shed labourers and named after their union man. Once, I believe, he had been a policeman, and he alternately used charm and truncheon to gain recognition for the diverse collection of people he represented. So the old

Stratford was traditionally a shanty town, for on the "outskirts" there were mess-rooms for the back-end shunters, for the brick-arch man, and for everybody else somewhere or other, even for the mythical chap whose job it was to clean out blast pipes. The Running Foreman who knew where to find his staff needed years of experience!

Such, in a few sentences, was the incredible Stratford depot – never to be confused with the works "over the uverside", as it was called. It was a depot surrounded by running lines, to and from Temple Mills, North Woolwich, the Fork sidings, Chobham Farm and Channelsea. Nothing at Stratford was conventional. The locomotives were booked to their workings by clerks in the same office as those in which the enginemen's rosters were compiled, so that not only were there fearful altercations about the rosters, but at the same time, trouble when a driver or fireman could not have their regular engine. The link structure was such that the suburban work was senior to the main line work, and this had the advantage that men in their early 50s took charge of "Britannia" Pacifics for several years before going into the L1 links and then on to the N7s. The Woolwich gang was the senior at the depot until the arrival of the railcars in 1956-57, and the last F6s were employed on the Woolwich trains. In their state of health, and with young and inexperienced firemen, it made life very difficult for the older drivers, many of whom longed to get back on the main line they had forsaken several years before.

Next door to the Timekeepers' office was the Running Foreman's office, the centre of the web, whilst along the corridor were the Maintenance Foremen, the Shedmaster and the Locomotive Inspectors. The Chief Locomotive Inspector, Arthur Weavers, was a remarkable character. He died suddenly in his early 60s and was greatly missed. He had the gift, not given to all locomotive inspectors, of being very much a management man and yet a good driver's man. He knew exactly how far he could go in dealing with incidents and irregularities, so that only those that were really serious reached the discipline stage. He picked his relief inspectors with the same qualities and so when my time came to take charge of the District, in 1959, I had support of the same high standard, although Arthur Weavers by then had passed on. Arthur sat at the end of the office, very pugnacious, and more than once I've walked in as somebody else more or less fell out through the door. On being questioned, Arthur would simply say; "He's had to be done, Force 10". But that would be the end of it.

And lastly, in this very general description of a huge concern employing more than 3,000 men, there were two establishments unique in my experience, not so much because of what they had but how they did it. The Crane Shop housed the turners, millwrights, carpenters (Classes 1, 2 and 3!) and the Breakdown Foreman, latterly the irrepressible Syd Casselton. The breakdown gang in bad weather would run up from 40 to

50 calls a month, and in addition was used at most week-ends for bridging work, as well, at that time, for work in connection with electrification. The turn-out time was normally 30 minutes, and the performance, under Syds' flamboyant, humorous, but very tough control, was the best in the country. The work was hard, the hours were long, and not many men could stand the constant interference with sleep for more than a few years. The pay was good and this, quite understandably, was the reason why many men came into the gang. Some notable jobs were undertaken, particularly with the small 36-ton crane, and the German hydraulic rerailing equipment, which was first used at Stratford in 1954.

When the Drewry 204hp shunting engines came in 1955, they were accompanied by a service engineer, James Davidson, whom we all liked very much. These little locomotives were good value, and we put them to work in the heavy yards such as Spitalfields. The text-book method of driving involved a pause between changes of speed, but on the late turn, with much work to do, the shunters would be pressing for results. So the drivers resorted to racing gear changes, and then there was trouble. There was a bell near the foreman's office and as James walked across the yard in the morning, Syd Casselton would start his tintinnabulation – one dingle for brakebands, two for a gearbox, three for a fire and so on!

The old Stores are no more; but so far as I know the ageless Charlie Lock is very much alive. The stores were no ordinary affair but a complete District supply-point, squeezed in under the main water tank. They were entirely home made, having gradually worked upwards as Charlie collected more items of equipment, built more wooden shelves, pockets and step-ladders, painted more labels and generally earned himself a reputation for storekeeping and hoarding in the old-fashioned tradition. He was a little man with a straggling moustache and a humpy back, a pencil behind his ear, *a Park Drive* smouldering in his mouth, and an indignant and frequently outraged voice; indeed he *was* the Stratford stores.

Add to all this an enormous staff of drivers and fireman, a staff of 500 artisans, the yard staff and shed staff, and often acute staff shortages, and you will begin to see the Stratford of the middle 1950s. There were bad working conditions, many locomotives in indifferent condition, restrictive practices and labour relations which were a curious mixture of excellence and explosiveness. And the responsibility for handling this turbulent and difficult organisation fell on the shoulders of the District Motive Power Superintendent, his Assistant and the Stratford Shedmaster.

The variety of the locomotive allocation was in keeping with every other aspect of the place. At one end of the scale, there were the "Britannia" Pacifics, nearly all manned by three crews and doing splendid work on the Norwich services. They were aided by the Class B1 4-6-0s, many also treble-manned and some spare, some of them splendid engines and some getting rough; in the Stratford vernacular they were all known as "Bleedin' old Bongos". There were a few B17 4-6-0s, and if you haven't travelled on a Stratford B17 in its last days you haven't lived. There were the B12 4-6-0s, the various tank classes, all the varieties of GE goods engines, some K3 2-6-0s, a few Class 4 BR standard 2-6-0 locomotives, a 2-4-2 "Gobbler" tank or two, one or two London Tilbury & Southend Class 3 4-4-2 tank engines, the little 0-4-0T "Pots" that worked at Devonshire Street, and our "week-end" locomotives for the summer season. These last were Class K1 2-6-0s, which strangely enough used to run into trouble with burnt fire-bars on a Friday evening, and yet would be thrashing down to Clacton or Lowestoft on a Saturday, while our own steam-brake-fitted goods engines had been presented to an ungrateful Traffic Department on the Friday night to run fast fitted freight trains – decelerated – back to Whitemoor. This happened every Friday. There was usually a post mortem, but geographically we were in a strong position, for we had to run our extra trains on the Saturday, and, of course, we were *Stratford*, regarded by the rest of the Eastern Section as a law unto ourselves.

The out-stations, closely related as they were to the parent depot, also had their individuality. Colchester, Parkeston, Clacton and Walton, not to mention Braintree and Maldon, had once belonged to the Ipswich District before it was dissolved in the late 1930s. The staff were largely East Anglians, and so they preferred, if they could, to get along without the help of the Londoners. Colchester had a mixed bag of locomotives, including a number of Thompson rebuilds of 4-6-0 class B2, which covered the stopping and semi-fast work on the main line and occasionally acted as cover engines at Clacton. Here there were eight distinguished gentlemen working four B17 4-6-0s between them, Nos 61650, 61651, 61662 and 61666, on the Clacton-London service, at that time so slackly timed that they were by no means immaculate as to punctuality.

Mention of the B2 class rebuilds reminds me of a short journey I made on one of these engines after I had been at Stratford about a year. They mostly got very rough in the high mileages but were required to cover about 65,000 to 70,000 miles between shoppings. The engine in question, No 61607, was particularly bad, but our proposal for shops was turned down by the Shopping Bureau at Doncaster. The reason was that the chief of the Bureau had known me since I was an apprentice at Doncaster and had learned, bless his heart, that Stratford "wanted watching". I reacted sharply to the instruction to re-propose this old tub in three months' time, and the Chief sent one of his Shops Inspectors to ride on No 61607. Luckily I got there first, so to speak, and the engine was put on the 2.27pm from Liverpool Street to Ipswich, fast to Shenfield. The driver was none other than Charlie Parr of Ipswich, who at first did not appreciate the reason for working a bone-shaking Stratford District engine on an Ipswich job. However I told him out of the side of my mouth that the faster he got to Shenfield – if necessary to Chelmsford – the sooner No 61607 would go in for a General Repair and he would be back at Ipswich. As a matter of fact the deed was done before we were through Stratford, and the old engine was in shops within the week!

Parkeston still worked the "Hook Continental", with a "Britannia" Pacific, and the staff still remembered the depot's glorious past by improving on it where ever possible, for Parkeston men would work any train anywhere, and the young men got plenty of opportunity in the summer time with special passenger and freight workings. The famous name of "Rocky" Chapman was very much alive, for his son was a driver in the main-line goods link. The passenger engines were Thompson Class B1 4-6-0s, regularly manned, Nos 61135, 61226, 61232, 61264 ("Frenchy's" engine) and 61149. In 1961, I was invited to the annual ASLE&F dinner at Parkeston, which was a great event for the retired enginemen. After dinner I was asked to say a few words. I was talking about the pride that enginemen had in their machines and mentioned each of the Parkeston engines except No 61149, which I forgot. I paused for a second and there was a loud voice from the bottom of the room. It was retired Driver Fleming – "What about my old engine, 1149", he called out. "She was better than any of them!"

At that same dinner I was waylaid by an old driver by the name of Lilley, who told me he had something of interest to show me and would I like to keep it? He slipped round the corner to his home and came back with a beautifully made "Jimmy", a relic of his firing days on the "Little Black Goods", which was the vernacular for the very simple steam-braked, small-boilered Great Eastern 0-6-0 which became LNER Class J15. Before the grouping, these little engines performed some amazing work at all the main line depots, and to keep time with heavy loads firemen often had to resort to unofficial methods of steam raising. This "Jimmy", which was a bar of triangular section to fit across the blast-pipe, with two distance-pieces, a thumbscrew and chain to lock it firmly in place, was made at March depot in 1915 for either 1s 6d or 2s 6d. The blacksmith used to turn out these extremely effective but quite irregular devices in the Great Eastern Railway Company's time, and from the Great Eastern Railway Company's materials, and then would pocket the proceeds!

But I suppose everybody was happy for the inducement *plus* the "false door" for reducing the size of the fire-hole door helped the locomotives to achieve regular work well beyond their normal capabilities. And so I treasure that little souvenir of the far-off days when the Parkeston men worked "45 of Bacon" from the Quay to Spitalfields with a 6-inch white-hot fire, using the little and often technique of firing to perfection.

Enfield and Wood Street were running their well-tried N7 0-6-2 tanks. The Wood Street engines were "Westinghouse only", the original Great Eastern design with right-hand drive; the Enfield engines, kept in a splendid state of cleanliness by their crews and in excellent mechanical condition by the chargeman and his fitter, were in fact the last short-travel valve N7s to be built, Nos 69658 to 69671 inclusive.

The N7 tanks bore the brunt of the Great Eastern London suburban work for many years, and worked faithfully and efficiently right up to the days of electrification and the introduction of railcars on the North Woolwich service. We thought of them merely in two categories – "short-travel" and "long travel"; additionally, one had to remember that the first twelve engines built were Westinghouse only and could therefore be used only on the "Jazz" Chingford and Enfield and the Woolwich – Palace Gates services. The Walschaerts gear was fitted and the valve-travel on the first 22 engines was the relatively short one of $3^{15}/_{32}$in. The N7 tank was a small engine and the valve gear had to be fitted into a confined space.

On the later short-travel engines, the valve-travel was very slightly reduced, but in 1927-28 a further 62 locomotives were built, and Gresley decided – against the advice of his technical staff – that long travel valves were to be fitted with a travel of $5^{23}/_{32}$in. The result was a marvellous little engine, of great character, capable of fine feats of haulage, speed, acceleration and economy. But at a price.

In a Walschaerts valve gear, the pin most heavily loaded is the bottom-end pin of the combination lever. The valve gear on the original engines was inevitably cramped and, with a short eccentric rod and long motion link, the forces in the gear were increased. Despite special methods of lubrication, the wear on the combination lever pins was considerable but in addition the anchor-bracket bolts were constantly trying to work loose. Sometimes the looseness led to shearing and then the combination lever would run free and tie itself into knots. Various methods of fixing the anchor-bracket bolts were tried, but none of them successful until they were welded to the crosshead.

This settled the short-travel engines, but on those with the long-travel, with their nine-and-a-half inch eccentric throw, the movement at the top of the combination lever was greater and the valve-spindles and guides were subject to very severe stress when the travel was below about 40 per cent. The weakest part of the gear was in the spindles and guides, which were of the same small dimension as on the short-travel engines, and the classic type of failure with the long-travel N7s was the bending of the valve-spindles at high speed. Even with the lever in mid-gear for coasting, failures were fairly common but it was not until 1956, when No 69721 fell to bits near Ingatestone on a passenger train at 60mph, and closed the down line for over an hour, that all the long-travel locomotives were fitted with much thicker valve-spindles, heavier valve-guides, reversing wheel-catches and toothed sectors, instead of the traditional chain hooked in the wheel spokes, and notices in the cab instructing enginemen to coast in mid-gear. Even then failures were not eliminated, which all goes to show that you cannot get a quart into a pint pot.

Otherwise these little engines were tailor-made for their work. Failures are a nuisance when they happen and cause delay, but luckily they are intermittent. The bunker of an N7 was just the right height for hand-coaling, they were light on coal and water, they did not slip, they did not often refuse to start a heavy train, the brake was good, everything was nicely to hand in the cab, they were clean to work on, and the valve gear, for all its faults, could be seen and easily got at. The tubes were relatively simple to clean, and the boilers straightforward and easy to wash out. They fitted nicely into the run-round loops at the terminal stations and, in short, they were ideal engines for onerous, hard-working duties and they reacted to the regular manning which helped so much to keep the suburban workings at Wood Street, Enfield and Hertford going well right up to the time of the electrification.

Now the Assistant District Motive Power Superintendent could very easily be the dogsbody, but it was never thus under Terry Miller. He trained me not only to be his Assistant but to stand on my own feet when he was away. My duties were many, but primarily I was responsible for co-ordination of the repairs function through the District, for casualties, for shopping, for availability of locomotives, and for discipline at the out-stations. Within the depot at Stratford also I had certain responsibilities, but with a man like Dick Robson as the Shedmaster, these presented no problem. Dick had been a contender for my job when I was appointed; moreover he was my senior and had been acting as the Assistant for several months. But we had the happiest of relationships and nobody was more delighted than I when, a year later, he went to King's Cross on the promotion he so richly deserved.

Before I came to Stratford one of my colleagues on the Southern, an old Stratford man, told me it would not be long before I met the famous "Mr Groom". And if I were to be asked to say what was one of the most rewarding and interesting tasks that I tackled during my time at Stratford, I would rate my work with Jimmy Groom very highly indeed.

Jimmy Groom was a boilermaker. He was also the Chairman of the Workshop Committee responsible for the representation of the artisan staff of all grades in Stratford depot, a man-sized job handled by an elected committee of four men, assisted by shop stewards. He had been in the depot since before the war, when he served his apprenticeship, and after the war he became the men's leader during some fairly difficult, not to say turbulent times, when local agreements, that ultimately became national, were thrashed out and not without some stoppages and working to rule. I inherited the Chairmanship of an informal committee which had been set up several years earlier to take action on the many minor points of dispute which, if allowed to grow unheeded, could become serious and thus make matters extremely difficult for the District Officer. We covered an immense amount of ground, meeting as we did fortnightly or monthly.

When one deals with such a man as Groom, one learns to do one's homework beforehand. His committee always did so and were intolerant of inefficient management. Hard though some of our meetings were, the final solution was clearcut and invariably honoured. And maybe Jimmy Groom's greatest strength was in his ability to reach agreement with management and then see that the agreement was honoured by his constituents, the 500 staff that he represented. This ability was never more valuable than in the period following the introduction of the diesel locomotives, when changes were not only inevitable but radical, not to say revolutionary.

One day Mr Miller and I were coming back from lunch, and as we crossed in front of the New Shed, we passed the bunker-end of an ex LT&SR Class 3 4-4-2 tank engine, standing on No 6 Road. Recently we had received an ultimatum from the Southend enginemen that they would not take these engines off the depot except on carriage pilot work in the Southend Victoria Station Yard. The reason was that the injectors, of the suction type, mounted on top of the tanks inside the cab, were difficult to operate, particularly as the overflow was passed back into the tanks and was not fixed below the footplate as was the usual practice elsewhere. As a result, the water in the tank would overheat if there was difficulty in operation and then neither injector would work. In fact, if kept airtight and in good order, they were good injectors and the method of operation was quite easy when you understood the knack.

But the Southend and Southminster tank engine work was in the hands of senior drivers and very junior firemen, and so there was always trouble. Mr Miller asked me to get some Great Eastern injectors fitted, and after a fortnight of trial and error the job was done and the engine was in steam for its trial. Back at Southend, it went into traffic after a little argy-bargy with its driver, and did some good work. I submitted a report for Mr Miller to put to our Motive Power Superintendent, Mr Trask. In fact, we had no right to interfere with the design of the locomotive and we were told so in no uncertain terms, but after getting that off his chest, E. D. Trask, in his letter, said he had had the same experience with similar engines elsewhere, and thanks to his recommendation, several of the class were modified and did useful work until the railcars made them redundant and they were withdrawn.

In 1956 and 1957 life got increasingly difficult. The writing was on the wall for steam, the number of diesel shunters increased, the diesel railcar sets – which we then thought were a remarkable advance on anything that had gone before – arrived, as did one or two German four-wheeled railbuses for the Maldon and Braintree branches. Life got difficult because it was increasingly clear that, at Stratford, we were losing our battle with steam maintenance and even the "Britannias" on the Norwich services began to deteriorate. Many of the younger firemen were not experienced enough to cope with inferior locomotives on the lesser services, and some of the running shed fitters had to be taken off steam work to learn and then cover the railcars. We were always critically short of power, and at times equally short of manpower, both enginemen and artisan. We collected odd engines from time to time to help us along for a few weeks. For example, we even had a Western Region 94XX pannier tank for shunting work: we tried her on a Hertford turn but she was very tight for water. Then we collected an old Fowler LMSR 2-6-4 tank engine in the last stages of decrepitude, and ran her into the ground.

These were times when a sense of humour was vital and sometimes the pressures were such that one became overstrained and tense. It must have been in 1956 that an L1 2-6-4 tank, working up from Bishop's Stortford, had to come off its train at Waltham Cross with broken firebox stays, caused by dirt in the firebox water spaces. No sooner had this occurred, than a K3 2-6-0 belonging to us was found in similar condition at March. I remember the latter incident particularly well as I was in charge of the District during the absence of my new Superintendent, who was on annual leave. This meant serious trouble for me, for I was largely responsible for locomotive repair throughout the District. Dirt in the water-spaces of a firebox is simply because washing out and examination have not been perfectly carried out. We were all in it, boilersmiths, boiler-washers, boiler-examiners, running foremen, and the District boiler foreman.

In the endless quest for locomotives to work freight trains, engines would come in late in the evening, the steam would be blown off, the boiler would be washed out blazing hot, the examination would be sketchily done, and by the early hours of the morning the same engines would have been boxed up, filled up, and lit up ready to go out in the early hours. This had to stop and, if necessary, the trains had to wait. But thanks again to the help of supervisors whose job was now being made more difficult than ever, and to the support of Jimmy Groom and the Workshop Committee, we evolved a satisfactory system that gradually pulled the boilers round to a reasonable condition. But not without a fair amount of trouble, for there were times when staff, who had rushed and sometimes skimped their work in the past, found that they had to work the full eight or ten hours and there were often refusals to complete a half washed out boiler or to box up after washout which led one to say things to men that one might well regret afterwards. However, performance gradually improved, and with it, punctuality, and we pulled away from the late starts from the shed of three, four and sometimes five hours, to a system of nominated freight engines to jobs. Provided we started right on Mondays, we could play the operators at their own game and demand the right time return of freight locomotives from traffic. As a result, for some months, we began to achieve what had seemed to be impossible, but with the beginning of the transition from steam to diesel, and consequent alteration to the workings, we gradually lost ground with our nominated engines, although we maintained a high standard of boiler efficiency right up to the end of steam.

In steam days, most depots had a "bottomless pit". It was from this "pit" that Stratford used to drag its last resources in times of acute shortage. In those days, before one fully understood the commercial implication of one's action, any old freight train could go to the wall and it was traditional that we covered all passenger trains whatever else happened. In these more enlightened days, there is no bottomless pit, no old crocks and the allocation of power is rightly restricted. At the bottom of the Stratford pit were the express engines restricted to local freight work until they could go into shops for a general repair, the L1 2-6-4 tanks which were so rough that the knock of the axleboxes set your teeth rattling, the old K2 and K3s which would not percolate and the many Buck-jumpers, Gobblers and J15s which covered various imaginary turns such as "6 A.M. in steam", the Ark Pilot, the Boilermakers shunt and the famous "Double Extra". I shall never forget one Saturday in August 1957 when the shedmaster was on leave and I kept an eye on things for him. The Running Foreman, Charlie Benton and I worked closely together for six solid hours and, despite the ruthless use of other depots' locomotives, it was touch and go to keep moving. All through the morning we were providing engines with five minutes to spare, doubling up on anything that came to hand, getting Liverpool Street to service incoming engines in 45 minutes and lifting out of the "pit" the worst of the worst such as Nos 61613 and 61815, almost unfit for local freight, machines that nobody had dared to run faster than 30mph for weeks past, and slipping them on to passenger trains. We usually chose trains worked by foreign crews for this sort of thing as they would not know too much about it until after they got started.

The extraordinary thing was that we were almost "right time" all day out of London and almost "right time" on the road, which was unbelievable. But then Stratford was like that. The staff as a whole were remarkably resilient and tough and included some of the best railwaymen in all grades that I have ever met. And lest it be thought that many of the things I have said about Stratford are unfair, or exaggerated, I would stress when one keeps no diary, that one relives and remembers the things that stand out one way or the other. And to get the right atmosphere over to those that do not know and have not been there, one must talk first of the eccentricity of the place, before later on coming to some of the real achievements.

Londoners are famous for their sense of humour and we were all given nicknames, used normally behind our backs. If you asked a certain foreman in the District how things were going, his reply would be "We're on the floor, Guv"; so naturally his nickname was "Lino". Three Running Foremen, all on the same shift at Stratford, wore black Homburgs, dark suits and in winter when going off duty and home to Romford, black railway macks, so not surprisingly they became "The Romford Undertakers".

One of the Running Foremen, Arthur Davey, was a great worker, well respected, who answered, why nobody knew, to the name of "Jake". One night in the winter of 1956, at the height of what I might call a brass famine, one of the chargemen came to Jake to tell him he had found an axlebox in a highly suspicious position near "the outskirts", ready for "off". The Police – who were used to this sort of thing – were called and a posse, including Jake and the chargeman, crept stealthily to the spot where the axlebox was standing. At that very moment two men were just about to lift the box with a long bar under the crown; they saw they had been spotted and made off across the field. The police and Jake, who ran fast for his age, gave chase and almost succeeded in catching the thieves. But their pace began to flag and one of the policemen in desperation threw his truncheon at the rearmost – and put poor old Jake in hospital for a fortnight! But he soon returned, his energy unimpaired, still chewing pencils, singing and communing to himself, always a sign that he was up against it and enjoying life.

These foremen, if they did their job properly, had a hard life, including the 12-hour turn-round at weekends, for it was not unusual to come on duty on Sunday evening for 12 hours (after a week of nights) to find you were short of steam-raisers, short of boiler-fillers, short of repair staff, short of power for the Monday and short of enginemen. Each winter there was one of the brass famines just mentioned. Engines would come into the shed with two steam-heater cocks, front and back. There might well be engines waiting for steam-heater cocks to be fitted. There were none in the stores, and the only hope of a punctual start was to rob the incoming engines. Not every fitter would work on the disposal pit, so the engines would go down the back end, get coal and so on, and an hour later come up through the Jubilee shed *without* steam-heater cocks! Where these cocks went, nobody could ever find out. In 1957 I decided to order enough cocks to keep any normal shed going for the next ten years or so, and we just got through the winter without any late starts from the depot. Luckily we were able to accumulate this massive stock without too much difficulty, and with a minimum of correspondence, for there were some realists about who understood our little problems.

Talking of nicknames brings its reminder of the Stratford Works Foreman Painter, Mick Gabbitas, always known to us as Michael Angelo. He was our great ally, and took immense trouble to paint any locomotives particularly well that we needed for Royal trains and other specials. It was he who did what was necessary on our beautifully-kept Liverpool Street Station pilots and he also was responsible for the painting of the little J69 0-6-0 tank, No 68619, when she was turned out in Great Eastern blue. For was he not, indeed, the last Great Eastern Railway painter – the only man who really understood the composition of Great Eastern blue or who could correctly paint the cast iron GER crests that were so popular in earlier days?

In late 1956 we began to get some Welsh coal from outlandish pits like Deep Navigation, Ocean Park and Penrhiwceiber. The suburban locomotives were making too much smoke, and the Welsh coal was used in an endeavour to eliminate the smoke nuisance. For a time it nearly eliminated the production of steam as well, but while some of it could have been reasonably classified as "anti-glow", it was mostly good coal, and in a week or two most engine-crews had mastered the art of handling it. Time was the essential, as with the Welsh product you could no longer blow up a thin fire to produce steam at short notice. You needed a body well burned through, and then very light firing; once this was understood all went well.

One evening there was trouble at Hackney Downs, where No 69620, coming up from Chingford, had stalled across the junction. The Acting District Motive Power Superintendent, a very good friend of mine, told me to get off up to London to find out what on earth was going on. When I reached Liverpool Street, No 69620 had managed to arrive, had taken water, and had been attached to a crowded ten-coach train which showed no sign of leaving. In the cab I found two very warm gentlemen in their shirtsleeves, the dart on the footboard, and the fire looking dirty and unpromising.

The driver was a man for whom I had considerable respect, but with whom I had disagreed on many occasions. After some forcible remarks about the situation, he paused for breath, and this enabled me to look at the fire and to suggest that it was time we started. Also I told him that he would have the pleasure of taking it out of me, as I was going to be his fireman. So we got a little more steam and then off we went. The harder George worked the engine (he loved that!) the easier life became for me. With fire hitting the roof and cascading along the train, the tunnel past Bishopsgate became like daylight, and soon the fire was a beautiful white-hot mass of immovable Welsh coal. Up the bank to Hoe Street George let the lever out, and she blew off against the injector; all I had to do was to close the bunker door, tidy up, and shake with silent laughter. In our differences sometimes George got the better of me, for he was cute, but this time I was one up on him!

In my eight years at Stratford I was involved in many disciplinary hearings in and around the Depot. Many cannot be mentioned here, for the decisions were personal to the men concerned. Some, however, were of great interest, involving as they did at times a complete difference of opinion, and of evidence also, which made it extremely difficult for the Officer hearing the appeal to come to the right decision. I have never believed in "the benefit of the doubt". Each case must rest on its merits, and it is up to the Officer to reach a fair and just conclusion, remembering that there are two sides to every argument.

There was one case, however, that bears repetition, because it shows that life can deal some hard blows, and because such an incident is very, very rare. In my Ipswich days a certain driver of a freight train had stopped at a red light in the loop at Shenfield. He and his mate were not doing too well for steam, so he had a look at the fire himself. While he was working away, he looked up through the front cab spectacle, and saw his signal as he thought, showing yellow. "Righto, mate, give her a stir up", he called out to his fireman; then he got up in his corner, whistled, opened the regulator – and went into the sand drag! I could not understand this case, and thought it was simply an error of judgment. In fact, when this driver was sent for a medical examination, he failed his colour vision test, and that move he had made at Shenfield was the last of his driving career.

Passing along the main line through Temple Mills yard one night, a freight train came to a stand at a red colour-light signal. After a while, the signal apparently turned to yellow and the train restarted. They were stopped at the next signal and told by the signalman that they had passed the preceding one at danger. This meant an enquiry and then a disciplinary charge. At the time I was acting as the District Motive Power Superintendent, and at the hearing had to decide the level of punishment. All the evidence was against the driver and yet I was convinced he was telling the truth. The medical examination proved that he was doing so. So the charge against him was withdrawn, but he never drove again. Imagine his feelings – life is hard indeed to some!

Stratford Works presented as big a problem to its management as our place did to us. Our working liaison was very close indeed, not only between managers but between foremen. Many were the times when the works helped us out with some sorely-needed spare part or with a repair "on the quiet", so to speak. We, in our turn, were not beyond a spot of camouflage here and there to help the works output figures and our disagreements were never serious or extensive. Should we be tempted to cause a spot of bother over a locomotive that ran hot, or would not steam when it came out of the shops, we were always faced with the spectre of the bent and mutilated parts that had been removed from Stratford engines when they went into the works for repairs. These were kept in the Engine Repair Shop against the day when we might cause trouble and so proved an excellent deterrent!

By the middle of 1957, it was clear that great changes were coming. The Railcar Shed was all but completed and new diesels were forecast for the near future. At the same time organisational changes were afoot, and by the end of the year, the old departments had ceased to exist as separate entities. No longer could I look forward to the day when I hoped to reach the top positions in the Motive Power Department. I was to be moved to Hamilton House, Liverpool Street Station, together with my District Motive Power Superintendent, away from the centre of affairs at Stratford at the very time when matters seemed to be most in need of our constant attention. This was most disappointing and when the time came to leave the excitement and difficulties of the District Office at Stratford for the relative calm of the new Traffic Headquarters, I felt depressed and unhappy about the present, and intensely disturbed about my own future.

Little did I know that the years ahead were to prove even more interesting, exciting and enjoyable than anything that had gone before!

XVI

District Motive Power Superintendent

THE YEAR 1958 was one of change. Not only were we facing the new year with an untried organisation, when many of the old departmental barriers were to be broken down, and yet where much of the old pride in achievement and standard of performance was to remain, but we faced the start of a complete revolution in Motive Power to be completed on the Great Eastern Line by the winter of 1962.

I disliked the environment of a Traffic Headquarters at Liverpool Street. It was alien to me to be away from the rough and tumble at the very time that I felt I was most needed. I was no longer the second-in-command of one of the most important Motive Power Districts in the country, and although I knew many of my new colleagues quite well, I still felt that I had to fight passionately for the Motive Power interests that had been my life for the last twelve-and-a-half years. So maybe for a start, I was not easy to live with and as my power and influence was markedly reduced, perhaps I might be forgiven.

It took all the assistants, whether they were Operators, Commercial or Terminal and Cartage men, a little while to shake down. This was not to be wondered at, for in every department there were barriers which were not easy to remove. There were times during 1958 when I despaired of personal success, but by the end of the year I knew I was a member of a great team, working together as we had never done before. And when, early in 1959, I was promoted to District Motive Power Superintendent under our Traffic Manager, Harold Few, I was a very happy man. Not only was this appointment to bring me into close association with a Traffic Manager in a thousand, but also it was to give me nearly four years in charge of the Motive Power side during this great revolution, when only the people intimately concerned could realise just what was being achieved.

At the beginning of 1958, the steam locomotive was beginning to tire and the new diesels began to roll in. I shall never forget the arrival of the first Brush 1250hp locomotive. She was brought in by Fred Griffin, who had been a well-known "Britannia" driver on No 70037, and was taken straight into the works – into the works, mark you, when she was available for traffic! We were not allowed to touch her. Next day, the Chief Mechanical Engineer's people handed her over. I don't think they liked that part of it, but away she went to Liverpool Street, to work the 10.36am from Liverpool Street to Clacton, which, to our surprise, she did without blemish. At first, however, every journey was an adventure, for you never quite knew whether that gently-opened controller would induce movement or whether the little red light on the driver's desk would pop up brightly, enough to stop your breathing as it heralded an ominous silence behind you!

Not only did we have to learn, but so also did the manufacturers. It must be realised that once the locomotives began to come into traffic, the race to keep abreast of change, of training, of modification and of maintenance was on with a vengeance but again this story belongs to 1959 rather than to the more experimental days of the diesels in 1958. It was no easy matter to learn how to maintain the locomotives even with the help of the firm's representatives, and in this period it was fascinating to see how men suddenly emerged as potential diesel specialists after a lifetime of steam. And they were sorely needed, for in the early days our methods in the depots by no means always pleased the manufacturers.

We realised, too, that the cramped and dirty conditions of steam maintenance would no longer do, but we could see ourselves heading for overcrowding in the new railcar shed, which was opened early in the year, as we tried to maintain in it not only the railcars, but the rapidly increasing fleet of locomotives. However, we were able to hold on until the new depot was opened in the summer of 1960, followed by electrification in North-East London in November of the same year.

In 1958 the English Electric Type 4 2000hp diesels began to take a share of the working on the Norwich services, and by the summer of 1958 the monopoly of the "Britannias" had come to an end. These wonderful engines had revolutionized the running on the Great Eastern main line, as had the timetables which had been introduced to exploit their ability. Now they were to share the Norwich trains and to cover most of the Clacton services as well. The new Clacton timetables were built around the "Britannias" and we knew our men very well. Most of the more elderly gentlemen at Clacton on the B17 4-6-0s had been content to keep time – but only just – on the very easy timings then in force, which were no inducement to sharp and smart running. Life being what it was, I knew perfectly well that when the new timetable came out, we could not guarantee a "Britannia" on every "Britannia" timing. And I knew perfectly well that time would be lost by the B1 or B17 substitute with some of the crews who were regularly on the Clacton–Liverpool Street services unless we did something about it.

The only way was to prove the point physically beforehand, and then to ensure that when a "Britannia" turn was covered by a less powerful engine a locomotive inspector would be on board to make sure time was kept. To prove the point, it was necessary to do it oneself, and I selected for the purpose the 5.36pm from Liverpool Street to Clacton, which ran non-stop to Colchester in about 64 minutes. It was difficult to lose time with this train if you tried, and yet time was sometimes lost. The engine selected for this self-imposed trial was No 61666, a "Footballer" B17, and the Driver, Stan Pittuck, was an outstanding engineman. Our aim was to get to Colchester as soon as possible, but I made no special arrangements as this was to be a personal trial. In firing I had to stand up to whatever treatment Stan Pittuck meted out, but our net running time was no more than 52½ minutes for the 51¾ miles. We filled the smokebox with ashes before the day was out, but we had proved the point that one of these locomotives could work to the Class 7 schedules. And having made this point, we were able to stick to it when the new timetable came into operation, for the word got round, and even when the less powerful locomotives were used time was generally kept.

Much has been written about the "Britannia" Pacifics, not least by E. S. Cox, who had so much to do with the design. So I will not dwell on the aspects that have been so ably covered. In 1958, because of the increase in the number of diesels at Stratford, the maintenance of all the 4-6-2 locomotives was transferred to Norwich, where they came under the care of my friend David William Harvey, certainly the finest practical steam locomotive engineer it has been my honour to know. He did one or two little things to his fleet to raise them in their last years on the Great Eastern to a tremendous peak and the week before they ceased to work into Liverpool Street I achieved my ambition to start Brentwood Bank at 60mph and finish at the same speed, with the standard eight-coach train on the 5.40pm from Liverpool Street to Clacton. The fact that all the "Britannias" were now maintained at Norwich interfered with the regular manning at Stratford, but this was inevitable in any case because diesels had to be quickly and effectively fitted into the workings, and the greatly improved standard of maintenance more than offset the loss of a regular engine.

Before the transfer, Stratford ran No 70000 and No 70042 as spare engines, No 70000 generally covering the "Hook Continental", whilst each of the remainder was allocated to three sets of men. For three weeks out of thirty-six, the crews lost their engines on most days while they covered the rest days of other men throughout the link, but otherwise they kept pretty regularly to their engines and this was reflected in the condition of many of the cab interiors.

Some men went to great lengths to clean, scour and polish the interior of their cabs, and one or two of them cleaned the boiler, cab and tender as well as the rods and motion – no mean task on a "Britannia" between trips. We helped by issuing "Derby" paste for polishing and tallow for the firebox front, tender end and cab roof, but not much external cleaning could be done in the depot. There was an acute shortage of firemen and as soon as boys passed their sixteenth birthdays, they went firing, to return but rarely to the hard and sometimes unrewarding work of cleaning steam locomotives.

The "Britannias" were in excellent hands when I came to Stratford in 1955.

The drivers working in the link had had some three years on the Norwich services and most of them remained on the expresses until well on into 1956 before going on to L1 2-6-4 tanks, or into the Temple Mills shunting link, where some people liked to spend their last years' service to avoid the rough-and-tumble of the Woolwich link and its decrepit "Gobblers". After 1956, the speed of retirement at the top was so great that some drivers were in and out of the Norwich link within twelve months, sometimes less, which was no good to them, or to us – a poor reward for forty years on the footplate and an expensive proposition in terms of road-learning, for all the Norwich work was covered within the one link.

The endless changes in the link and the gradual deterioration in the condition of the main line engines until the transfer to Norwich was effected, coupled with an acceleration of the schedules, brought an instability into the running that had to be challenged. And fight it we did by all the means at our disposal at Liverpool Street in 1959, and effectively too, so that the last days of the "Britannias" on the Eastern section were as brilliant – and some would say even more so – as when they came to us new in 1951.

There were many stories told of the work of the "Britannias". During 1955 and 1956 No 70037 was certainly the smartest in appearance run close by Nos 70001, 70002 and 70038. In time, as always, there were changes, and the palm passed to Nos 70036 and 70039. There was never much to choose between them in performance, however, and every driver preferred his own engine to anybody else's.

No 70037 was run by Drivers Griffin, Rolstone and Philpott, although the latter emigrated sometime in 1955. She was used for many of the special jobs that came along, including such Royal or semi-Royal workings as came our way. Fred Griffin was a great personality. For a start he was a good engineman; he was always dressed in spotless overalls, with a white collar, tie, clean shirt, and his cap carrying the highly polished batswings, the insignia of the Great Eastern Railway. He did a very great deal of cleaning himself. One evening he and his mate set to work to clean the paintwork on No 70037, which was standing at the front of the Jubilee Shed opposite the messroom. "Griff" was generally impervious to criticism, and as they worked away they were given plenty of gratuitous advice from men waiting for work, sitting on the bench outside the messroom. Fred and his mate carried steadily on, and having finished one side, he walked across to the group of amused watchers, and said to them, "Now, boys, bring your bench along and watch us do the other side!".

The "Britannias" went through Crewe shops for their intermediate and general repairs and the London Midland Region would use them for several weeks before sending them back to us. The cabs were always filthy and the brass and copper work was always painted over. This we tried to get put right, but letters and telephone calls were of no avail. When No 70037 went to Crewe, we decided to try sterner stuff, backed with a spot of publicity. In addition to correspondence with Crewe Works, with a formal request for them to leave all the polished work unpainted, we had the engine cleaned to perfection on the day of her departure for general overhaul. Fred Griffin was booked to take her to Camden and, on arrival there, offered her for the "Midday Scot". I doubt at that time if there were many engines in the country in her state of sartorial excellence and to say she caused a sensation at Camden was putting it mildly. So she worked down to Crewe (but not on the "Midday Scot") and disappeared into the works. Crewe kept their promise and gave her a thoroughly good going over, leaving the bright work untouched, but when we began to receive repair cards from Carlisle, Perth, Polmadie, and such like places, we knew that the worst had happened, that the London Midland and Scottish Regions were using her, that she was nobody's baby, and that she would come home with her steelwork red with rust. So she did, but at least it had not been painted over, and after the initial wailing and gnashing of teeth she was soon round again in her old state of perfection.

No 70036 was an excellent performer and was always kept clean. In 1957 and 58 she was run by Ted Whitehead, who was Ipswich born and bred and who very rarely lost time with any engine, let alone his No 70036. Incidentally, this engine must have been the first "Britannia" ever to be driven by a Frenchman. In Ted's last week in the link, in September, 1958, he had André Duteil of La Chapelle, Paris, with him on the 3.45pm from Norwich. André was the driver from Ipswich to Stratford, under careful guidance, needless to say. Certainly this was some change from his enormous 4-6-4 No 232.S.002 of the SNCF, but he did his work beautifully. When he ran into Chelmsford with the station lights shining on his goggles, unknown in this country, somebody in the cab said that he looked like "the man from Mars!".

As for No 70001, she was run by Dick Brock, A. Snell and W. Cardy. They always burnished the top of the vacuum brake pedestal in the cab. Then either Snell or Cardy moved out of the gang and the vacant place was taken by A. E. Page, who came off B1 4-6-0 No 61399. I happened to be at Liverpool Street one day as the "East Anglian" ran into No 11 platform with Driver Page in charge. As the engine came to a stand, the relief crew, Dick Brock and his mate, Billy Hart, got into the cab. A teacup had obviously been standing on the top of the brake pedestal, leaving its trade mark behind, and Dick, having looked at it critically, said with much emphasis "Look here, Pagey, if you're coming to work opposite us, we're not having any of your so-and-so 1399 habits on *our* engine – and don't forget it!" Perfectly true and he meant it.

Albert Page was a character in his own right. He and his mate Charlie Gunner were neither of them silent men but they could do their job extremely well. I had more than my share of journeys on their footplate, testing doubtful locomotives if theirs happened to be laid off, and on one occasion proving to a visitor from the Southern that the old B1, No 61280, could do anything – or nearly anything – that a "Britannia" could do. Albert had a smile that was hidden as he spoke but came out like the sun immediately afterwards. If a man said to you – and not many would! – "'Ere, guv, I bleedin' want you" at the top of his voice in the depot yard, you would tell him what to do about it. Not so with Albert, for as you drew breath, that smile would come out and rob the words of all offence.

But in 1958, the Norwich service punctuality had begun to deteriorate for the reasons already stated. I decided to challenge this trend and the only way I could think of, other than by improving the condition of the engines, was by introducing an additional Inspector to ride on the 9.30am from Liverpool Street to Norwich and the hard 1.45pm return day after day, week after week, month after month, so that all the crews were ridden with. In this way my views on the importance of punctuality were not only fully understood but acted on without fail.

I was lucky enough to pick a unique man for this job. In fact, he never became an Inspector and up to the time of his tragic death in January, 1960, he remained a driver, although paid the rate for the Inspector's job. Had he lived, what an inspector he would have become! W. J. Mason was the man, conscientious, a strong and bold engineman, a little fellow, skin and bone, who had the great gift of being able to communicate at the right level, so to speak, with any driver or fireman with whom he was travelling. Wally could swear, if it was necessary; he could coax or act "diplomatic"; he instinctively knew how to handle a difficult situation or a clash of personalities, and there is no doubt at all that the impression he made on Norwich punctuality was remarkable. My instructions to him were to get to Norwich right time, never mind the condition of the engine or the coal or the weather or the fireman's back. Wally got there right time most days and through him – and the regular Inspectors who did an amazing job in the last years of steam, and through my own observation and work – we pulled the running back, not only on the Norwich trains but on the other services. So we got the best out of steam in its last days, and not only so but in this way I got to know the capabilities of the many crews employed on main line work.

Although a very sick man, his dedication pulled him through day after day. The desire for right time running made him do things that his body sometimes resisted, for even a flagging steam pressure needle would demand his attention if he thought fit. When the French drivers André Duteil and Henri Dutertre (from Calais) came over in 1959, he looked forward to taking them to Norwich on the 9.30am. This he did, and the journey was a great success. But he never worked again, having steeled himself to hide his inner feelings from everybody else, and he died a few months afterwards, well under 60 years of age. Over the years my own wife has had to hear a lot about railways and so it is with most railwaymen, for ours is work that must by its very nature come into the home. Similarly everything that happened during every day was relived in the Mason household, where, in pre-war days, Jesse Mason had catered every week for more than twenty up-and-coming firemen who were being put through a course of "improvement" in rules and enginemanship by Wally and the other spare time instructors. He was a great little man!

Welsh coal had come and gone by 1958 on the Jazz services. However, it became increasingly clear that we were going to have great difficulty in making the N7 tanks stay the course on the Chingford, Enfield and Hertford services, until the electrification in November 1960. A decision had been taken to withdraw engines of this class as they reached the normal mileage for a general overhaul and by the middle of 1958, with no cover of engines for the shops, we began to get into real difficulty, We had pointed out in no uncertain terms just what sort of a service we should be giving, but to no avail until the decision was reversed in 1959 by the newly-appointed Chief Mechanical Engineer. There is no doubt that this change saved the day for us. A limited number of engines were "half soled and heeled", in other words they had their axleboxes, crossheads, large and small ends, side-rods and valve gear re-fitted, and a certain amount of boiler repair also was carried out. This kind of work gave little trouble to the main Stratford Works, whereas in our state of transition it would have been quite impossible for us to effect repairs on this scale at Stratford locomotive depot.

As a result, the "Jazz" services went out in a blaze of glory, with no time lost by locomotives in the last week of steam working in November 1960, and many of the crews kept their own engines right up to the end. That reversal of the original decision was a triumph for commonsense.

The Westinghouse brake when properly maintained and handled, indeed when handled with courage and confidence, has no equal. But it also used to have an unknown quantity that was missing in the solid, dependable, but dull old vacuum brake. The Americans have a wonderful railway folk song called "The Wreck of the old 97". In the second verse come the lines.

"It was on this grade that he lost his air-brake
So you see what a jump he made".

Now and again, suburban trains on the "Jazz" used to lose their air-brake, but we never got as far as making any jump off the trestle! But many men, if such a failure occurred and the remainder of the journey was on the level or uphill, would get to their destination on the hand-brake with the help of the guard.

One day I was going to Wood Street depot, and just before 2.15pm, stepped on the footplate of a long-travel Stratford N7 0-6-2 tank with Driver Harry Hibbert. He asked me if I was going to take her down to Wood Street and I said, "Yes, I will". Our journey was quite uneventful in the early stage; my stops at the stations to Hackney Downs were perfectly orthodox "service" applications, but running into Clapton on a falling gradient, I had to make a sharp application to stop. Looking up at the gauge, we saw that our air needles had dropped back, so the fireman went out along the side of the engine to give the top cover of the pump a tap (or more). This had no effect at all; the pump would not restart and the guard was urging us to get a move on. Brake or no brake we couldn't stay where we were, as there was a Cambridge express behind us which we would have delayed. So we set off hell-for-leather past Clapton Junction and on to the branch. It is uphill all the way from there to Chingford, so there was no cause for alarm. We had a good guard, a good hand-brake, and we only lost three minutes with our slow stops. Wood Street found another engine, took the top-head off the pump, and soon sent our N7 back into her working again.

The two pilot engines at Liverpool Street Station were kept beautifully clean. The day shift men were paid an hour overtime a day to keep things right and the London Running Foreman used to see that the work was done. Occasionally enginemen used to say that their cleaning days were over and refused to do this work – which was not compulsory – and other men devoted hours between shunts and trips to the cleaning. The result was two locomotives cleaned to pre-1914 standards, admired by all, and inspected by all manner of city people from time to time. Two pilots certainly were a luxury, for the little 0-6-0 tank simply did a spot of shunting, covered the New Cross freight trips if the booked engine failed, and ran through the East London Line tunnel with a night freight trip. The bigger engine, the West Side pilot, was an N7 0-6-2 tank, No 69614. Her job was to cover the entire steam suburban services and she was a useful machine, stepping into the breach on many occasions in the event of last-minute failure.

When the office moved to Liverpool Street, I used to make my last port of call on the way home the Running Foreman's office at the country end of No 11 Platform, usually calling in at about 6.0pm. One evening during the bus strike of May, 1958, I called in early, as we had friends coming for dinner and I wanted to get home by 6.30. The Foreman told me that the West Side pilot had gone out on the 4.10pm to Hertford, and that the little pilot, No 68619 of Class J69 would have to cover the 5.29pm to Chingford. It was now 5.20, the engine had just moved from the centre of the station, the fire had recently been cleaned and had been re-made with good Welsh coal. These pilots were the only engines left in London burning the Welsh product, and soft coal was the last thing we wanted to use to build up a recently-cleaned fire.

I walked over to No 2 platform where 68619 was just backing on, with Driver Alf Peeling, 60lb/sq. in. steam pressure, and an inch of water showing in the glass. The load was ten coaches, packed to bursting point; the platform was packed with those who could not get in. At 5.29, we had 75lb/sq in pressure but still very little water in the boiler. Five minutes later we were still there but the longer we stayed, the more we shut the station up, and by the time we had 90lb/sq in and half-a-glass of water, I said "Away we go", thinking we might not get very far, but at least we should let another train into the terminus.

The little engine, in full gear and with the regulator full open, climbed Bethnal Green bank, and just got round the curve beyond Bethnal Green, but she had enough steam to keep the Westinghouse pump beating and the train running. Half-way to Hackney Downs (we were on the fast lines) we shut off to get some water in the boiler, but with pressure back to 60lb/sq in neither injector would work properly, and we were in desperate trouble. While driver and fireman battled with the injectors, I kept the train moving and watched out for signals. Somehow or other we got to St. James Street none the worse, and released some of the passengers from their purgatory. I got off to ask the Control to get a Wood Street engine on the front to help us up Highams Park Bank, but when I got back to the engine I found that good Welsh coal had well and truly caught alight, so away we went again, full of hope. By Hoe Street, we had 100lb/sq in and could begin to thrash the little engine.

My order to Wood Street was carried out promptly and an N7 0-6-2 tank backed on to help us. But unfortunately this engine had been put away for the night and the boiler filled right up, with the steam pressure has been knocked back to below 100lb/sq in. So he was not much use. For a start he smothered us with water, and we had to push him till he got warm. Eventually he muscled in, however, and by that time our No 68619 was in tremendous form. So we drew into Chingford blowing off like fury from the safety valves, and with the fire in perfect steam-making condition. But we were 25 minutes late. The passenger's comments luckily could not be heard too clearly, because of the noise of escaping steam, and we went home light engine as quickly as we could. Everybody had done his best, but that was no compensation for infuriated commuters.

Although I realised by the end of the year that the new traffic organisation was a winner, I had had enough of 1958. For the first time in my career, I had doubts as to my own future prospects in the railway service. We were not doing well and we seemed to be gradually losing ground.

But 1959 was different: indeed 1959 to 1962 were great years. Complete dieselisation on the non-electrified Great Eastern Line, a major suburban electrification scheme introduced, conversion from 1,500 V dc to 25kV ac on the Shenfield, Southend and Chelmsford services, electrification to Clacton, all this was part of a complete revolution in traction. This was enough in itself, but all the time the standard of performance on the Great Eastern Line was rising despite the early difficulties with the high voltage electrification, and despite the immense amount of bridge-building, track relaying, week-end work, resignalling, station rebuilding that involved many people in hours of planning and careful thought and gave no respite at week-ends to operating men who were driven hard enough during the week.

XVII

Stratford MPD – the Transformation

To DO JUSTICE to the events in the years 1959 to 1962 in a single chapter, or even two, would be an impossibility. A book would be needed, and even then not everything would be included. To many of those involved, whether in the Division or in the Great Eastern Line Headquarters, it was without doubt the most rewarding, unforgettable experience. We were up against time, against problems we had never faced before, managerial and human problems, a whole way of life changing for every man on the locomotive side of the Great Eastern traffic organisation.

The operating people worked tirelessly, together with the Civil Engineers and Signal Engineers and our own breakdown gangs, relaying, resignalling, rebuilding, putting in new bridges – endless week-end hours of toil, fighting to be ready in time for each Monday morning's services. There were overhead structures to be erected for the 25kV ac electrification, motorcoaches to be converted from 1,500V dc, new stock to be tested, and a hundred and one other tasks to be achieved by these Departments and also that of the Chief Mechanical and Electrical Engineer.

It was essential to think ahead, to plan and scheme, and to keep the staff in good heart, for I was determined to keep on top, in advance of requirements, and never to get out of my depth in the gathering wave of dieselisation and electrification. It may seem ironical to my readers when I say that no one tried harder than I did to bring about complete dieselisation of the non-electrified lines in the Liverpool Street Division, but equally no one made greater efforts to get the best out of the remaining steam before it became a thing of the past in November, 1962.

Electrification had come to North-East London in November, 1960, and relieved me of a maintenance responsibility for the ageing steam fleet. Then the electrification gap between Chelmsford and Colchester was to be closed, so that the electrics might run through from Liverpool Street to Clacton.

It was on January 26, 1959, that I became District Motive Power Superintendent, temporarily at first and then permanently from the following August, when we all became Running and Maintenance Engineers (Designate). In due course we were to take responsibility not only for locomotive running and maintenance, but also for carriages and wagons, road motor vehicles, stand-by sets for electric signalling in the event of power failure, station lighting and even station barrows. Indeed, there was precious little as to which we could absolve ourselves from responsibility, except, of course, electric traction, which came under the Chief Mechanical and Electrical Engineer!

In our Division, and at the Great Eastern Line Headquarters, the Commercial men, the Operators and the Motive Power people had become drawn together as a team as never before. Naturally, we had our disagreements from time to time, for each of us had to fight for what he considered to be essential for the good of the job, but such an organisation (from which there evolved the Divisional set-up of to-day) could do no other than thrive in such conditions of challenge and excitement, and under the type of leadership that we were fortunate enough to have.

My own Divisional Manager, Harold Few, has now retired. Bent as I was on achieving my own targets, I did not always appreciate the immense strain to which he was subjected, the type of decisions he had to take, the compromises he sometimes had to make. The rougher life was, however, the calmer he became; if we really were up against very difficult problems, as when the new electric services were running badly and public opinion was very critical, he was as solid as a rock. He let me run my side of the job, gave me freedom to act and to take decisions, and yet was always there if need be to help and to encourage. He supported us in what we were trying to achieve, and drew my colleagues and myself together in one common purpose. In other words, Harold Few was a Divisional Manager of great standing and ability, and as I have just said we were fortunate indeed to have him in charge.

When I was moved up, the post of Motive Power Assistant also became vacant, and the appointment made was for me another stroke of luck. Or, maybe, it was a stroke of genius on the part of other people who knew their man better than I did. By now I was 35 years old, which in those days was quite young for a senior officer, even if I was getting pretty experienced; but my Assistant was to be a Bert Webster, a man of 57 years. This, was however, the happiest of partnerships. "Webby's" experience was of the right sort. He would try anything new, but occasionally, when I was pushing ahead too fast, his advice proved timely in getting me to pause and think. So our temperaments produced just about the right blend of management for those particular times.

"Webby" was a London & North Western man. After the 1914-1918 war he had been a Crewe apprentice, later working as a fitter at Crewe, at Bow Works of the North London, and for some years at Stoke-on-Trent, where he had been the only "Nor'-West" man among a crowd of "Knotties", as the North Stafford men were known. He had come up the hard LMS way, as Shed Foreman at Swansea, Widnes, Preston, Speke Junction and Trafford Park, at the last of which he had had an illness. After this he got away to March in the Eastern Counties. Merely to say that he was respected at March as Shedmaster would be an understatement; here he might well have stayed until retirement but for the need to find an Assistant for me at short notice.

And so Bert Webster and I started operations together. He was very shrewd and an excellent judge of men. Above all, he was one of the most amusing men I have ever worked with, and this wonderful gift of the ridiculous kept us both in tune with one another and with our work, even when things were at their most difficult. I remember one evening, just before he went home, when he looked round the door between our two offices. I was feeling jaded, and he knew it. He was wearing what had been a bowler hat, but a section had been taken out of it, turning it into a low-crowned clerical hat of the kind worn by the more elevated dignitaries of the church. Lower down was a white cardboard collar back to front, and lower still some BR issue gaiters round his trousers. The arrival of this apparition, and the remarks that came from it, made me laugh till I cried, and so I went home in good spirits and ready to tackle whatever the morrow might bring, or, for that matter, the telephone during the night.

On another occasion we escaped for an hour one day when we had been at Bishop's Stortford, and looked in on some friends who lived at Audley End.

They had a steam-roller for work on their land, and neither "Webby" nor I had ever been on one in our lives. I was offered the wheel and the regulator, but he would not come on board and stayed on the ground. Here he acted as a policeman, running backwards while I steered the machine drunkenly from one side of the lane to the other in an endeavour to keep a straight course. My shirt remained clean for rather less than a minute, and as the big-ends began to fly round under my nose I soon learned my lesson! "Webby" needless to say, was highly amused.

The year 1959 opened with about 30 main line diesel locomotives allocated to Stratford, and accommodated in the railcar depot. Since its opening this small shed had been overcrowded, and as a result the pits were dirty and there was a general air of slackness. The locomotives were dirty, too, as we had no youngsters to clean them, but many of the staff had gained a grounding in the mechanical side of diesel maintenance at the works of English Electric, Brush, Mirrlees and other builders, while a start had been made also in driver training. Two main developments were in view. One was the opening of the new diesel depot at Stratford in the summer of 1960, and the other was the inauguration of electric traction in North-East London in the following November.

The training of electric drivers had to be carried out at Clacton, as the only stretch of the Great Eastern Line electrified up till then at 25kV ac was between there and Colchester. Some 140 men had to be trained, which meant that they had to lodge at Clacton and their places had to be filled by men from Stratford; but these were the least of our problems. One tended to forget the task that faced drivers at the smaller depots, such as Colchester. Here a senior driver in his late 50s and early 60s had to learn not only how to handle the electrics, but also five or six different types of diesel locomotive as well as assorted shunters, while in 1960 he might still from time to time be working a steam locomotive. However much variety may by the spice of life, these men certainly went through a strenuous and very difficult time, though they emerged from it with great credit.

At Stratford Depot on any one day there must have been about 700 drivers and passed firemen. As men retired, others were promoted from link to link, so being required to work on locomotives different from those on which they had been trained previously, or on the railcars newly arriving from Derby and powered by Rolls-Royce engines. Approximately 200 men were on regular shunting duties or on shed work, while the other 600 were spread over the fourteen suburban passenger, main line passenger and freight, and local freight gangs. It was my Utopian aim to have every driver capable of handling all the various diesel types, but I found that such an ideal was virtually impossible of attainment. It has to be remembered that in steam days there was no recognised training in the handling of particular locomotive types and no corresponding examination. A driver was expected if necessary to get on a strange engine and get cracking, finding out about his charge as he went. In consequence, Running Foremen had considerable flexibility in the deployment of engine-crews. But with the numbers of diesel locomotives increasing every week, and faster than we were training, the men, something drastic had to be done, and quickly.

After the English Electric Type 4 2,000hp 1Co–Co1 diesels had arrived a day's work for one might be two double trips from Liverpool Street to Norwich and back, or one Norwich trip and a freight turn. Apart from the trainmen in the Norwich link, sets of men were required in London to relieve the long-distance men on arrival, and if the locomotive then was required for a freight working, one of the goods links would be involved. So, in 24 hours, three or four different gangs might be concerned in the working of one locomotive. And the only way to avoid endless chopping and changing of the men's rosters was to get on as fast as possible with the training, which went on day after day, month after month, sometimes involving up to 80 men each day throughout the Division, including the electric training.

Each diesel type required a basic training of three weeks or a single week's conversion course. By 1960 the different diesel types comprised the Type 2 Brush A1A–A1As of 1,250 and 1,365hp, the Type 3 and Type 4 English Electric Co–Cos and 1Co–Co1s of 1,750 and 2,000hp respectively, the North British Locomotive Type 2 Bo–Bos of 1,100hp, and the Type 1 North British and British Thomson-Houston 800hp Bo–Bos, as well as the many smaller shunters. For all these varieties training was needed, as well as simple conversion courses for electric drivers about to change from 1,500V dc to 25kV ac traction. That this vast programme was carried through in so short a time was an extraordinary achievement; it may have been comparatively straightforward to organise, but the speed at which it was carried out put a very considerable strain on all those who were involved, and few of us who were intimately concerned will ever forget the problems that the foremen, the list clerks and the control faced in finding enough men to keep the normal services going.

Some time in late 1959 or early 1960 a decision was taken by my Divisional Manager, together with our Line Manager, that in my experience was unique. We had our targets for training, but the normal train service had to run, and in the absence of drivers under training the Motive Power side would cancel this or that freight train to find temporary replacements for these men. But we were a traffic organisation and my commercial and operating colleagues were understandably critical of my insistence on reaching the ultimate goal at all costs, especially as freight receipts were being seriously threatened.

So the bold decision was taken severely to pare down the suburban steam passenger services during the off-peak hours, so that the small depots could provide their own relief to cover their men under training. This saved the need for Stratford to send drivers, say, to Enfield to cover these vacancies. This move gave us more men at Stratford, and with the decision that certain freight trains *must* run on certain days, come hell or high water, the situation was put as nearly right as it was possible to get it. It is unlikely that in the old departmental days a decision of this radical nature could have been put into operation so effectively and with such enthusiasm; indeed, I doubt if it would ever have been taken.

So much for the difficulties of the footplate men. But what of the artisan staff? Here the problem to be faced was even greater. In the summer of 1960 the new diesel depot was to be opened, providing us with space, facilities and accommodation the like of which we had never even dreamed of in steam days. Between steam conditions and the as yet uncompleted depot there yawned a great gap, for most of the Stratford steam locomotives by now were in poor shape, and the steam sheds were riddled with restrictive practices which bade fair to stultify progress from the start. On top of all this, we were short of fitters and had had next to no electrical experience.

During 1958 the depot had been responsible for the mechanical side of the diesel locomotives, and the main works for their electrical maintenance, but with this arrangement there were all manner of complications, and it was a blessing when both sides came under our control early in 1959. With this transfer, however, very few men came across to us from the main works as the running shed pay was lower than that in the works. So we had to make a start both untrained and understaffed, and knowing that it would be almost impossible to attract traction electricians from outside industry.

So we took a decision that might well have been criticised in later years, but was the only one possible at the time. It was quite unconventional, and it would never have been possible at all but for the ability of the Workshop Committee, under Jimmy Groom, to foresee the problems of the future; and it worked. As we had next to no electricians and no training school, and time was not on our side, the decision was to start a small school of our own in the depot – a homespun unofficial outfit run by Groom who, though a boilermaker, had studied electricity seriously as a hobby. So we worked out a four weeks' course in the school, followed by a much longer practical course in the depot, and were able to convert certain of the boilermakers and fitters, who had the right aptitude, to the electrical side. Those available for training were chiefly men who were bent on learning a new trade, and some of them, like Groom, already had made a study of electricity.

This was not done without opposition, and there were many obstacles to overcome, but the initial group of men whom we trained completely justified our faith in what we felt to have been a right decision. Even then, such a transition is not easy for a man who has been trained in so totally different a craft as boilermaking. It takes months, and sometimes, indeed, years to achieve the highest standards in a new craft, but this was done all the same. In time we were able to introduce dual-trained men working on both the mechanical and electrical sides for a bonus of so much a week, first on the railcars and later, tentatively because of the complexity of the machines, on the main line diesel-electric locomotives. Thus we had the fascination of being in the lead. We had to take the decisions, for nobody else was "in the know"; it was just a matter of going forward and learning as we went.

Possibly the greatest hurdle we had to surmount was the move into the new depot. This meant the closure of scores of messrooms and the elimination by mutual agreement of many restrictive practices. Having got thus far, there had to be insistence on the highest possible standards of discipline in maintenance, and, perhaps even more important, in the cleanliness of the depot and of the machines. In brief, we had to create an entirely new attitude of mind.

One of the men who had to take a large share of the responsibility for the transition was the Shedmaster himself, Norman Micklethwaite. He was trained at Doncaster, and was both forthright and determined. We used to laugh at "Mick", who never pulled his punches. It was said that if he got a strong note or a telephone call of a critical nature, he would bound out of his chair, bang through the office door, and storm across the front of the shed "full regulator and 60 per cent", with his head down and his temper up, bent on the immediate task of putting matters right. But we had the best possible man for the job, and for several weeks before the transfer he and several others, including members of the Workshop Committee, spent a great deal of time working out the staffing plans and the many small points of detail which needed to be covered before the day arrived.

When we moved into the railcar depot in 1958, the staff had to work among a clutter of contractors' ladders, with some of the pits not available, and the stores only half finished. In 1960 there seemed every likelihood that history would repeat itself, and this would have spelt disaster to our plans. However, commonsense prevailed, with the decision to put back the date of opening until we were really ready. I had been sent to Winterthur in Switzerland in October, 1959, to visit the Sulzer Works there, but we had also been shown round the running shed of the Swiss Federal Railways at Zurich, where I had never before seen a standard of cleanliness such as I have just referred to as a new "attitude of mind". This is how one of my colleagues described what we witnessed at Zurich, and this is what we hoped to introduce when the new Stratford depot was opened in August, 1960.

The decision to defer the opening was not an easy one to take, for it is a responsibility of management to push ahead with developments with the maximum speed possible; this meant delaying the implementation of a big scheme in which many different aspects of the railway business were bound up with a specific target date. But to us that deferment of two months meant all the difference between weeks of muddle and mediocrity and success from the start.

Well the day came, we were ready and everything went as planned. To start right with the cleaning programme we used no fewer than 36 cleaners, working round the clock, and adults, not youngsters. This was an expensive business, but it was worthwhile. The depot floor and the pits were given constant attention. Some of us remember the little Pole who used to clean the back end of the Jubilee shed in steam days; he took on the back end of the new Diesel depot, and like so many of his countrymen turned in a wonderful job. If such a thing as a cigarette end was seen on the depot floor, the chances were that whoever dropped it there in the first place later had to pick it up and put it in the waste bin.

The main line locomotives were cleaned on arrival at Liverpool Street by a small gang of adult cleaners under the guidance of a driver whose indifferent health prevented him from working trains any longer. We were determined that "handsome is as handsome does" should be the motto for our fleet, and this meant that the locomotives must be clean. Up above in Broad Street one could see the beautifully-kept Finsbury Park diesels, and we could never stomach the thought of playing second fiddle to the Great Northern. So our standards improved, and despite the many difficulties we still faced and any imperfections in our performance, we felt that Stratford had made a fresh start, and that the past had been well and truly buried.

One day I remember infuriating Micklethwaite by pointing out that his locomotives still were not clean enough. He disputed this, so I sent him a photograph which had been taken by a railway enthusiast from the top of the Stratford water tank. This showed a line of his excellent diesels standing below, clean and shining, including their roof tops, except for a black line along the length of each roof which the cleaners had not quite been able to reach!

Perhaps our greatest day after the opening was February 15, 1962, when the Queen visited Stratford Works and the new Diesel Depot. Thanks to the selflessness of my superiors, it fell to me to take Her Majesty on a tour of the depot. About a year previously, the Great Eastern had earned a black mark because a Type 2 Brush diesel *(not* a Stratford engine!) had failed on a journey made by the Queen from Liverpool Street to King's Lynn. Not only was I able to take her into the cab of a similar diesel, No D5694, but also to show her the actual component that had caused the trouble. Then I introduced members of the staff to her, and she left us all with the feeling that the sun never set at Stratford!

An hour or so before the Queen's arrival, a member of the Workshop Committee handed me a poem that had been composed by a fitter's mate in the depot, by name Bill Mankin. It was a simple and genuinely sincere piece welcoming Her Majesty to the depot, and comparing the new conditions with the old. I managed to arrange for the poem to be handed over by Norman Micklethwaite to the Queen's Private Secretary, Sir Michael Adeane. Next day I received a note from him saying that the Queen had read the poem, and asking me to convey her warm thanks to Mr Mankin for his kind and loyal good wishes. Bill Mankin certainly will not forget that day for as long as he lives.

Neither shall I forget the departure of the Royal Train from alongside the new depot. Standing outside what was left of the Jubilee shed were four locomotives – two spotless diesels, a clean B1 4-6-0, and a dirty little old Westinghouse goods, one of the last of the Class J15 0-6-0s still in service – which we had so arranged in order to portray the change that had taken place over a period of sixty years. Each locomotive was packed with men, standing on the gangways, on the cabs, roof tops and nose ends of the diesels, and on the coal of the J15's tender, cheering themselves hoarse. The diesel horns were sounding, but clear above all was the shrill Great Eastern whistle, clear as a bell, of the little J15 goods engine, bidding *au revoir* to Her Majesty and signalising the end of steam at Stratford, which took place very shortly afterwards.

Such was the end of an era. It had been a great era, but there is not a shadow of doubt that the arrival of the diesels and the building of the new depot were the salvation of Stratford. Although we were working the new depot with many of the old staff, they were refreshed in mind and outlook by new machines and good working conditions, and this gave them a new relish for the future.

XVIII
Last Years of Steam

A GREAT DEAL had to be got out of steam in its last years, for we were determined to see that matters did not deteriorate as they had done prior to the first electrification. This meant that our Locomotive Inspectors had plenty on their hands; but they were the right men in the right place at the right time. All five of them had done their main line work on the B1s and the "Britannia" Pacifics; they had tried their hand also on the Southern "West Country" 4-6-2s when we had borrowed a few of these to replace "Britannias" which had developed a temporary defect; with many years' experience of firing and driving behind them they were essentially practical men, knowing how to get the best out of steam while at the same time able to absorb the new diesel techniques. Wally Mason had died, but we selected to take his place on the Norwich service as an acting Inspector another driver, who we knew would get the trains to Norwich on time. "Hit 'em hard" was our motto on these tightly-timed expresses.

In 1960 an interesting phenomenon began to show itself. The more junior passenger links covered the semi-fast trains on the Cambridge and Colchester main lines with steam locomotives, though they also spent much of their time on main line diesels. Quite a few of these men had started on the railways in the early 1920s, a very bad time for those hit by the depression who had temporarily to go back to cleaning or labouring for years instead of firing.

The 1918 and earlier 1919 drivers at Stratford – and indeed on the Great Eastern Line as a whole – had had years of main line firing, and mostly had become passed for driving before the last war, but those who were not so lucky eventually went firing on shunters and suburban work, and did not get on to express trains until well into the war period.

By this time the main line passenger work had been curtailed considerably, whereas the slow and heavy freight work had proportionately increased, and when many of these men eventually became drivers, they had had comparatively little main line firing experience. So through the 1950s these drivers shouldered the brunt of the local freight work, including the trains over the East London Line to Hither Green and elsewhere, and then suddenly found themselves, at 60 years of age and over, on semi-fast trains demanding high speeds, with rough-riding steam locomotives one week and easy-riding diesels with comfortable working conditions and speedometers the next.

It needs little imagination to see that some men would not take kindly to this see-saw existence, and so we used to watch carefully the working of such trains as the 5.57 and 8.50pm from Liverpool Street to the

Cambridge line, on a Monday evening particularly. If time was lost, the driver concerned would probably have as companion on the following evening one of our inspectors, who very likely already had done a useful day's work. Such action was necessary to give confidence to men who, through no fault of their own, had not had the experience that other more fortunate of their colleagues had had in abundance; and as each of our Inspectors had been first rate drivers with all the confidence in the world, they could handle this situation most competently.

Stratford men did not always see eye-to-eye with the management, but they were very good to each other, and particularly to men coming to Stratford from other parts. There have been depots in my experience where men coming from elsewhere on promotion have had a very frosty reception from the residents, and particularly from those whose own promotion may have been deferred as a result. But this was never so at Stratford in my day.

I well remember the closing-down of the Midland & Great Northern line, which brought a number of men from South Lynn and Melton Constable depots to Stratford. Their arrival caused a considerable setback to the promotion chances of men junior to them, and yet the "Joint" men had a great reception.

This meant a great deal to the strangers, as it is no easy matter to be uprooted from the far end of Norfolk and suddenly to find yourself working in a huge London depot, and living in a hostel. I myself also did what I could by way of welcome to these men, nearly all of whom I knew from my South Lynn days. But the real reason why these Norfolkmen wanted to stay in London rather than go back to East Anglia if they had the chance, was the fact that the London engine-crews were splendid mates in every way, helping with road knowledge and with the complexities of such workings as those over the East London line.

It is a pity that much of what took place in my world of those days cannot be set down in print. I had many ups-and-downs with the men, over lost time, absenteeism, and all sorts of queer things that happened from time to time. I could fill many pages with the cases with which I had to deal, some interesting, some sad, some in which it was impossible to feel any sympathy with the man concerned, some in which decisions were really hard to reach and harder still to take. There were cases in which an injustice had to be remedied; others which proved to be the turning point in a man's career, when he rounded a corner and from then on became a first rate railwayman. But many of these matters must remain confidential. They are part of the essential structure of railway life in a large Division, where so much must be left to the initiative of so many people.

One example comes back to mind, of an afternoon when I was going back from the Stratford depot to the Liverpool Street office. I have excellent sight, and as I walked towards the tunnel under the station I could see, from several carriage sidings away, a man sitting on a seat on the Cambridge line platform, waiting for one of the railcars for the Lea valley. The same instant that I saw him he spotted me, and promptly covered his face. "Now why should he do that?", I thought. On getting to my office I rang Stratford to have this man's time checked, and found that he was still – nominally – on duty! The result – two people "in the cart" instead of one. But they both should have known better.

When one becomes the officer-in-charge, life inevitably takes on a sterner look, and because of this it is less easy to set out the lighter sides of one's work. The latter often came in short bursts, when one least expected them, and thank goodness they did come, for they eased the intense pressure under which we were often working. One evening in 1961, not so long before the end of steam on the Norwich trains, I took an old friend to Ipswich on the footplate. During the day he had ridden to Norwich on a "Britannia", and I suggested that we might go down to Colchester with the "Essex Coast Express" on another "Britannia", and on to Ipswich with a Type 2 diesel, returning to London on the 6.45pm Mail from Norwich on a B1 4-6-0, provided that I could arrange this last with Bill Harvey, the Norwich Shedmaster. All his Bis were in first-class form, and could time the hardest "Britannia" turns if worked really hard.

When we reached Ipswich, the Stratford crew for the Mail were waiting at the London end of the up platform. We walked across, and just before No 61045 ran in the fireman said very sharply to me that we had no right to roster a B1 to this particular job. I did not like his tone of voice, and told him that we put on to any train the type of locomotive which we knew could keep time with it. The arrival of the express put a stop to any further arguments. Nevertheless, when we started there was an atmosphere on the footplate which, though it would not have worried me, was not of the most propitious with a visitor on the footplate.

We started off gently through the tunnel, and just before we came out into the daylight the driver turned to me, saying, "Would you like to take her up?". I thought I would wait a bit, and replied, "No, Cyril. I'll bend my back for some of the way, and you can knock hell out of this old cart and pick up all the time you can". So he opened her out, fixing the cut-off at 40 per cent as we swung past Ipswich Loco ("There goes another Cockney", no doubt they said as they heard us), while I began to fire. I found the tender full of lovely dry dust, but fortunately there was nothing that a B1 liked better than small coal and dust when she was working hard, and it was light and easy to handle. No 61045 was running like a bird; it needed six shovelfuls, then another round, and then another round, until we got to the top of Belstead bank, with a full boiler and a full head of steam. And so on to Colchester, by which point we had picked up a minute or two, despite encountering a speed order at Ardleigh.

When we stopped at Colchester I said to the driver, "I'll have that little drive now, Cyril, if you're agreeable", and so off we went again, with myself at the regulator. By now I had proved that No 61045 would take all the knocking about you cared to give her, so we got the Stratford fireman going nicely, and then warmed him up a bit more, then a bit more still, until he had a really wet shirt by the time we were passing Shenfield. But the B1 had gained four minutes on a "Britannia" timing, and although the fireman may have felt that next time he would be more careful in what he said to the boss, he had proved himself to be a *real* fireman. The other three of us had a quiet chuckle about that journey, and certainly it was an evening for our visitor to remember! I had a great respect for the Thompson B1, and always felt that the 61040-61059 series were the best of the bunch.

We must have sent quite a few people to Norwich on "Britannia" Pacifics in the last days of steam; it was good that the privileged few should see how railwaymen have to work, and learn a bit about the railway tradition and pride that are deeply ingrained in many railwaymen. One such traveller was W. O. Bentley, an old Doncaster apprentice who designed the famous Bentley car and founded the firm of that name in 1919. One evening he went with me down to Colchester on the "Essex Coast Express", and on another to Enfield Town on the heavily loaded 5.31pm from Liverpool Street. He was astonished by the work which we got out of N7 0-6-2 tank No 69719, despite the fact that we developed a hot big-end in the process. She was past the first bloom of her youth, but notwithstanding overtime at many of the stops because the train was so crowded we kept time, though we had to flay the N7 to do it. He was fascinated, too, by the comparison between the Type 4 diesels and the steam "Britannias", going down to Norwich on No D202, and coming up on the 4.45pm flyer on a hard-riding Pacific, for most of the way in pitch darkness. But W. O. used to fascinate me too. He would sit in the corner of the cab, saying nothing, looking almost bored, though with occasional flashes of apparent surprise. But at the end of the journey it would be clear that he had not missed a single trick.

From what has just been written it will be clear that the 5.27pm "Essex Coast Express", non-stop from Liverpool Street to Colchester and after that calling only at Thorpe-le-Soken to Clacton-on-Sea, was one of our "show" trains for footplate guests. One evening in 1960 I joined Cecil J. Allen on the footplate of No 70011 for what, with his age well into the seventies, was to be his last footplate journey on a steam locomotive. We had promised him that we would take him up the length of Brentwood bank without the speed going below a mile-a-minute. In his Autobiography C. J. has described this trip, and how we were going grandly on 35 per cent cut-off with full regulator up the final pitch of the bank, beyond Brentwood station, at over 60mph when suddenly a double yellow appeared ahead. This

Shenfield distant was sighted for considerably lower speeds than that at which we were travelling, and had steam not been shut off immediately and braking begun, we might have overrun the home signal to which it referred. But *Hotspur* gave a fine demonstration, for all that; despite *seven* speed restrictions, two to 5mph, five to 20mph, and one more to 33mph, we came to a stand at Clacton 3 minutes ahead of the published arrival time, and so delighted some of the regular commuters that they came up to the engine to congratulate the crew on a fine achievement.

In 1959 there came to England two French drivers of my acquaintance, André Duteil and Henri Dutertre of the La Chapelle (Paris) and Calais Depots of the French National Railways. They went down to Norwich with a "Britannia" on the 9.30am, and returned to Ipswich on diesel No D200. On this journey they were introduced for the first time to a bottle of cold tea, and I cannot pretend that they enjoyed this as much as I have enjoyed a drink of their *vin ordinaire!* From Ipswich we took a diesel multiple-unit to Colchester and from there an electric local to Thorpe-le-Soken, where we waited for the 2.0pm from Clacton.

In due course "Britannia" No 70007 ran in, with Stan Pittuck in charge as the engine reached me, I could see that the drive for the axlebox lubricators on the right-hand side had come adrift; indeed, the drive arm was fractured. Immediately the two Frenchmen were interested spectators of what they expected would be a quick job of dismantling. Instead, the crew tied the gear up with some signal wire, while I got on the telephone to Colchester for a fitter to strip us down as soon as we got there. Henri was very critical about the absence of tools on the engine, and my French was too limited for me to be able to tell him why we never (in recent years) carried pin punches and handy little spanners in the tool equipment.

Well, Colchester squared us up, and we prepared to leave for Liverpool Street without the lubricator drive – no problem this for a journey of only 51¾ miles. A quick word with the stationmaster to try and ensure that we got a clear road, and then we set out to show the Frenchmen that they are not the only people who can pull back lost time. Approaching Romford we were nearly back on schedule, but only to be stopped by signal as the Romford pilot was being put across the road into the yard just in front of us. So we were five minutes late into London after all. Later on I looked in at the Great Eastern Line Control to see what had been reported as causing the delay, and was amazed to see the entry, "2.0pm Clacton LV Engine 70007 Fitters Colchester and slow running Shenfield to Stratford". Like a shot I hurried upstairs and put that one right with our Divisional Control!

If I had been an Operating and not a Motive Power man by training, I might well have written a good deal more about the Control. They are the boys who get most of the criticism, who wake you up in the middle of the night and give you bad news, but who have a job that is relentless except, perhaps, on a winter Sunday afternoon. This particular Liverpool Street Control was on top of its job, and played a great part in the 1959-1962 revolution – a part calling for initiative, inventiveness, and boldness and courage in making decisions.

After our arrival from Colchester that afternoon, I took the two Frenchmen to Chingford on an N7 0-6-2 tank; at least, I travelled in the front brakevan while our two guests were on the footplate. It was the engine's last trip of the day before the fire was cleaned, and I could hear all the sounds of shortness of steam, the laboured starts, and the infrequent use of the injectors. No 9608 finished the climb exhausted, with 90lb of steam on the clock and half an-inch of water in the glass, together with two paralysed Frenchmen who had never seen anything else like it in their lives!

One complete change that took place with the arrival of the diesels was on Bank Holidays. It had always been my practice to spend the afternoon and evening of the Easter, Whitsun and August Bank Holidays at Liverpool Street with the Locomotive Foreman, keeping in touch with all that was going on in the Division. Sometimes nothing would go right, and nearly always there was a pandemonium of smoke and steam as relief crews struggled to prepare incoming engines for outgoing trains in less than no time. At other times I would be congratulating myself that all had been going well, and would be on the point of going home at 9.30 or 10 o'clock when suddenly the 'phone would ring with the joyful news that No 76031 or some other engine had come unstuck, say, at Witham with the tail-end of its train out on the main line.

So I should have to wait until eventually the casualty arrived, having delayed various other trains some 20 or 30 minutes; I had then to get on to the engine, look at the fire, listen to a tale of woe, and reply with a few home truths about firemen who let their fires run down the slope of the grate. It was not always the fireman, of course; it might be the engine, the coal, bad maintenance, and sometimes sheer bad luck, but the fact remains that the diesels and the electrics brought such consistency into the working that after Whitsun, 1962, I gave up the practice of Bank Holiday appearances. Up till then I had not had a single August Bank Holiday off duty since 1948, but from 1962 my presence was no longer needed.

Earlier I have spoken of our problems in finding sufficient locomotives for freight trains, and of the fact that in a Divisional organisation one really did begin to understand the other man's problems, and the necessity to do one's best to meet the demands of the Commercial people, particularly when the reasons for those demands became known. Just before Christmas, 1959, we were in deep trouble. Freight traffic was being left in the yards and we were critically short of both power and men. Harold Few asked me to come to his office, and there I found a council of war in progress. One of the operators, actually a good friend, said something which annoyed me, and after Harold had laid it down that certain trains had got to run that evening I banged out of his office the worse for a ragged temper, but with the determination that those trains *would* run! If my memory is correct, we had to find power and men for six trains, two to New Cross, one to Hither Green and (always a difficult one) another to Norwood Junction, as well as two to Acton on the Western Region.

Without any clear idea as to how we should manage, I rang straight through to the Running Foreman at Stratford. He happened to be Bill McMeakin, the Rest Day Cover man, and I told him that somehow or other those trains had got to run, and without cancelling anything else. The first thing in our favour was that McMeakin was a very calm customer and knew his men; the second was that Christmas was getting near and some of the crews would have no objection to earning a little extra at the end of an eight-hour day. I asked him to ring me back in ten minutes or so to tell me what he could do!

A minute or two later McMeakin was on the 'phone. "We can do the New Crosses", he said, "and we can do the Hither Green and the Actons, but we've nobody who knows the way to Norwood. We can only manage to Latchmere, and Redhill Control say that Stewarts Lane have no conductors". "Ah ha", I thought. "This is where the Old Pals' Act must be applied". So I went down to Line Control, asked the Deputy to get Stewarts Lane Loco on the 'phone, and waited. An answer soon came from the Foreman at Stewarts Lane. "Newman's the name", he said; this was Harry Newman's signature tune. Could he send a pilotman to Latchmere for the 8.50pm Temple Mills to Norwood? "Of course, Chief", was the reply. "We'll have Alf Long going home at about that time. He'll look after your driver to Norwood". That the 8.50pm to Norwood was able to run was much to the surprise of the Southern Redhill Control, who could not understand where the pilotman had come from or by what means. So all the trains *did* run, except for one of the Actons, which had to be cancelled because there was no guard. Bill McMeakin had done a great job, and I had to spend a lot of time on the 'phone to the depot during the evening. My operating friend, who had made the comment which upset me, had to be informed of the good news, but I waited in the hope that he would have gone to bed before I rang him. He had!

It is no good being a Manager with an engineering background if you cannot get yourself out of trouble in times of breakdown and failure. Most experienced Managers know how to do this, what chances can be taken with complete safety, and how to avoid delays to trains by on-the-spot decisions. The young qualified engineer, on the other hand, has to learn these things before he can be regarded as a competent Manager, for it is not the slightest use saying that this or that piece of equipment has failed to the detriment of the service when the art rests in the further knowledge gained in the rough-and-tumble of daily work.

This reminds me of an amusing little incident in the last days of steam on the Woolwich line. The Westinghouse-fitted stock had been replaced by four-coach vacuum-fitted sets, and one day when I had an hour to spare I joined an Enfield driver by the name of Bob Baker, a splendid man, on one of these trains at Stratford Low Level. Our journey through Dockland to North Woolwich was without incident, and on arrival there I got off the engine and walked out of the station to spend a few minutes watching the traffic on the great river, with the ferry-boats performing their antics. After ten minutes I sauntered back, and found our engine attached to the other end of the train by its coupling, but without a vacuum pipe. Our young fireman had coupled the vacuum pipes but had omitted the necessary precaution of coupling the locomotive to the train as well; Bob Baker had then drawn forward to get water, leaving the train behind, and one of the vacuum pipes also.

As only three minutes were left before departure time, we had to do some quick thinking. The two alternatives were to turn out the passengers and cancel the train, or to use our knowledge and intelligence to keep the service going. The guard was a grand old boy past 60 years of age, and he agreed to do his stuff at the various stations, while we had an excellent Westinghouse brake on the engine. The timings were quite easy, so that we could take our time running into each station as far as Lea Bridge, while from Tottenham South Junction onwards the gradients were uphill all the way to Palace Gates. So, with splendid co-operation between driver and guard, the job was done with no more loss of time than a couple of minutes. At Bounds Green Carriage Shops, near Palace Gates station, the driver got hold of a new vacuum pipe, fitted it in double-quick time, and was back in service with very little delay. We knew exactly what we were doing, and that given skill and co-operation at both ends of the train we could run it quite safely without the continuous brake.

This was an isolated and relatively unimportant incident, but it was an example of the importance of knowing what to do in a rare kind of emergency, without causing delay and inconvenience to those who have paid their fares and expect to arrive at the advertised time. This applied particularly to London dockers, who use these trains at this particular time of day, and would tell the train crews exactly what they thought of them if they were kept waiting!

And now for a railway happening which was no concern of British Railways On the North Downs in Kent there is a wonderful miniature railway belonging to some friends of mine. It is hidden away in the depths of the country, and runs through woods and sunny glades for some two-and-three-quarter miles before returning to its starting point. It was built before the war by the owner of the land, and now is kept going by his family. From time to time I was honoured with an invitation to spend an afternoon there with my family, not only to picnic in that lovely setting but also to help run the railway.

On one never-to-be-forgotten occasion, I took with me six young Stratford drivers, all of them leading lights in the enginemen's Mutual Improvement Class, down there for the day, and we got there in two cars by lunch time. On arrival we were asked by the owner if we would be prepared to operate the line for some 15 to 20 children who had come over during the school holidays to ride on the trains while their parents took it easy. This suited us very well, and if all had gone well would have kept us busy until dark. But fate decided otherwise.

As I knew the road, on the first journey I took charge of the engine which was a Great Northern Atlantic, while my six companions were fascinated by the engine, the scenery, the tunnels, the single line tablets and the signalling. On the second trip, Charlie Gunner took over the engine, but whoever was looking after the tablet dropped it as we were passing on to the single line. I jumped off, picked the tablet up, and ran back to my place on the now accelerating train and got aboard just as there was an almighty bang from the locomotive. On stopping we found that a small-end pin had worked out and had caused enough damage to put an end to the proceedings as far as that engine was concerned.

So I hurried back to the terminus to find the owner, feeling rather bad about the incident, as I had been in charge, while the train-load of kids looked as though they were in for a disappointing day. But there was one ray of hope. There was another engine, a little 4-4-0, and she was in steam, though in rather a doubtful condition; however, I got the owner's consent to use her when we had got the Atlantic back home. There would be only one way to succeed with this little machine, and that was a method that sometimes has to be used with the real thing – "Fill her up and shut her up". So we packed the firebox full, shut the firehole door, and gave her the lot, full regulator and full cut-off. Going through the tunnel was a hair-raising experience, with red-hot cinders in one's eyes and hair. But our diminutive charge worked unfalteringly for hours, pitching and rocketing provided she was belted good and hard all the time.

So, on a glorious evening, we came to the last train of the day, with the children now all away home. As we approached the start of the big loop I got off the moving train to pay attention to the call of nature. After this, I sat on a tree stump to wait for the train's return, reflecting on the simple pleasures of life, and then suddenly realising, with everything completely quiet, that the train should be back by now. What could have happened? Six experienced enginemen would not be likely to run too hard round corners and over crossings, but could it be possible that they had become derailed and had damaged the locomotive? In a panic I set off up the track. With not a sound to be heard nor any sign of movement I felt that my worst fears were being confirmed. Then suddenly I saw the train, with my six Stratfordians, convulsed with laughter, pushing a dead locomotive and several coaches.

Soon after I had jumped off, the fire had well and truly died on them, and the little boiler had very quickly lost steam though they had managed to reach the tunnel. They thought they could get through, not realising that a small locomotive on no bigger than a gin gauge has no reserve; so they stalled in the tunnel, which fitted the train so like a glove that nobody could get off to push. The only way out was to clamber over the three or four empty open coaches behind the locomotive, and so to get to the back of the train and shove. So we pushed the cavalcade to the top of the hill and coasted down nearly to the terminus; then Bert Pell dropped the tablet for the last time and chased down the bank in the beech wood to pick it up. Finally we emptied the 4-4-0's firebox, put her to bed, thanked our host for his kindness, and went home exhausted, amused and thoroughly contented.

And now we must take leave of Stratford and of the Liverpool Street Division. So far from saying that things are not as good as they were, I would claim that the Stratford of 1962 was an infinitely better place in which to work than the Stratford of 1958. There have also been tremendous changes throughout the industry since 1962, and no doubt many of the decisions that we had to take in those pioneer days of dieselisation will have been found wanting in the light of later experience. Or have they not? We had to come to decisions when guided only by commonsense and managerial *expertise*, together with a knowledge of men and of their capabilities, and it was an experience that I would not have missed for anything in the world. Indeed, my life over the years 1941 to 1962 provided all that any man could desire. The work was of great interest, tough and exacting at times, but never dull or boring. It was a job that entailed close and regular contact with men striving to earn a living in their own individual and sometimes independent way in the service of the railway.

We were none of us particularly well paid for what we had to do or for the responsibility we had to bear, but although this meant that we never had much to play with at home, the work was such that there were few days when I did not look forward to the excitement and the challenges that lay ahead, as well as to the comradeship that exists among railwaymen. And above all, serious though the task was, one's spirit was constantly infused by the inherent humour around one on all sides. Where else could one find a Mechanical Foreman wearing a Palm Beach suit, or a coalman-in-charge of a little depot who sported a kilt, who wrote his monthly report in alternate lines of red-and-blue ink, and who concluded with his signature and "God save the Queen"?

XIX

Firing and Driving in France

IN MAY, 1958, James Colyer-Ferguson persuaded me to go to France with him, for what was to be my first introduction to French steam locomotive power. We travelled from Calais to Paris on a PO Pacific No 231E16, on the Saturday, and returned the next day. Since then I have been across 55 times, always with a railwayman or a close friend who will benefit from the journey, while French railwaymen, many of whom never would have dreamed of doing so, have visited this country to see what we can do. Groups of railwaymen from Stratford, Parkeston Quay, King's Cross and Liverpool have been given a wonderful welcome on the Nord Region of the SNCF, and every man who goes across, comes back with the happiest and warmest thoughts for the rank-and-file French railwaymen who have made him feel so much at home.

For three week-ends in each year I go across, and never for more than three days at a time. My holidays however, are spent with my wife and family, with as little thought for and contact with the railway as is possible, and in the same way that a holiday puts one right for the task ahead so do these quick visits to France. They give me a chance to do the same physical work that delighted me as a younger man, working on steam locomotives. But although steam has nearly gone, replaced by diesels and the wonderful SNCF electric locomotives, it is to meeting one's friends again that one looks forward, and even when steam eventually goes, these same ordinary French railwaymen will still draw me across the Channel.

These week-ends are a mental refreshment. There is work to be done, and there are charming straightforward people to meet. There is no red carpet; if there were, I should not want to go. There is simply the warmth and friendship of people who welcome me in their homes, who speak no English, who help me with my French and laugh at my mistakes, and who have the common international bond of being professional railwaymen. I have been able to see how people live and eat and work, and as I have had the immense pleasure of working on the unique French locomotives, not only steam but electric, I have learnt much of the work and actions of others, experienced at first-hand methods of railway operation different from our own. Every journey shows me something new, something fresh, or brings me into contact with somebody I have not met before, be he a porter at Rang-du-Fliers Verton, a Fivois electric driver, or a Calaisian steam man whose name I have known but whom I have never met in person before. Indeed for a few minutes on a recent journey, and whilst still in my overalls and unwashed, I met the legendary M. André Chapelon himself. When other people go across, it is my pleasure to work out the details of what they will do and whom they will meet, and it is an even greater pleasure to hear in the greatest detail on their return an account of their experiences and their impressions. If they have enjoyed themselves, and broadened their knowledge of railways and railwaymen, that is good enough for me.

The man responsible for all this is Philippe Leroy who retired from the position of *Ingénieur en Chef, Traction, SNCF Nord* in July, 1969. He was the Motive Power Superintendent and spent a lifetime on the Nord. As a young man, he had been a Polytechnician and joined the railway at Dunkerque: he had done some firing and driving there, and then worked his way up in exactly the same way as I had done on the locomotive side. During the war, he was active in the Resistance, and from the time I first met him in May, 1958, by the side of No 231E16 in the Gare du Nord at Paris, he went to endless lengths to see that the railwaymen of Great Britain who went across at my behest not only had the experience they wanted, but were welcomed on arrival and made to feel at home.

When Philippe Leroy retired, he came across here with Madame and it was our pleasure to entertain him in London. We gave them some lovely presents to which over fifty people contributed, and at an evening gathering we thanked him for all he had done for us. The company included drivers from both the Eastern and Southern Regions, Motive Power and Traffic Inspectors from the London Midland and Eastern Regions, Divisional Managers, Running and Maintenance Engineers, Divisional Movement Managers, people like old Sammy Gingell, Group-Captain Basil de Iongh who had been in digs with me in Doncaster days, my old Assistant from Liverpool Street, Bert Webster, and many others, some of whom were not connected with British Railways but who had a deep interest in, and understanding for, railwaymen and what they are striving to achieve.

James and I got on that old Chapelon Pacific No E16, when she backed on to the "Golden Arrow" at Calais. The driver was Henri Dutertre, then 35 years of age, while the fireman, Henri Meyns, was over 47 years old. Dutertre had been running No E16 since he was 27, and in case this might surprise people, that particular generation of Calaisians went through the workshops, spent a short time firing, and then became drivers at about 24 years of age. At between 25 and 27, they passed their examinations to drive the Chapelon Pacific. Indeed, Henri was by no means the youngest: I believe that the palm went to René Gauchet, who had No E14 at 25 years of age, whilst his fireman, Gilbert Sueur, was 23, a total of 48 years to take a Pacific on a boat train from Calais to Paris "*à la perfection!*"

Henri was quiet and charming. I could hardly speak a word of French, so that I was reduced to shaking hands, smiling, and offering cigarettes. But after a hundred miles or so of watching every move, I asked him, through James, if I could be his fireman to Paris. Henri had no objection, and although naturally unacquainted with the road, I had little difficulty in giving him all the steam he wanted for this beautiful engine had enough in hand to make the work relatively light. Both driver and fireman were amused and surprised to have an Englishman on their footplate prepared to have a go for 80 miles on a strange engine in an unknown country. Our journey was quite uneventful, but I shall never forget it. The rhythm of that locomotive, so easy and smooth, the precision of operation, the compound working and complications of the independent control of high and low pressure valve gears, the little cab and the tender end stacked with briquettes, also the idea of a 47-year-old fireman working with a driver over 10 years his junior in age, the fact that the blower was never used in the Boulogne tunnels or anywhere else so that flames came round the sides of the fire-hole door and up in front of the highly polished faceplate fittings through the roof of the cab, these and a hundred other impressions are still fresh and clear in my mind.

In Paris, Philippe Leroy was waiting for us, as indeed he has done on every single visit I have made to Paris, and so was Basil de Iongh, by now Assistant Air Attaché in Paris. Next morning we were at the Gare du Nord by 8.0am, seen off by M. Leroy, and our driver was André Duteil of La Chapelle, with his own engine, the De Caso Baltic 4-6-4 No 232S002. Away we went to Lille in relative silence with this huge engine and heavy train. This was the first stoker-fired engine on which I had ever travelled, and we left Paris with a couple of inches of dull red fire and an inch of water in the gauge-glass. I had never seen anything like it. In next to no time, she was forging up the Survilliers bank with the same inch of water in the glass, with the steam pressure exactly on the red, and with André and his mate, François Demeurant, standing in their respective corners as I was to see them many times until September, 1962, when I made my last journey with André on a steam engine.

André was a little man, so small, indeed, that he would never have been allowed to start on the footplate on some railways in this country. François was a bit taller, say 5ft in and more solidly built. André has always been to me the classic Frenchman of the more volatile type. Fresh-complexioned, lively and quick of movement, with gold teeth flashing, he stood slightly crouching at his work, every movement of the controls made with delicacy and understanding and his use of the Westinghouse brake a joy to behold. Both men were wearing the usual French *bleus* with scarves streaming in the wind, both had their caps on back to front, held firmly in place with the standard SNCF goggles, giving them the same dashing appearance that one associated with the earlier racing motorists. They gave us the most wonderful welcome and after Arras I was put in charge of the firing. I had to maintain an inch of water in the glass as before, but I was not permitted to shut off the ACFI pump to bring the level down, for it had to be delicately eased, as also did the rate of firing, to achieve the required result. I was amazed at the delicacy of the control in the steam-valve to the stoker, which had to be minutely adjusted by tapping with a light spanner.

At Lille, we said goodbye and joined another Calaisian, Emile Lefebvre, on No 231E36. Here again we were received as though Emile and his mate had been welcoming us back after a 20-year absence, whereas they did not even know we were coming when we stepped on to the footplate. This journey was comparatively easy. I did the work and this earned me a mixture of Emile's bottle of Bordeaux and Jacques Deseignes' coffee. We got to Calais Ville, they ran us on to the Gare Maritime, and we boarded the *Invicta*, 28 hours after leaving it, overjoyed with what we had seen and with the people we had met, and knowing that we *must* come back again.

Since then, Philippe Leroy has visited us nearly every year, André has made four visits to this country and Henri five; they have worked on our engines, seen our country, stayed in our homes and gone away feeling that there is no country where they could have had such a welcome as in that strange island across the Channel where, reputedly, all is reserve and tradition, and foreigners are never welcome.

I do not need to describe in any detail the French locomotives on which I have travelled. Books have been written, and articles too, by various authorities on the design features that have made possible the most extraordinary performances by quite moderately dimensioned machines. But I would rather write of the incidents and experiences that have come my way, the kind of thing that is not often recalled or indeed experienced.

André and I used to go to Jeumont on the Belgian border, nearly always with James Colyer-Ferguson. André would tell me to get on with the job as soon as we had left Paris, and after things had settled down and we had taken that tremendous junction at Creil at 75mph, swinging one way and then the other, he would go to the cupboard, get out four little glasses wrapped in tissue paper, open a bottle of Pelure d'Oignon with due ceremony and fill the glasses with the same discrimination as if we were dining at home. And all the while, the great machine would be tearing through the northern countryside while we were toasting each others' health with our eyes on the road ahead. Soon it would be time to stop at Compiegne and the procedure was to make one firm but by no means hard, brake application, put the Westinghouse brake-handle in "block" position, ease it once or twice to reduce slightly the pressure in the brake cylinders, and, if possible, stop without a further application. In the early days, André's instructions in his very quick French were not easy to follow, but over the years, I learnt the road and got the better of those great 4-6-4s, including the famous 232U1.

To start away from a station required three pairs of hands (until you had got the knack) on No 232S002, and as for a Chapelon, you needed three hands and an optional foot to deal with the by-pass. The initial difficulty, to someone accustomed to the response of our locomotives on which movement usually began at once, was that on the regulator being opened no movement was felt until a second or two later, during which time the receiver safety-valve had lifted and frightened me no end. So much did this shake me the first time that I instinctively closed the regulator, with the result that we stopped dead and I then had to start all over again. Fairly sophisticated performer as I was on our own engines, that first start from Compiègne makes my cheeks burn to this day. But practice makes perfect, or nearly so, when coupled with thought, care and understanding.

Just before the end of the S class Baltic I was to make an amazing journey from Jeumont to Paris. I was then the Running and Maintenance Engineer, at Liverpool Street, and under my control in Zeebrugge I had a Wagon Inspector responsible for checking the condition of the wagons going to Harwich *via* Zeebrugge and the Train Ferry. After spending the earlier part of the day at Zeebrugge, Harry Noden, our Carriage and Wagon Engineer and I got to Jeumont *via* Brussels and Namur, and picked up André with the heavy evening train to Paris. It was only two months before the end of steam; the regular manning of the engines had broken down and we had No 232S003. She was no longer in perfect order, but nevertheless was not too bad. So we set off from Jeumont with a comparatively light train of 380 tons, which was made up to 750 tons at Aulnoye. We had some speed orders and other checks, but we were into Paris no more than a few minutes late. It was done this way.

André told me to run right up to the maximum permitted speed of 120kph (75mph) uphill and down dale, and never to come off the figure. André's mate was one of the very best of firemen, Renè de Jonghe, now retired and living near Calais. So we ran mile after mile with steam absolutely immovable on the red line, with water out of sight in the bottom of the gauge glass, and now and again drawing the level up by opening the top cock. The pyrometer showed steam at a tremendous temperature and every so often the feed-pump, cut fine, was reinforced by an injector to keep that bubble of water in the bottom nut. I kept my eye both on the road and the working of the engine, and pondered from time to time on the extraordinary confidence of this 49-year-old fireman who could control matters with such precision on a locomotive that was past its prime. Naturally we had "un *p'tit coup de rouge*" on the rising gradients, and we were diverted round the station at Creil so that we passed that point at about 30kph instead of the usual 120. This gave us the chance to open out this great compound locomotive on the 1 in 200 climb to Survilliers. I cannot remember the speed to which we accelerated, but it was close on 60mph and I shall never forget her beautifully clear-cut exhaust, with the occasional rocket thrown sky-high, as she forged her way up the bank to *"Le Pont de Soupirs"* the "Bridge of Sighs" – before starting to spin silently down towards Paris. Steam has all gone now; the "Bridge of Sighs" remains, but has lost its significance. What good fortune I have had!

Up in the driver's corner on most French locomotives, or across on the fireman's side, is *"L'Espion"*, the "spy" or the Flaman speed recorder. As is well known, the speed is recorded on a card which is attached to a revolving cylinder and which is checked at the end of the day. So if a driver exceeds a speed restriction by 3 to 5kph, or the maximum for the road or locomotive by roughly the same amount, he receives a *"lettre de reprimand"* or a summons to see authority. On this card, other equally important matters are recorded.

If, for example, the driver has come down Caffiers bank, hell for leather, at 120kph, and is getting near Calais, he must look out for the "80" board as he passes the allotments before Les Fontinettes. As soon as he sees the "80", he presses a button above the speed recorder which marks the card. Thus his vigilance is checked. As he passes the "80" warning board, his engine strikes the *"crocodile"*, the air whistle in the cab sounds long and loud, and he marks the card again and then comes down to the "80" at the appropriate point.

If he has been advised that he will encounter a temporary speed order at some point on the journey, he will be watching out for the warning board. As soon as he sees it, he marks the card, and as he passes the board, a temporary *crocodile* fixed for the purpose sounds the air whistle and he marks the card again. And so, each time that warning whistle sounds, whether for a speed restriction or an adverse signal, that card is marked and later on checked. Thus, this particular discipline is tight and if there are discrepancies, there is a little matter of an interview with the *"chef méc"*, or locomotive inspector.

Now these *"chef mécs"* are men of authority who have come from the ranks of the drivers and who have gained their positions by competitive examination. From my earliest days in France, no inspector has been instructed to travel with me on the Nord. But not long ago, I was on a type 141R 2-8-2 on Train 16. Climbing Caffiers bank, the "Flaman" stopped recording and we changed to the *"machine de reserve"* at Boulogne. Here we were joined by a man of 42 who was one of the *"chef mécs"*, a remarkable character who was not satisfied until all was moving well on the new locomotive. He had been a driver at Lille at 24 years of age, and an Inspector at 28 – by competitive examination. But I have often had the Calaisian Inspectors with me, for we knew each other well, and they often came along "to instruct and advise me in the performance of my duty", bless their hearts. And as you can read all about French locomotives elsewhere, but not about personalities, a word about two of these people would not come amiss.

Until he retired in 1969, Marcel Dewevre had been the Chief Inspector at Calais for a number of years, working closely with the Ingénieur, Eugéne Lavieville, and his Assistant, Georges Chatillon. The former had come from the footplate grades, doing his early work at Amiens and, as an Inspector is job was to keep a close watch on the activities of the footplatemen at Calais, sharing some of the work with the Assistant Inspector, Edmond Godry, who also covered Boulogne. In many ways, their work was exactly the same as that of their counterparts in this country, but as they had fewer men and fewer engines to follow, they were able to go into matters in greater depth.

When Marcel was a fireman at Amiens, during the occupation of France by the Germans, he and his driver were working some kind of special train which was derailed by the activities of the Resistance. The engine went over on its side, and the driver was badly injured, though not fatally, whilst Marcel was thrown clear. It was very dark but he sorted himself out, found a flare in the wreckage, and set off up the line to carry out such protection as he could. He hadn't gone more than two hundred metres when he ran into a couple of German soldiers and a sergeant who stuck their weapons in his ribs, accused him of sabotage, gave him the old "one-two" across the face, kicked him in the stomach and threw him in the ditch, from which he eventually emerged, sore but still capable of fulfilling his task of protection. And this, of course, is not the sort of thing that one normally expects when going to protect a train! But it is no more than what happened to countless Frenchmen at one time or another in widely varying circumstances and it makes one think deeply on what life could have been like in an occupied country no longer free. Many of the men I know have had experiences which they relate now with some amusement, but which were the reverse of funny at the time. Many were sent to Germany to work in factories under forced labour conditions, and many of those left behind played their part in the course of their duty in making life difficult for the Germans, even in small ways, such as the odd slip which threw dirty water from the chimney, or a sudden movement of the locomotive with the cylinder cocks open.

Edmond, now the Chief Inspector, is a remarkable man. Born in Calais, he started at the depot, did his time in the workshops, and then after a short time firing became a young driver of the same generation as Henri Dutertre. As a driver, he stands on the gangway of the PLM Pacific in Cuneo's great portrait of Calais Gare Maritime. To say that he is a perfectionist is putting it mildly. His knowledge is profound, his charm is great, and the respect in which he is held by the Calais staff is there for all to see. Yet like most perfectionists, he still seeks for improvement in his own performance and those of others.

The Nord enginemen work to a laid-down system in almost every aspect of their handling of the locomotive and it is Edmond's duty to see that the standards were maintained. For example, if one were running into Boulogne with train 19 and a Pacific, the speed would have dwindled normally to about 100kph. The driver would make a single *"dépression"* or application of the Westinghouse brake, allowing just under one atmosphere to pass to the brake cylinders. He would make this application at the second *"mirliton"* before the last signal which heralds the approach to the station. On this basis, it is more than likely that he would make a fast stop, and he should need neither to apply nor to release the brake again, all adjustments being between the block and service positions to reduce the pressure slightly in the cylinders.

Or, if he were driving a 141R limited to a maximum of 100kph, or a double-headed Pacific-hauled train with the same speed limit, he would delay his application until he reached the actual signal, for the speed might well have fallen away already to under 100kph. These instructions and countless others have been worked out and are rigidly applied, having been taught at the onset in the instruction schools when drivers were under steam training.

Edmond has paid two visits to this country, the last in charge of a party of Calais drivers and firemen, and including Fred White, the BR Interpreter from the Gare Maritime, a Frenchman born in Bleriot Plage and who was deported during the war whilst his mother, an Englishwoman, continued by some miracle to live within 10 miles of Calais through the four years of occupation without being apprehended. I took the party of Calais men to Liverpool Town Hall to meet the Lord Mayor, who had fought in France during the First World War. Very simply, he spoke of his experiences and deeply impressed his guests, and at the end of the visit Edmond replied with all the charm, depth of feeling and perfection of which he was capable, having anticipated this moment and thought about it for days. Not only was I proud to be showing visitors from abroad a part of our heritage so very British, but I was proud of the way my friends from across the Channel conducted themselves on an occasion that was unique in their experience.

On that particular visit, we arranged for the party to travel on some of the ore trains run for the Summers iron and steel plant, between Bidston and Shotwick. It must be remembered that no French driver has had experience of working heavy loose-coupled, unfitted trains such as these ore trains, which consist of a Type 4 Brush-Sulzer diesel and brakevan with 11 unbraked bogie hopper vehicles of iron ore in between. Two of the visitors were put on each engine and two in each brakevan, and all were fascinated by the vital importance of co-ordination between driver and guard without which the heavy trains could not be rigorously controlled on the falling gradients between Heswall Hills and Shotwick. The visitors had never experienced the feeling of a heavy train behind beginning to push the locomotive down the gradient, and appreciated that this sensation had to be checked instantly by enginemen who did not have a continuous brake at their command.

In the days of the Chapelon Pacifics, pride in the machine was taken to the same lengths as it was in this country by our men who ran their own engines or (as was, of course, more common over here) shared one. I doubt very much if the Paris – Orleans type engine was ever shared in the interests of improved utilisation in France, for it was an extremely involved and delicate machine from which it was possible to draw power out of all proportion to its size, and with an economy of coal and water which had to be *experienced* to be fully understood. I found these engines very easy and comfortable to fire, and by keeping the section of firehole next to one's legs closed it was possible to keep the heat down. The only difficulty that could arise was really due to the freedom of steaming.

With a heavy train, and working hard, one might be tempted to let the fire down under the firehole door and in the back corners of the firebox, to reduce the chance of blowing off steam from the safety-valves which used to be forbidden. But one did this at one's peril for two reasons – hard work was needed to rebuild the fire and it was possible during the rebuilding process that steam pressure might fall, which again was not allowed. But if it did happen, one had to work quickly and effectively with the excellent briquettes stacked round the end of the tender, using some small coal and dust "to fill in the gaps". That Pas de Calais dust was splendid stuff which, if properly used, could get the pyrometer round to 380°C. And what an aid that little gauge was to a driver and fireman! It would tell you if you had a hole developing in the fire, if priming was on the way, and whether the fire was in any way imperfect. The Chapelon Pacific carried a gauge showing the smokebox vacuum, and the triplex brake gauges, together with a similar one showing boiler pressure and the pressure in high and low steam-chests, gave the driver very complete information on the performance of his locomotive. These engines were beautifully proportioned, and had a personality of their own. There was a marked *esprit de corps* among their drivers, the Nord *senateurs*, who were in a class apart, and special skill was needed in the maintenance of the engines.

I had only one bad journey on a Chapelon Pacific. This was on 231E40, with Maurice Saison of Calais; it was his own engine. She had developed a blow from a steam-pipe joint in the smokebox, but she rallied quickly enough with the regulator closed. Between Etaples and Amiens, however, we gradually lost ground and I was very glad that the days of through Calais-Paris running were over, and that we were coming off at Amiens. The fireman was Saison's own mate, the 50-year-old Auguste Beauchamps, and nothing he or I could do had any real effect. But we kept time all the same.

Auguste remained as a fireman on the main line until the Pacifics were withdrawn, retiring at the age of 54 after he had completed 25 years' service. He was one of the men who had not started on the railway until he was 29, the upper limit unless you happened to be a family man. I can never really believe it, but I am told that the limit is raised for men with large families, so, with an allowance of one per year, a man could start as late as 33 years, therefore, if he had four children. Good old France, like us they have some very humorous ideas! My last journey with Auguste, which also was my last on a French Pacific, was on No 231K8 on the *"Flèche"* from

Amiens. When I got on at that station, he gave me a packet of Gauloise cigarettes in return for my services, and then kept me going for a while on an excellent Bordeaux, until our driver decided he would change places with me. *"Le Vieux"* worked very slowly, smoked endlessly, and smothered everything in dust, but he always had steam and plenty to say.

The PLM Pacifics survived the Paris-Orleans type because they were more numerous and of a much simpler design. They did not have the great reserve of power of the Chapelons, but they were extraordinarily efficient and powerful locomotives, in fact some fireman preferred them to the Chapelons except on the very heaviest work. They could survive the comparative rigours of *"banalisation"* or the sharing of engines between crews and the falling off in maintenance standards. After the post-war recovery, main line steam failures were virtually unknown at Calais, for the engines were watched very closely indeed, but in 1967 the odd failure began to occur, and the end of the world seemed to have come when a K class connecting-rod broke coming down Caffiers bank at speed. Nevertheless, these old engines were maintained at a high standard, and only by comparison with what had gone before could one say that there had been a deterioration.

The PO type rode so superbly that the traditional glass of water (or wine) would not spill, whilst the movement was as near silent as it is possible to achieve with a steam locomotive. The PLM locomotives were almost as smooth. To fire the PLM Pacific was again a relatively simple task physically, although the heat thrown back by the fire could be tremendous. It was even more necessary to keep the firebed heavy at the back end of the firebox and to feed the sides and front very sparingly indeed. Unlike many engines, the fire also needed feeding in the centre of the firebox, not only immediately in front of the firehole door but further forward, and preferably with small coal or dust spread with a quick but comprehensive flick of the firing shovel. If this was not done, the engine would come off the boil at once.

A number of these engines lifted the water in the boiler surprisingly far, and if you were not very careful you could be involved in priming, but this was dealt with by having the confidence to run with a very low level of water in the boiler. I remember leaving Amiens on No 231K73 with a train of over 650 tons. Our driver, François Joly, marked the level of water he required with a piece of chalk, on the gauge glass protector one inch above the bottom nut. His handling of the engine was so gentle, with advancements in cut-off as the speed fell on the banks keeping the rate of steam consumption surprisingly even, that it was possible to maintain that inch of water throughout the run without the slightest difficulty, because every move he made was predictable, and there was no need to worry about having a full boiler to guard against emergencies.

One day I had gone across in November, and by the time we arrived in Calais it was blowing a full gale. We had No 231G266; our driver was the smallest Frenchman I have ever seen, André Desmolliens, and our fireman the robust Aimé Deloison, one of the very best fireman I have ever encountered, also well on in years. The force of the wind going up Caffiers was tremendous, and it was hard work to keep time but we were through Etaples on the dot and then settled down to run in the usual consistent manner across the Northern Plain. Soon, however, we swung round a corner to see a stop signal at danger. We reduced speed quickly and came to a stop at the next station to enquire what was wrong. Here we were told *"C'est le vent"*, and off we went again. This happened twice more and despite much hard work we were late into Amiens. This was the first and only time I had ever known a train to be stopped by "signals blown to danger", but with the French chessboard type of signal it is possible for the force of the wind to strike the board and partially turn it to the danger or "half-cock" position, which means a stop so far as the driver is concerned. Incidentally, André Desmolliens reduced the orifice of the variable blast-pipes on this journey on two occasions, as it was necessary to work the engine hard to regain time. The effect on the fire and the consumption of coal was immediate. This sort of thing was frequently done with the Paris-Calais boat trains in bad weather conditions. In the teeth of a westerly gale, it was extremely hard going with the heaviest trains and discreet use of the "fireman's friend" was sometimes necessary if time was to be kept with a PLM pacific.

On another occasion, Colin Morris from King's Cross and I were making our last journey of a splendid weekend from Amiens with Train 19. Our driver was a spare man *(mécanicien de remplacement)* called Michel Dutrieux, one of the trades union representatives, a splendid man who died in his very early forties. Our journey was quite straightforward until, on the last length of Caffiers bank, about a mile-and-a-half short of the summit, we suddenly lost speed and despite all Michel could do with the change-valve and high pressure steam on the low pressure side, No 231G42 came to a stand. We were on an up gradient of 1 in 125 with a train of over 600 tons. I said to Colin that we should never restart again without assistance, but I spoke too soon.

Michel and his mate were off down the train in a flash and found a brake dragging some six coaches back. There were two *chefs de train,* one of whom went back to protect the train. In seven minutes Michel had cured the trouble and was back on the engine. Sand was applied, the regulator was opened, she moved, and after two or three exhausts we went over to full compound working and we were on our way again, without a slip, gradually accelerating round the curve and up the hill.

To see what to me had seemed to be impossible achieved without hesitation or difficulty was thrilling enough, but to see two men deal with a failure so quickly and get away in such circumstances with an overall delay of no more than just 15 minutes was something I shall never forget. Just before we started, the whistle was sounded. This no doubt arrested the progress of *M. Le Chef de Train,* who was down on the track at the rear doing his work of train protection. Quickly he turned round and obeyed the instinct so often seen in the old silent films of staring and then of running after the train. But it was too late. I'm sorry to have to record that he had to walk home and we got the most appalling rocket from the other *chef de train* when he found at Calais that his mate was up in the hills over 10 miles away!

Colin Morris will never forget this week-end. The previous day, we had travelled on Train 34 to the little station of Rang-du-Fliers, giving us about 25 minutes there before returning to Calais. We knew that there was to be a dinner for us that evening, so we kept out of the *buvette* and walked down the platform to while away the time. Colin saw one of the long-handled Continental bass brooms leaning against the wall and just as a member of the staff sauntered along to talk to us, he said he thought he could just do with that broom at home. I mentioned this to the porter who fetched Marcel Gille, the Stationmaster; the latter then went to the store, got out a new broom and presented it to Colin with the compliments of the SNCF. For the next two days wherever we went, so did the broom, up the gangway on to the ship and into the *coupe* on the "Golden Arrow" at Dover. And the next time I was at Rang, I received a similar present, and this time with red, white and blue ribbons pinned on the handle and a bunch of flowers for my wife.

That same evening we were back at Calais. Eugene Lavieville always organised a dinner for about ten of us, perhaps six drivers, the two inspectors, Georges Chatillon, his assistant, Fred White the interpreter, and ourselves. The evening started by a visit to somebody's house for an *aperitif* and then at about nine o'clock, we would start a very heavy meal, with much talk and jollification until we were too tired to talk much more. Everybody would say it was time he went home and so in the end he did, after endless handshaking. The "deux chevaux", Simcas, and Renaults roared off into the night, but one of them always returned for a final *Calva* – Maurice Vasseur.

"Monsieur le President" used to drive No 231E7, and he worked No 231E22 on the last journey from Paris of a Chapelon Pacific. He is president of the *Amicale*, the Social club of the Calais depot, a fine engineman and a wonderful character. Maurice is very tall, very thin, has gigantic feet, and is immensely French in gesture and appearance, whilst his speech is very Calaisian. As President of the Social Club he visited England with the first group that came across in 1963, and when I took them on the last day of the tour to the Romney, Hythe & Dymchurch Railway, he excelled himself. No SNCF *mécanicien* had ever seen anything like the *Green Goddess.* After his journey with George Barlow of

New Romney, he got off the engine saying again and again *"Inimaginable, in-im-ag-inable"*. No visit of a Calaisian group to this country is complete without Maurice; no visit would be complete without his little holdall full of bottles of *Chateauneuf du Pape* to be opened whenever possible, and at unexpected times; and no visit would be complete without a visit to *"Le p'tit Chemin de Fer Inimaginable"*.

I shall never forget travelling to York in 1966 with the French party. We were in the compartment and Maurice was just opening a bottle when a ticket inspector appeared at the door, saying "All your tickets, please". At the same moment that the Inspector saw and recognised me, Maurice offered him a glass of wine. *"Coup de rouge, Chef"*, he remarked. And so our inspector, a little doubtfully at first, shook hands all round and sat down and toasted the SNCF before continuing on his rounds. And it is perhaps so typical of these rank-and-file French railwaymen that they *fit in*. It is natural for them to be friendly and to appreciate what they see and the people they meet, and because of that, they have the world at their feet when they come over here.

XX

French Locomotive Men

Most of my French footplate journeys have been on the Nord, but I have been on the Est on three occasions, and once on the Ouest from Le Havre to Paris. This last was on an early morning train, when James and I had an uncomfortable crossing on the *St Patrick* from Southampton. Our engine was No 231G763, a modified Pacific with poppet-valves, very little inferior in performance to a PO Pacific. I was fairly tired and had never been over the road before, but I did the work from Rouen and found that I had plenty on my hands. The train was booked very tightly indeed and I was kept busy almost to the outskirts of Paris. Our driver, by the name of Dupont, had been given the alternative during the war of joining the railway or going to Germany, and rather naturally plumped for the former.

When we were on the Est, we had an excellent trip on an Est 4-8-2 No 241A8, driven by a very competent engineman called Delaborde, who had been a prisoner of war throughout the war, and whose fireman had had a very rough time indeed in a forced labour camp. Altogether, I had four runs on these very fine engines. Our trains were round about 700 tons in weight, and the coal was mostly very small indeed, but the work was comfortable and the cab seemed roomier than on most French engines. The Est Mountains were quiet and purposeful machines, which rode smoothly and seemed to be complete masters of their work. Each cab was scoured and polished and the crews from Chaumont were uniformly high in standard.

We had travelled to Vesoul on the 7.40am from the Gare de l'Est, another very fast train. On the first occasion, we were running right up to the 120kph limit through thickish fog, keeping time well, and it was not until we were near Troyes that we ran out into the sunshine. There we changed engines and stepped on to a beautifully clean PLM Pacific belonging to Troyes, No 231G21. We got on uncommonly well with this crew, shared the work and the wine, and by the time we left them at Vesoul we had been given an invitation to spend an evening with them next time we came to Troyes. I have had many memorable meals in the homes of railwaymen in France but there are perhaps two which stand out because of their sheer size. The first was at André Duteil's home, and this was the first I had ever had in a French house; it ran to eight courses and taught me the essential lesson of not filling up with bread between courses. The second was at Troyes. We had first been invited to the home of Jacques Vidal, the fireman. There we had the formal but friendly welcome so common in a French home, with the family sitting round the long table in the living room whilst we drank our Pernod or Ricard. And then we all went on to the driver's home and sat down to our meal at eight o'clock in the evening. After some eleven courses and many different wines, *Grandmère* decided to call it a day at about one-thirty in the morning, but Alfred Bouchard and his wife said we were now to go to a night club. I longed for my bed, but the cabaret near the station entertained us until four a.m. After more hand-shaking and goodbyes, we were in bed by four-thirty. But by six we were up, by no means in the best of form, and by six-fifty we were on the engine again, with M. Vroux of Noisy-le-Sec; so soon after seven o'clock I was busy firing an excellent little Class 141P 2-8-2 on which we were travelling to Paris. André met us at the Gare de l'Est and took us home for lunch, another marathon which proved to be my undoing, for by the time we had got to the dessert I was fast asleep!

Jacques Vidal had had an interesting career. He was workshop-trained and had been an examining fitter at Troyes. He had then gone off to French Somaliland with the family as a Mechanical Inspector, but something had gone wrong and they had come back home again. He had then chosen to transfer to the footplate grade, becoming a fireman in his late 30s. His technical training earning him quick promotion, but although he subsequently became a driver on the diesels, he had to come off the footplate through ill health a short time ago. It was he who sent me a short letter when Winston Churchill died, a note so beautifully expressed as to be very moving. It was a testimony not only to what the great man had done for his own country, but to what he did for so many Frenchmen also who were living under the domination of Nazi Germany.

In the same way that tea is drunk on nearly every footplate in this country, so wine seems to find itself tucked away in the tender cupboard of most French locomotives or perhaps alongside the firebox to get the right temperature in cold weather. I am not quite sure what the rule book says about all this and certainly it is unknown on the electric locomotives. I have heard that wine is *"absolument interdit"*, which sounds to me like the enforcement of rule to forbid the drinking of tea in this country. Yet not so long ago, I was travelling on a 141R, and we were accompanied by a very senior Inspector who had been to Calais and was on his way home to Paris. Now the driver was *"Monsieur Le Président"*, and soon after we left Calais *M. L'Inspecteur*, who had stood in the centre of the footplate for a few minutes surveying the scene, came across and altered the setting of the regulator and reversing gear, issued instructions here and there, and generally dominated the proceedings. But no conflagration came as a result of what in this country might well have been interpreted as unnecessary interference.

We reached Boulogne without further incident, and *M. L'Inspecteur* then took over the driving as far as Abbeville in superlative style. However, we had not been on the move more than ten minutes when he was tapped on the shoulder by Maurice and asked to stand up, whilst *M. Le President* withdrew from the locker under the seat an excellent bottle of Bordeaux and some glass tumblers. *M. L'Inspecteur* resumed his seat with his eyes on the road ahead, accepted a glass of wine, toasted us, drained the glass in one gulp and threw the dregs out of the window.

Both on the footplate and in the messrooms I have had some splendid wine. Henri Dutertre, a connoisseur with a cave under his house full of excellent claret and white wines, likes the white stuff in the morning and this has to stand in a bucket of water on the tender end or over the cab-side to keep cool. André Duteil has served Muscadet, Julienas and Côte de Brouilly at Jeumont to balance the five-course cold luncheon he had brought in his gigantic Gladstone bag. These lunches were memorable in that they took place on my earlier visits to France when everything that happened was new to me. Indeed, I had a strange feeling of being in a dream and that any minute I should wake up back in England. The Jeumont coal crane-driver (Mimile, *"Roi des Cornichons")* and the shunting drivers used to come in while we were eating to examine these extraordinary English people who had come out of the blue to their little border town. Afterwards, we would take a walk across the canal bridge to a little *auberge,* and after a cognac and coffee, we were ready for the long journey back to Paris.

Bébert Bethune, a gay Parisian, who used to run No 231E26, taught me that red wine may taste good, but as a thirst-quencher it is a very poor substitute for tea. On my third visit to France, in 1959, we left Calais 34 minutes late with a light train of 440 tons. Bébert stood up in his corner on this very hot day and ruthlessly pulled the time back so that we were on time at the Gare de Nord. By then I had 184 miles in my back and arms and the best part of a bottle of red wine inside me, together with a raging thirst, a headache and the blackest of faces.

If one travels with the driver of a French locomotive whom one does not know, one often gets him talking about his career on the railway. With pride, he may say: *"Monsieur, j'étais chauffeur Super"* (meaning a Nord Super-Pacific), but never with more pride than the man who says: *"Ah, Monsieur, j'étais mécanicien PO six ans"* or *"j'etais chauffeur de route PO quatorze ans, neuf ans avec Gauchet", – or* Vasseur or Dutertre or whoever else it might be. Men worked together for years on the PO engines for this kind of life, despite the continuous lodging away from home, was very worthwhile and these men were financially at the top of their profession. This was particularly so in the days of coal premiums, bonuses for regaining lost time and for cleaning their engines.

Henri Dutertre drove No 231E16 from 1951 until she was withdrawn in 1962. He was then given No 231E9, transferred from La Chapelle. He got this engine to his liking and ran it until late in in 1965, when it was withdrawn from service and he was given a part share in a PLM Pacific. He complained bitterly to me in his letters that to share an engine even with René Gauchet would do neither them nor the engine any good, and he refused to be consoled when I told him that many of our engines had got a good deal more dilapidated than his shining No 231G42. When No 231E9 was withdrawn, he sent me some colour photographs of the engine in all her glory, and on the back of one were the words. *"Dernier train de cette locomotive – Que de Souvenirs!"* As for No 231E16, I have the SNCF brass plaque from the smokebox door, given to me by Henri, and now mounted on polished wood, in pride of place in my office.

The railway enthusiasts of France were drawn to the PO Pacifics in the same way as our own fans were to the A4 Pacifics. When No 231E22 had finally been withdrawn after her last journey, a remarkable man called Oscar Pardo, living in Bidart, with a tremendous knowledge and understanding of railwaymen as well as engines, gave to each driver and fireman at Calais who had been *"titulaire"* on one of these locomotives, a beautiful model of No 231E13, a Paris engine. Not only this, but he included the depot officials and, as if this was not enough, he gave me a similar model, perfect in every detail, which is a permanent reminder to me that I have worked with the men who have been responsible for the running and upkeep of what could be described as one of the finest, perhaps the finest, steam locomotive design ever conceived, built and operated in the history of railways.

But the life of the French engineman is not all PO and PLM Pacifics and glamour. There is plenty of far less rewarding work and there are plenty of rough patches. Lodging away is the rule rather than the exception even at suburban depots. There are some long days and some extremely difficult duties. Promotion is slow at Calais, for the depot has lost work and suffers from the usual winter recession as did the depots just across the Channel. Men have passed the examinations for driving but are still working as firemen, having waited patiently for some years to go into the training school to become *"élève mécaniciens"*. Not until they have passed through the school are they permitted to handle a locomotive. Compare this with what was happening twenty years ago at the same depot, when men were being promoted to the Pacifics in their mid-twenties!

Some of the conditions are excellent. Four weeks holiday, travel concessions, salaried status, retirement at 50 (or after 25 years' service for those that started at 29 years of age or over) with an excellent pension. There are no youths and no old men on the footplate in France and one notices this in a messroom when one sees drivers and firemen of similar ages eating, drinking, talking, looking at the television and so on. In this country I have had my battles with men, particularly young men, on the question of early morning or week-end absenteeism. Because timekeeping is so vital to the efficiency of a railway and because the absentee's work has to be done by some other man with a job of his own to do already, I have always felt it right to be tough on this issue. But the French shedmaster, at a depot such as Calais, knows of no such thing as absenteeism nor lateness for duty, for it simply does not happen!

The French driver spends several weeks, indeed some months under training, whether for steam, diesel or electric traction. He works to a set pattern for each type of locomotive, train and route. In this country, each driver was an individualist with his own ideas, methods, and theories to be discussed and argued. Certainly the British driver has been allowed a degree of latitude in the way he does his work which is denied to the Frenchman, whose mistakes, when he makes them, are subject to a severe and unrelenting discipline. A Frenchman may come on to the footplate grades in his early twenties after a short time in the workshops, or after many years if he so wishes. There are junior drivers at Calais who saw the red light of redundancy in the workshops and became firemen by application over-night. There are drivers like Maurice Vasseur and Gilbert Sueur who have never been in the workshops, and who spent many years firing as in this country. But they had to pass the same examinations and were therefore entitled to the same promotional ladder as the workshop-trained man.

Our individualists each has his own little practical dodges and idiosyncrasies for getting over this or that problem, whether with steam or diesel traction, and so does the Frenchman. I have seen with pleasure the expression on the face of a well-known Southern Region driver as he watched Henri Dutertre using sand to keep the speed of No 231G42 down to exactly 50kph over a speed restriction on a slightly falling gradient without the use of the brake, and again, later in the journey, to keep the locomotive, working on a light rein, to exactly 120kph. No electrically-hauled main line train is brought to a stand at the Gare du Nord by the use of the continuous brake. The straight air-brake, with sand, and very carefully used, does all that is necessary to arrest the gentle crawl along the platform necessitated by the speed limit at the entrance to the station.

The electric locomotive in France continues to be developed, and indeed electrification is being pushed forward rapidly. The results in terms of speed, reliability and consistency of performance are remarkable, at any rate with the more orthodox locomotives, whose power, when experienced for the first time, is something that one cannot forget. The control of the Type 16000 BB locomotive is very much in the drivers' hands, including the three stages of field weakening, operation of circuit-breakers at neutral sections and a notched controller. The 16000 BB is a rugged, reliable, very fast and very powerful locomotive, and when pulling back time, which is not easy owing to the tightness of the schedules, an electric driver is well worth watching. In these conditions he rarely sits down, and bringing his speed right up to the maximum with the quickest possible acceleration he holds it there up hill and down dale by the most delicate attention to controller and field weakening. He has the usual "dead man" pedal, the vigilance detector, and the Flaman speed recorder, with the standard recording arrangements for adverse signals, and both temporary and permanent speed restrictions. He may be working from Paris to Brussels and back in an eight-hour day, on a single-manned multi-current locomotive, or from Paris to Lille and back to Amiens, to lodge there, and one thing is certain; a high speed electric locomotive requires the highest level of concentration and attention to detail.

These locomotives may not have the glamour of steam, but there is no comparison in their performance, and their very brilliance is an education to those who have never before experienced modern high speed and high-powered electric traction. In these days, a Paris – Calais boat train will run to Amiens for mile after mile at 87mph, and from Amiens onwards, when one of the 141Rs takes over, the fall to a maximum of 62mph is most marked. But the American 2-8-2 locomotive as used on the SNCF, and modified to the extent of some refinements in design, is a great machine, ideal for a very heavy train with frequent stops.

In the North of France, these locomotives are coal-burning and stoker-fired and so, provided that the fireman works closely with the driver, there is no need to worry about steam. If it is necessary to thrash the locomotive uphill to regain time lost on the level or down hill because of the 100kph limit, one does not have to think of the fireman's back. But it must not be thought that the use of the mechanical stoker necessarily makes life easier for the fireman. Physically it most certainly does but he must concentrate hard. The coal is fed to the distributor plate inside the firehole door by a screw driven by a small two-cylinder steam engine in the tender: the coal is then carried by jets of steam to the appropriate parts of the firebox. But the regulation of the feed has to match exactly the requirements of the driver, and the power of the jets has to be regulated to meet whatever type of situation is being dealt with.

Because I like to experiment, I have twice run things too tight on a stopping train in that I have arrived at Etaples on Train No 27 with the fire so low as to be alight on only half the grate. The fact is that one's confidence in these locomotives is such that one tries every dodge to economise in fuel. As I had over 600 tons and was facing the stiff Neufchatel bank, people may think that I was too clever by half to run my fire so thin. However, I should not have been in difficulty had I restricted very carefully the supply of coal to the firebox on starting away. But I adopted the hand-firing technique of firing heavily and relying on the blast to bring the fire to life. With a very thin "stoker" fire bed, however, this treatment can charge the grate too heavily under the firehole door. Suddenly I have realised that all is not well; the pressure gauge has begun to hang, the smoke at the chimney top has disappeared, I have looked into the firebox and seen the little heap of black coal which tells me the worst and which means that I have got to get busy with the poker and quickly at that. Once I had a letter from René Gauchet, the next driver to retire at Calais, in which he said that whilst you needed to be a real driver to understand the French Pacifics, anybody could drive one of the American locomotives.

Be that as it may, there is plenty to do and think about. Unless they are getting rough, the 141Rs ride quite comfortably and they can accelerate very quickly. I have had hold of them on many occasions, and to start away with over 600 tons from the Tintelleries station at Boulogne on a 1 in 100 gradient in a tunnel, and without an atom of slip, gives one a feeling of great satisfaction. But I shall never forget the occasion on Train No 34 with No 141R428 when, with a load of 658 tons, I had dropped a minute from Calais to Boulogne by being a little too gentlemanly. The train was already behind time and overtime at Boulogne put us more than four minutes late away. Edmond Godry was on the engine and I was told that I must get to 100kph well before Hesdigneul and hold it as far as I could up the 1 in 133 to Neufchatel. Never had I heard a locomotive exhaust like it in all my life. I kept dropping the little reversing lever over tooth by tooth until it must have been at about 50 per cent travelling at 55mph up 1 in 133. A great column of black smoke was going high up into the sky, the steam pressure and water level were remaining constant and we gradually pulled back the time so that with a sharp stop at Etaples, we were only a minute behind. The noise, the vibration of this living machine, our concentration on the job in hand and then the pleasure of achievement made a wonderful combination.

The exhaust of the Kylchap locomotives is even sharper than on such an engine as No 141R428, and the contrast between the Pacific and the American was never more marked than on Train 19 when double-headed by a PO Pacific and a 141R. In the tunnel after Boulogne, the stately movement and gentle beat of the compound would be lost in the uproar from the chimney a tender-length behind us. But they respond to delicate treatment both from the fireman and the driver. The Worthington pump feed can be infinitely adjusted and the travel of the valves is controlled by a compressed air servo-operated reversing gear that is not only positive but sensitive. L. P. Parker used to insist on the minimum travel of the piston-valves when coasting; John G. Robinson used to insist that full gear and then half-a-turn up were needed to take the block away from the top of the link; while the LMSR said that a drifting position half-way down was just right. On a French 141R 2-8-2, however, one keeps the regulator just open until the train enters the platform. And then only does one close the regulator and put the lever into full travel.

When Henri Dutertre first saw a regulator fully closed at over 80mph on a "Britannia", he gave me a serious little lecture, but he did not realise that we were coasting with the reversing lever at about 15 per cent cut-off and that everything was nicely cushioned. The French have always followed the practice of easing the regulator back notch by notch until the speed begins to dwindle and the brake does the rest. Even in this there is no variation in technique, as coasting steam-chest pressures, and the speeds at which the regulator may be closed are laid down for each type of locomotive and rigidly observed.

As will be realised, the French railwayman has given me endless pleasure. I have been invited to social evenings in the Calais depot organised by the *Amicale* to honour the retirement of some person who has come to the end of his career. Husbands, wives and families begin to arrive about 8 o'clock on the Saturday evening, and soon the big instruction room has filled and the participants are seated at long tables, set lengthways down the room and headed by M. l'Ingénieur *Chef de Depot* and his Assistant, their wives, the *retraités* and their wives and families, and any visitors there may be from England.

Everybody starts eating, and drinking wine from bottles provided by the committee of the *Amicale,* until somebody decides to sing. There is no piano, but we have Arnaud Flament, who was once Henri's fireman before he became *Moniteur de Chauffe,* (a sort of super fireman) and who, besides being champion billiard-player of the Nord, has a splendid voice. Then come others, including Marcel Dewevre. Although for myself so far I have resisted all requests to sing a solo on the grounds that nobody would understand what I was singing! However, before I know what I'm doing, I do find myself singing "Goodbye Tipperary", which the French sing on every possible occasion, pronouncing "Tipper*ary*" as it is spelt, and immediately afterwards "Mademoiselle from Armentières," in broken English.

Then come the speeches led by the *Chef de Depot* and *M. Le President* and if I am involved in the retirement, I also say a word. *M. Le President* is in his element, and by this time his collar has worked its way outside his coat to give him a rakish appearance, enhanced by his thinness and expansive gestures.' There is a pause whilst champagne is poured out and the health of the *"retraités"* is drunk, and then the band, which has slipped quietly into the adjourning room, starts a deafening and vigorous polka. Now I am not a good performer on the dance floor, but before I can do anything about slipping out for a breath of air, I am ordered to dance with Madame "Dunstable" (her husband is a driver who comes over here to see his sister in Dunstable every year and speaks comic English). Madame D. does not speak English, and whirls me round the floor, whilst I try to compose myself sufficiently to carry on a conversation in French. The dance is being held in another instruction room, with great charts of the Westinghouse brake and various valve gears on the wall. The music stops but before I can retire, it starts again, and I am now ordered to dance with Madame Gauchet. So off we go once more, the brake systems flashing by again and again until I feel that the whole thing is a fantasy of unreality.

Luckily, before my flying feet have broken anybody's ankle, the music stops again, and steaming hard I return Madame Gauchet to her husband who has craftily dodged both dances. We go back to the other room where Maurice Vasseur, now in his shirt-sleeves, is standing on one of the tables ready to announce the results of the "Tombola". I win a cushion for my wife and a bottle of champagne for myself, and at one o'clock in the morning, after shaking hands with everybody many times over, I stagger off to bed to sleep like a log. But not so the contestants, for most of them dance the night through, the *"retraités"* are well and truly honoured and the function closes down at about 6.0am. The members of the committee, however, wash the pots, tidy the rooms and get everything ship-shape before going home to breakfast. Great fun – but not too often!

No doubt I shall keep going to France and I hope the Frenchmen will come here too, for it gives me pleasure and amusement to see their reaction to what to them is a different world. I have seen them stare with unbelief at the London Transport buses, so unlike their own, and their praise for the Underground is lavish indeed. They have seen a little of the great city, through the kindness of Thomas Cook, and they have absorbed the fascination and friendliness of Liverpool, visited its two Cathedrals, sailed down the Mersey, heard all the stories about the Liver birds, and seen enough of British Railways to convince them that we are no second-raters when it comes to running a railway business. My little friend in the Amiens Control, André Corbier, who has seen me so many times when I have passed through, has even achieved his life ambition of going to watch Liverpool play at Anfield and hearing the Kop in full cry.

I recall with amusement the visit of Henri and André to this country in 1959 when they travelled from Derby to Birmingham on a Stanier "Black 5." This old engine got over the ground at tremendous speed, but threw its unsuspecting visitors about and demonstrated all the party tricks of a really rough engine. One of the injector steam-valve wheels fell off and disappeared over the side, the flame scoop fell into the fire, the reversing gear flew into forward gear, and at one stage both driver and fireman were desperately trying to get their injectors to work, the water-valve having worked round to the closed position. Both the Frenchmen were good actors and their demonstration of the journey was most amusing but did not mask their appreciation of a lively if rough machine.

And now we must take our leave of these Frenchmen. *M. Le President* has finished his last bottle of wine, each of them has had his spell on the footplate of the Romney, Hythe and Dymchurch *Green Goddess* with George Barlow, and each of them is going home tired but happy having had four days in England he will never forget. Cars take us from Hythe to Folkestone, we shake hands for the last time and then they go aboard the *Côte d'Azur.* We walk along the jetty to wave our farewell and in a few minutes the ship slides silently across the Harbour. Just over an hour later, our guests are back once more on French soil whilst we also are on our way home, tired but happy. As Edmond Godry said in his letter after the last visit, *"Que de souvenirs imperissables!"*

RAILWAYS IN THE BLOOD

RHN Hardy

Contents

Acknowledgements

To Ian Allan for his acceptance of a sequel to *Steam in the Blood*, inexpert though it may be: to his editorial staff, who have guided me and kept me on the right lines.

Rhoda Powell, once my secretary at King's Cross, has typed the entire story. She had done a great job.

The choice of photographs and the task of writing the captions is equally onerous. I have used some of my own pictures but those that I have been loaned for publication highlight the fact that I am but an amateur at the game.

Finally, to Gwenda, my wife. Whenever I say in this book that I did this, that or the other, it is not me alone. Without her help, encouragement and support, without her willingness to upsticks and move her home a number of times at the behest of the railway, I could have done none of these things.

Foreword
by Sir Peter Parker cvo

HE IS A HAPPY MAN who finds his work; the work with the scope that can shape and be shaped by all his enthusiasm and energies all his life. A member of that happy breed is always recognisable – even on the murkiest, darkest day. It doesn't really matter what he is doing – steeple-jacking or holding the service in the church below, gardening or deep-sea exploring, conducting an orchestra or children in class or a bus, driving a business bargain or a train – what matters is that as he does it, he is all there, concentrated, every bit naturally devoted to the doing, and loving it. Richard Hardy is just such a man, a happy railway man who has the passion that makes all the difference to the performance of a job and to the writing of this splendid book.

It abounds with good stories and wisdom, with admiration for trains and for people, and with that unmistakable happiness of a life-time.

Dick Hardy's career – 1941-82 – shows the transformation spans of railways: nationalisation, steam to diesel, the competitive growth of roads and cars, the contraction of the rail network and manpower. But in this autobiography of railways none of this registers in that familiar moaning litany of disappointment over decline. Change there is, and deeply felt. It is not just steam that disappeared but a culture, he tells us. Yet this Hardy is a philosopher who lets cheerfulness keep breaking in. Another Hardy wrote, 'If a way to Better there be/It exacts a full look at the worst'. This Hardy would add, 'And the best'. His apprenticeship rounded off in the Running Shed, tough, demanding work, 'The roughest of conditions, no mess rooms, no changing rooms, no washing facilities, no heating and under locomotives just an acetylene lamp: and yet this was the life I loved and nothing ever disillusioned me'. His strength of spirit is in his friendships, in the art and craft of men and women who knew their difficult jobs, and in an indestructible readiness to enjoy things and the variety of human nature.

This career spans not only a crucial time of transition on the rails: it extends easily over the widest range of working relationships of any colleague I knew in BR, from footplate to Board room, in this country and France, inside railways as well as outside. There is one memorable journey, with the late Poet Laureate. The picnic party on the train had reversed in a siding at St Helens Junction. Sir John Betjeman thoroughly enjoyed the view – a car scrap yard: they were watching a mechanical monster devouring old cars...

There is pleasure and deep purpose in the memories. Two themes make their connections throughout the narrative – good management and leadership. For too long it has been fashionable to keep too often these themes academically separate. It is a relief to know that nowadays the business schools here and in the States are troubled by the widening fissures between their tests and the realities of organisational success, between the teaching of techniques and the development of will and purpose to make things happen. There is a difference between ideas and belief – belief makes things happen. Richard Hardy is a believer who has made things happen. He himself has managed on the basis of the humane thesis, as good tempered as it is convincing that authority must be firm and accessible, listening and learning, and straightforwardly, unsqueamishly responsible for leading. 'To have hold of an engine', to drive, is a loco man's phrase that he learnt on the footplate at his start, in 1941. Dick Hardy has 'had hold' from the earliest command, as a very young Divisional Manager, to his final success at HQ (and he seems to have been more than a touch reluctant to go there).

In his last phase he had the senior opportunity to work out his philosophy into the practice of developing managers, of all ages – certainly including, thank goodness, the Chairman.

I was at the farewell party for him at '222'. I never attended a celebration of retirement which reflected such a wide variation of jobs of all levels and all ages in the organisation. I suppose there is just a possibility that some railwayman or woman may know more about how to run a railway than Dick – there is certainly nobody who cares more or who has a better feel for the network, human and technical, that make up the great service. He boldly admits at the outset to the need for romance on the rails. Agreed.

One of the great Romantic poets agreed too: 'A man must travel, and turmoil, or there is no existence'. I am sure, given time, he would have been more specific and said 'travel by train', but Byron died a bit early.

Sir Peter Parker CVO

Preface

A S A BOY IN AMERSHAM, I had a happy time. Certainly I was an only child but I had some good friends, most of them sharing my passion for railways. With Michael Kerry, I visited the station to watch Great Central, LNER and Metropolitan engines working their magic, returning, in summer, to carry on equally passionately at cricket or tennis. By and by, I began to learn something of railway work from booking clerks, porters, signalmen and not least from the Neasden drivers and firemen, some of whom took a personal interest in the start of my education in life and human relations. When Bill Collins, one of these drivers, welcomed us back to Amersham in 1973, it was a joy to me that a man so deep in retirement should want to see again the man whom he had tutored as a schoolboy. Indeed, I have told the young engineer many a time to maintain his friendship for he who has helped him in the formative years. So often that man will become a lifelong friend.

In those precious days, Michael and I avidly read Cecil J. Allen's *British Locomotive Practice & Performance.* We timed the 'Directors', the 'Faringdons', the four-cylinder 'B7s' and more reluctantly the 'B17s' and Pacifics. The great man, however, was not impressed and our work was not published. We were what is loosely called railway enthusiasts and I had hoped to include a chapter in this book, even two, about my friends whose hobby is the railway for I owe much to many of them. Some encouraged me right from the outset of my career, others guided and advised me and now and again I have been able to repay their kindness by showing them the railway life from the inside.

To name but three: there was Ian Allen, East Anglian medical practitioner, railway photographer and one-time honorary shedmaster at Framlingham and Laxfield; Harry Mosse, Major Royal Artillery, 'that bloody little army chap' (so said my much loved Assistant, Bert Webster) who thrust his way into my Liverpool Street office to demand 'Are you in charge he-ah'. Somehow, he finished up on a 'Britannia' going to Norwich and back. And then there was Henry Maxwell, railway romantic, who made famous our Stewarts Lane 'King Arthur', *Sir Balin*, No 30768. Henry never lets me forget those days. Recently he wrote to me and here is a little of what he said.

> 'To myself, your whole personality was epitomised in No 768: she embodied the spirit which would keep the flag flying whatever the odds, round which the fond and the faithful could rally and she was as much the "heart" as the "pride" of Stewarts Lane'.

After my retirement, many of these kind people asked Gwenda and I to join them for an evening. Three of our closest friends, Basil de Iongh,

James Colyer-Fergusson and Bill Thomas did the honours. We had a wonderful time and we shall always treasure the wine glasses and the decanter that will see the Hardy family down the years. On the decanter was a perfect engraving of a 'Britannia' and the words:

RICHARD HARDY
Railwayman
1941-1982
from his Friends

The inspiration of those interested but not professionally connected with our great industry has helped me to write this book. I knew that *Steam in the Blood* gave pleasure to more than a few and Sir Peter Parker told me that it was high time for a sequel. I have retraced the same path as in my first book but I have also written of the last 20 years of service when my railway life widened and I became more involved in public life.

Locomotives, technical, operating and commercial matters have interested me but I would hate to write a textbook. But when it comes to the deeds of railwaymen, it is different, for one can write from experience and from the heart for we are drawn together by instinctive brotherhood. Sir Peter Parker's chauffeur, Tony, was once a Stratford fireman whom I had known as a cleaner boy and this created a bond between us. My 'engineer' through the Rockies from Calgary to Field in 1983 welcomed me not because I was an Englishman riding on the locomotive but because I was a railwayman. It is a brotherhood that transcends the barriers of position and which maintains a mutual respect. Never could I have stood a desk-bound drudgery for I have drawn strength all my working life from the practical railwayman, from my colleagues in management and from the enthusiasm of the young. Life has been tough at times but I have had great happiness and fulfilment.

And so my story is of railwaymen and women, and I must have known very well indeed some 25,000 in my time. Whoever they are and whatever they have achieved, in some way, they have enriched my life.

I

Magic to a Young Man

JUST BEFORE my retirement in December 1982, I visited the plant works at Doncaster where I had started my apprenticeship in January 1941. Life is not worth living without a degree of sentiment and romance and George Cox – at that time Chief Inspector in the works – and I walked round some of the shops, for we had been apprentices together. We turned into the new Erecting Shop, George and I, to find Denis Branton, one of the apprentices with whom I had done my first day's work. Now a Chargehand Erector, he had achieved his ambition. Our recognition was instantaneous and, for half an hour, work was suspended and the clock put back, for we had not seen each other since I left the works for ever in 1944.

It is strange that I should have made this particular pilgrimage in my last weeks of service and yet not found time to visit the Carr loco where I spent 11 marvellous months to round off my apprenticeship. The running shed life was what I wanted. The comradeship of the running shed artisans and staff, none of whom were on the piecework that drove the Plant men relentlessly onward: live engines, smoke and steam, excitement, the constant battle with the demands of the timetable, 24 hours a day, seven days a week, breakdowns, derailments, heat and bitter cold, football played in clogs, in the lunch hour – or cricket – sometimes extended until our foreman came after us, the roughest of conditions, no messrooms, no changing rooms, no washing facilities, no heating and under locomotives just a hand acetylene lamp: and yet this was the life I loved and nothing ever disillusioned me. We were a happy crowd, and now and again there was the chance to relax round the fitting shop fire or in summer on a bench in the sun while we waited for work or perhaps the chance for me to join Driver Harry Frith on the fitting shop pilot. Harry was excellent and instructive company and I used to do his shunting for him when nobody was looking. His engine was an old Ivatt 'J52' 0-6-0 Saddle Tank, also confined to the yard on medical grounds, but, unlike Harry, her days were numbered. If you blew the large ejector to release the brake quickly, it went on instead. An engine was needed for the job, this one would do until her days were up, and there were many more important repairs to tackle. And in any case in those days a shunting engine was credited with 7mph so it would knock up a fair mileage even if it was standing still.

In my Doncaster days my footplate experience was nearly all achieved, in my spare time, through the kindness of many men who had nothing to gain from teaching their art and who did so without reservation. I was always welcomed although I certainly learned when and where to appear (and disappear), for these friendships meant so much to me that to overplay my hand or to abuse this warmth of heart was unthinkable. After a year or so, it was clear to me that I was not going to become a gifted engineer, but it was obvious to Edward Thompson who gave me much encouragement that, given the right development, I could make my way in the Locomotive Running Department in the management of depots and districts, of artisans and of enginemen – that most independent breed of men who, paradoxically, require, indeed demand management which is strong, fair and utterly involved. He felt, given experience and firm directive, that I could become what is today known as a manager of men, a leader of men. It is for others to say whether I achieved that objective but I am certain that I gained a basic knowledge and understanding of what was required of me from the men who trained me, whether in the shops, the running shed or on the footplate.

Amongst the drivers and firemen of the West Riding, Ted Hailstone of Bradford was probably my great mentor. Much has been written of this remarkable man but one had to work with him really to know and understand his ways. He was a Great Central man, with a splendid memory, vast experience and gifts of expression. I could listen for hours to his splendid voice – 'when I was firing to Willoughby Lee', he said one day, which sounded tremendously grand. A strong yet kind personality, he had a dedication to the job, but with a hard, ruthless and sometimes unreasonable streak that made him heartily disliked by those who did not measure up to his requirements. The bosses did not escape his strictures, and his contempt for weak management and lack of discipline was profound. On the other hand, his admiration of a Chief such as John Blundell, his prewar Shedmaster, knew no bounds. Thus, he bred in me an understanding of what was right and wrong, that management had to be fearless, that it had to lead, to keep its promises, and maintain law and order – even in the difficult days of the war – but that men would respond if they were known, understood, encouraged and, if necessary, disciplined.

One evening, when I had been at work the best part of 14 hours, we stood at Halifax waiting to work our last trip to Bradford. Everything was set on our 'N1', a good body of fire made ready for departure for we had to attack the long 1 in 45 to Queensbury after we left the North Bridge station; and I was thirsty. We had 15 minutes before we were due away, Ted was examining the engine and I walked across to the refreshment room for a cup of tea (OCS we used to say, and it cost us 1d). I was back by the engine well before departure time but there, barring entry to the cab, was Ted, as I had never seen him before. He brought his face menacingly close to mine, his first finger pointing forbiddingly at my chest – 'Now, lewk here, young man, I'm responsible for you and don't you ever leave my engine again without my authority'. That was all. I daren't excuse myself, I simply did my work to perfection for the rest of the day, and we parted as if nothing had happened. Years afterwards, he asked me if I remembered the incident and felt sure that it was a lesson learned the hard way: never again, from that summer evening in 1944, did I leave a footplate for whatever reason, without a clear understanding of where I was going.

In Ted's last week of railway service at King's Cross, I went with him on No 60014 *Silver Link* his own engine – to Leeds and back. By that time I was Assistant District Motive Power Superintendent at Stratford but this did not stop us doing the job together. He knew he would miss his work and was in a fairly emotional state but all the lessons I had learned from him I applied faithfully to his great satisfaction. We were a team with no need to speak – the occasional gesture to indicate a clear signal, a smile and a nod, my old mentor and I were completely absorbed in our world apart where everything depended on the ability and temperament of two men, charged with a great responsibility, working together to find steam and power, to run to time and in complete safety.

And let us pause to remember this point. We are none of us perfect, life does not begin at 9.0am and finish at 5.0pm, we all have moods and queer and difficult ways, and yet discord on the unique little world of the footplate could turn the job from a joy to a misery.

There was only one Ted Hailstone; and perhaps it was for the good of the West Riding that Driver Benny Faux of Ardsley was unique. He was another of my mentors, the complete antithesis of Hailstone, except that he, too, was a splendid railwayman although he loved to be regarded as a daredevil. In the same way that Ted had made me keep to the straight and narrow, so Ben showed me how rules, though not necessarily the rule book, could be bent to advantage. I do not think that Ben went out of his way to influence me but I am glad he did for it is very necessary to begin to know a few of the tricks of the trade!

We would go to Thorpe Arch on the late turn with a Robinson 'B4' 4-6-0, probably No 6101; we took 10 coaches of factory workers, many of them girls, from Wakefield, Normanton and Castleford. We usually had an excellent night's work which meant that Percy Thorpe (his trusty mate) and I shared the work whilst Ben occupied himself with some dilettante supervision from the left-hand corner of the cab. In the early hours we would be back in Westgate, and having taken the empty coaches to Balne Lane Ben would consult his watch: 'we'll stay here a few minutes, Percy', he would say: and then he would arrive at Ardsley a few minutes too late to put the engine away without making overtime. There would be a set of men in the lobby waiting for disposals and Ben and his mate would slip off home that bit earlier, without having to dispose of the engine. Now, this sort of thing is as old as the hills and a good foreman knows how to deal with it but the point is that it opened my eyes to some of the manoeuvres that others, more serious minded, would never let me see.

Not for Ben a technical discussion on locomotives, not for him the Mutual Improvement Class on a Sunday morning or even railway talk in the home. He was quite amazed to hear that I had written about him in *Steam in the Blood.* A few years ago a Copley Hill driver, long retired, was looking at that book and at Ben's photograph. After a minute or so of silence, he said emphatically and with great finality 'That man was mad'! Ah, no – an adventurer, sometimes a daredevil but never mad: a man who knew what he and others could achieve but who never allowed latitude amongst those he could not trust, although this too had been a hard-learned lesson before I had the good fortune to meet him.

One last memory; New Year's Eve, 1944, a Sunday night and Ardsley men on the Bradford Mail via Batley. We had a very heavy train, six buckeyes and a luggage van, and a 'J39' 0-6-0 freight engine, No 2707. Just right for the job with 1 in 40 gradients. Ben bought us cigars, we were full of good cheer and we had Alistair Kerr with us, Deputy Borough Surveyor at Wakefield and as good a railwayman as you could find. 'Isiah then' says Ben, 'Dick will do the driving and Mr Kerr the firing and no messing about, Dick, you've got to knock her about to keep time with this load, and you'll have to run hard down Batley 'Oil.' From Drighlington down to Batley was 1 in 38, pitch dark, blackout, of course, no signals and a nice sharp curve near the bottom. As this was the LNER and not the LMSR, the West Riding and a Sunday, there were no 'Big Berthas' to give us assistance even if we had thought about a banker. Away we went, Alistair firing for all he was worth, Ben driving me on until, on Ossett bank, for the only time in my life, I drove an engine absolutely flat out with a passenger train at what must have been nearly 30mph. The noise, pure thrill and primeval excitement of this journey will always be with me. Our arrival at Wakefield, at the double so to speak, with No 2707, big ends thumping, negotiating the crossovers in a series of straight lines, was hilarious. This I am unlikely to forget but never the man who made it possible, 'Mad Benny' Faux whose motto was 'We never bother, it'll be reight'!

Let his daughter, Elsie, have the last word. In due course she married Alistair Kerr, and years later we were talking about her father. When she was a girl she worked in Leeds and travelled every day from Ardsley to Leeds Central. She had the choice of a train from Castleford or one from Wakefield, both worked by the Ardsley passenger men. Came the day when her father was driving. Eventually the train proceeded decorously into Leeds Central where Ben waved to a white-faced daughter as she passed the engine. 'From then on, when my Dad was on the Wakefield, I went on the Castleford. He never got another chance to frighten me!'

Bob Foster, of Copley Hill, was my first West Riding mentor – kind, friendly, a man who understood my very youthful enthusiasm. Through him I met many Copley Hill crews and, bit by bit, improved my firing technique on slow trains to Leeds with 'K3s' and 'C1' Atlantics, but, by 1943, I was gradually becoming master of the situation and so had the opportunity to fire on the ex-Great Central four-cylinder 4-6-0 No 6165, *Valour*, which spent several months at Copley Hill in 1943. The experience taught me that, given understanding handling, this engine could do an excellent job with very heavy trains. It was not the collier's friend that history would have us believe but it could be if things were not done properly.

Indeed, I can honestly say that, in my fairly extensive experience of the un-rebuilt 'B3s' and the numerically much larger class of mixed-traffic 'B7s', I have never had a rough or uncomfortable trip. On the other hand the Copley Hill Atlantics, once on the move and on a dry rail, were in a class of their own for performance matched with economy: furthermore, their maintenance costs were lower than the Robinson engines. But these considerations never altered my vision of the huge, gutsy, handsome machine with the classic name of *Valour.*

Late in 1941 Bob and I and Percy Carline, a Copley Hill top link fireman, were sitting on our 'K3' No 231 in the Garden Sidings at Doncaster. Alongside us there came to a stand a Great Central Atlantic, with a difference. It was one of the four 'C5' Compounds, No 5364 *Lady Faringdon.* Now Percy Carline originated from Grimsby and Immingham and had fired, years before, for Grimsby Driver Bill Richardson who was in charge of No 5364. I was introduced and was transferred to the footplate of the Atlantic.

Bill was one of the old school of Great Central drivers, at any rate in appearance. Much of my work with him was on Grimsby Doncaster slow trains worked by the Compounds and the Pom Pom Bogies', Class D9 4-4-0s. This was not high speed work but it was interesting and Bill was splendid company. At that time, he had a mate called Frank Vessey. There may have been better firemen than Frank and the old man used to lecture and encourage him patiently and with the odd wink in my direction. Most LNER engines were fitted with the Davies & Metcalfe Dreadnought vacuum ejector: the handle placed in the vertical position opened the large ejector and could then pass through perhaps 30deg without applying the brake. On hot days, when his mate was getting into difficulties, Bill would very solemnly draw imaginary pints using the handle of the vacuum ejector as a beer pull and pass them to Frank to keep him going.

The working of the Compound Atlantics was interesting and quite straightforward. It must be remembered that they were built on the Smith principle with one high pressure and two outside low pressure cylinders. Three sets of inside Stephenson's valve gear were used, controlled by a single reversing screw, the exhaust from the high pressure side passing directly into a large steam chest common to both low pressure cylinders. The engines could be worked simple expansion, any degree of semi-compound and, of course, full compound.

When starting a train, boiler steam entered the high pressure steam chest directly and also the low pressure side through a reducing valve controlled by a spring loaded regulating valve, adjusted by the driver who could therefore vary the pressure in the low pressure steam chest. And so the start would be made in simple and gradually, by turning the change valve wheel in a clockwise direction, full compound working would be achieved. This wheel was situated in the right-hand corner of the cab and connected to the regulating or change valve on the side of the smokebox. Once well on the way to full compound working, the regulator would be opened wide where it would stay when there was work to be done. From start to finish, the valve gear would remain at full travel and all adjustments necessary made on the change valve and regulator. If time was not being kept uphill, a touch of live steam would soon sharpen things up. I should dearly have loved to have worked a fast train with one of these Compounds which had had a splendid reputation on the Great Central main line, but I had to wait until 1958 to sample another Compound of a very different and more complex sort, the famous Chapelon Pacific.

Kindly, generous Bill Richardson finished up on the Grimsby Dock pilots but another great friend, this time at Grantham, stayed on the main line until his retirement in the 1950s. This was J. A. Thompson, always known as an endearing character with a lovely sense of humour and turn of phrase. We were working a Doncaster Grantham Tarty' with a GN Atlantic, No 4401, making frequent stops – for there were many wayside stations in those days – with quite a heavy train of 10 corridor coaches. This meant some hard work for us, particularly starting on the uphill stretches above Newark – we even stopped at Barkston!

The water scoops fitted to the GN tenders were lowered by a long lever which the fireman, facing the tender, pulled towards him through some 30deg from the vertical. Primitive but not dangerous unless, on returning to the vertical after struggling to get water, one trapped one's fingers on the tender end. Once the scoop was down in the water, it stayed there and there wasn't a snowball in hell's chance of getting it out until the end of the trough, with the result that, if lowered too soon, not only would the front of the train get a soaking but, more important, we would find a miniature Niagara bringing coal on to the footplate level with the firehole door. Until some sort of clearance had been made, one had to stand on the coal one was shovelling and that, as everybody knows, is a difficult business!

Shortly before the Muskham troughs, old Bill, standing tranquilly in his corner, pockets as ever stuffed with fruit, said: 'Right, Richard, we'll fill up at the "trawvs" and don't put the scoop down until I tell you'. We came up to the troughs at about 50mph and I waited and waited and then, of course, impatient and youthful, I pulled the lever down. In no time the footplate was deep in coal, lumps, small coal and water. Chastened and about to be faced with the hardest part of the journey I began the task of cleaning up while Bill's kindly eyes looked quizzically at me over his distinguished moustache: 'Never mind, Richard', he said, 'You'll be wiser next time, have another pear'! And this reminds me of the Edge Hill, Liverpool crew on a 'Lizzie' working the Merseyside express one evening. The fireman dropped the scoop and couldn't wind it out. Once again, coal and water everywhere, sheets of spray finding their way into the driver's corner, soaking him to the skin. He turned to his mate and said with perfect Scouse timing and simplicity – 'Did we gerrany'?

Not everybody by a long chalk has worked on a Sentinel car. There was one stationed at Ardsley but worked almost entirely by the Copley Hill No 2 Tankie link. I cannot remember the name of this particular unit nor its type but it was strong enough to spend much of its time slogging up Batley 'Oil. I wonder whether firing a Sentinel lorry is quite as hair-raising as it was on the 'Car'. We would attack the 1 in 38 out of Batley with everything just right, the fireman at the back end with his boiler, chimney, injectors and engine, the driver away from it all, in peace and solitude, up front. We were the first passenger train of a winter's morning and sparks streamed from our red hot chimney, which shared the compartment, while steam at 300lb/sq in from a well-filled boiler drove us hissingly upwards. To put coal on the fire one lifted, with the left hand, a lid on the top of the vertical boiler and with a domestic shovel popped the coal from a small bunker down the vertical shoot on to the fire below. The injectors were delicate but effective little things operated by minute handles and if the rail was dry all would be well for us in our warm little compartment. But if we started to slip, all hell was instantaneously let loose and 300lb of steam and a boilerful of water had become 150 and the water playing 'Bobby Bingo' as we struggled slowly up what seemed to be a never-ending treadmill.

Nevertheless, the car did a very fair job, having to stop twice, at Upper Batley and Howden Clough, in the middle of that fearful bank and it nearly always reached Drighlington to connect with the train for Bradford via Morley.

My journey on the Sentinel car was made with my dear friend, Stan Hodgson. He, with the extrovert nature, had invited me up on to the Copley Hill footplate of the old Robinson 'B4' 4-6-0 over 43 years ago and, by introducing me to his driver Bob Foster, had opened for me so many friendships. The great majority of those I knew have passed on but Stan Hodgson (on that evening in 1941 promoted to the top link to cover leave of absence) is still with us. When I retired, Gwenda and I gave a party at Headquarters. Not only did many of my colleagues of recent years accept our invitation but so did some of those with whom and for whom I had worked over nearly 42 years. Stan has never changed in manner and personality and little in appearance; he still lives up above Armley, Leeds in the same house that I visited as an apprentice, and he always remembers that first night when Bob Foster, Stan and I sat on old No 6100 in the Garden Sidings, Doncaster and he said after listening to my deep, gravelly voice: 'Dick, owd lad, you've missed your vocation; with that voice, you should have been a parson'! And so it was a great joy to me that he came to our party and enjoyed it so much for his friendship bridges, bar a few months, my entire railway career.

II

'And Mr Hardy will give us the answer'

The story has already been told of how I came to be sent to work under the great L. P. Parker, Locomotive Running Superintendent of the Eastern Section of the LNER. I have written of his character, his ability and the effect that he had on the lives and careers of us young people, because to work for such a Chief was an education and extended far beyond the normal relationship of those days. It has been said that he ruled by fear and there was an element of truth in this. Certainly we feared him for he had the power and intention to make or, if one let him down, to break but, nevertheless, it was the older men, the more senior people who really went through the hoop.

Before the war, when he was District Locomotive Superintendent, Stratford, the bell in the Mechanical Foreman's Office would sound. Charlie Greenwood, Chief Mechanical Foreman, held in high esteem by L.P., with a lifetime of experience as a running shed fitter and fitter's foreman and an artist in Stratford connivance, would rise far from gratefully from the paper work that he despised, square his shoulders and his chin, mutter something about an 'old barstard', remove his bowler hat and, clamping it under his arm, walk purposefully towards the great man's office. A knock, immediate entry without a pause and Charles would walk the mats which protected the carpets from coal and oil, coming to a stand at attention, in front of the desk, to face the music. If it was dusk the office would be lit only by a powerful desk lamp trained to shine in the eyes, whilst the little man, in bow tie, smart suit and no doubt wearing spats after the fashion of the day, sat behind the light, ready and waiting to toy with his victim. 'Tell me, Greenwood, how many axleboxes have we lubricated with water rather than with oil, in the last six months?' The impossible question, for nobody, not even Charlie Greenwood, dared hazard a guess with L.P. A simple question, the answer could soon be found but L.P. would never allow reference to a notebook in his presence and any reply had to be thought out, memorised, qualified and then dissected. One could never try anything on with that remarkable man and the penalty for doing so was likely to be systematic dismemberment.

Ah, you might say, the sadistic old devil. Of course he was: he was ahead of almost everybody and he knew it. But how it made one think, concentrate, control one's emotions and one's temper, tell the truth, maintain a courageous and civil fairness if in the right. How it made one try hard, how he liked one to have the courage to *do* something (even if it was wrong) rather than always play for safety and, above all, how it trained one for the years ahead. L.P.'s lessons in administration and management have been applied time and again when I have had to cope with particularly difficult problems. I have never used his methods, for they were unique to him but he drilled principles into me on which I have been able to build as I gained experience and confidence.

L.P. certainly had a sense of humour and, of course, the Londoners could see the comic side and liked to have the last word and although they rarely got the chance, made up for it subsequently with a wealth of picturesque and inventive detail.

Bill Morris, driver, was appointed by L.P. as an oil inspector to reduce the number of heated bearings on the Great Eastern section. He had a colourful turn of phrase, not bad language but a choice of word and idiom often denied to the better educated. Morris had been very concerned with coal dust from the large coaling plants which found its way into bearings. These plants were known as 'cenotaphs' as there was a certain similarity of shape. None of us would have dreamed of using the word to L.P. but Bill had no scruples and I heard the following conversation which gave me great joy but which also rewarded the two contestants!

Inspector Morris (in his usual loud voice) was called to explain a report: 'Nah, Mr Parker, the trouble is they go under them cenotaphs and get coal dust in the oil.'

L.P. (quietly, inscrutably but with the merest twinkle): 'The Cenotaph, Morris?'

Inspector Morris: 'Yes, Mr Par-ker, the cenotaphs, you know what I mean, don't you sir?'

L.P. (even more quietly): 'Morris, I was always under the impression that the Cenotaph was in Whitehall. I find your remarks interesting but you have not yet told me why the dust is getting in the oil and what you are going to do about it.'

L.P. would twinkle, if in the right mood, but he always asked the one question that had to be answered which made Bill Morris go away, think of ways and means, and then get things done.

Early in 1949 both the Operating Superintendent and L.P. were anxious to obtain some more powerful express locomotives for the GE section. We had the splendid 'B1s', the GE 'B12s' for the less important main line work, together with 'B17s' whose capabilities varied according to the standards of the depot to which they were allocated. The majority of these locomotives were allocated to two or three sets of men and the day had but recently passed when some of the Norwich 'B1s' were allocated to only one set of men. Something bigger was needed and I remember seeing correspondence about the possible allocation of 'A3' Pacifics. By this time, it must be realised that L.P. Parker had taken over the Western section and so had command of Motive Power affairs for the whole of the

Southern Area. It would have been simple for him to order a reallocation of power but I think there were Route Availability restrictions that vetoed this particular move.

And so in the early spring of 1949, for several weeks, we were host to the Southern Region 4-6-2 No 34059 *Sir Archibald Sinclair.* How we enjoyed that machine! In the first place it was in good order and secondly it was handled by only one crew – the driver being Burritt of Stratford. I had an excellent journey to Parkeston Quay and enjoyed firing the locomotive on the return trip where we terminated at Stratford having taken no coal at Stratford before departure nor at Parkeston – quite deliberately in order to clean the tender for a high speed trial to Norwich and back the next day.

This was to be its swansong before returning to the Southern and, indeed, No 34059 went out in a typical Bulleid blaze of glory, though not quite in the manner intended. I hasten to say that it did not catch fire (a little trick to which I would become accustomed 3½ years later when I joined the Southern Region) but it had something better up its sleeve, after five weeks of conscious innocence, that probably no other engine could have achieved.

The load was 12 corridor coaches to Norwich, starting from Liverpool Street, returning with 13, terminating at Stratford. The 13th coach was the General Manager's Saloon, almost certainly the Great Northern coach that I used years later as Divisional Manager, King's Cross. I was to sit in the front coach to be a lesser Cecil J. Allen. Chief Inspector Len Theobald was on the footplate to see that the engine was worked to the L. P. Parker dictum – with a full open regulator and short cutoff where practicable. Our running was the fastest I had ever known on the GE section and I remember that our speed at Ingrave Summit, up the 1 in 84, was 48mph.

We left Norwich with our extra coach in the early afternoon. L.P. was there and had had a conversation on the platform with Len Theobald. Something outstanding was to be achieved and it must be remembered that No 34059 had no speedometer at that time. L.P. and Theobald had always got on well, and Len had an infinite regard for his Chief to whom he was both civil and outspoken.

We climbed the 1 in 84 to Trowse Upper as if it were 1 in 300 and were accelerating brilliantly when, somewhere near Swainsthorpe, the first station out of Norwich, there was a terrifying hissing roar from the locomotive, steam was shut off and we came to an abrupt halt near the signalbox. Jumping up on the footplate, I learned the worst. A steam pipe supplying the reversing gear had fractured and here we were, 107 miles from Stratford with an engine that could neither reverse nor be notched up, and here we were faced with 13 coaches, a journey to London in full forward gear, a formidable General Manager and a far more formidable Motive Power Superintendent! Needless to say, to come off the train was never seriously considered. 'Dick, get back and tell L.P. we're going to

Stratford in full gear and tell the Guard to give us rightaway as soon as you are back in the train'. I ran to the back of the train, wondering what would be my fate. It must be remembered that I was but 24, that I had never set foot in an inspection saloon and I most certainly was not in the same league as any of the occupants.

The steps had been lowered by the attendant and after knocking I went into the dining room. Eight people were sitting round the table, the air was heavy with cigar smoke, the port looked as if it had been round more than once and I was asked to explain what had happened. Everybody was smiling and nodding and seemed prepared to stay at Swainsthorpe for the rest of the afternoon. L.P., however, had not yet spoken. Suddenly, quietly and twinkling dangerously but with port and cigar to hand, he said: 'Hardy, I am quite sure that Theobald will be able to keep to time with that extraordinary machine even if it will neither start nor reverse.' That was all. I beat it quickly before things had a chance to get nasty, hopped in with the Guard and mercifully we began to move. The story is soon told. We ran to London in full gear and 601b of steam or less in the steam chest most of the way. We kept time having stopped for water at Ipswich where we also passed the word for a clear road to avoid the chance of being stopped. The engine burned a fair amount of coal but I have often wondered whether any other type of locomotive could have run 107 miles at express speed in full gear and kept to an exacting schedule. But the Bulleid Pacifics, wayward, difficult, brilliant, fascinating, had that very human trait of rising to the occasion after a display of petulance, and so it was that day.

We never got the Bulleid Pacifics on a regular basis, for which Stratford depot later on had cause to be profoundly grateful. In 1951 the service to Norwich was completely recast and the 'Britannias' came, brand new and in tremendous form to make an unforgettable name for themselves.

Between 1947 and 1949 I worked very closely with the Operating Headquarters at Shenfield. Whilst the Motive Power Department existed, so the Operating Department was a separate organisation, The 'traffic'! Whilst we were 'traffic' and 'loco' there was a great battle to be fought with no side giving any quarter in their passionate desire to run the railway. It was at South Lynn on the M&GN section, where with a single line to Yarmouth one was constantly having to improvise with Cambridge and Norwich controls and our own South Lynn traffic people. I loved that sort of work and co-operation and, equally, it was joy to work with Shenfield for we had an understanding of each other's responsibilities whilst some of the finest operators that I have ever met were in charge at Shenfield in those years. The engine workings were simple, easy to read, straightforward and always took into account, even up to the end of steam, the need to book engines to regular crews. The Parker influence, of course, but the Shenfield people could see the wisdom therein.

We shall return to Stratford later on this story: L.P. will not only have retired quietly from the scene but sadly will have passed on before he ever completed the book that I believe he intended to write. So let me leave you with a memory of this extraordinary little man. He decided to take me to a meeting of senior Motive Power and Operating Officers of which he was Chairman. I was sent for, told to get my coat and whisked off with him from Hamilton House to one of the hotels. I was to take notes which I did with vigour and enjoyed my lunch although I found the exalted company a bit terrifying. The debate continued after lunch, with the Chairman ruling this, that and the other. I was thinking how nice a cup of tea would be, and paused with my notes, dropped into a momentary trance in which I heard, as in a dream, the terrifying words: 'And Mr Hardy will give us the answer'. Mr Hardy came to his senses, stammered out a reply which mercifully must have had some relevance as it was accepted with indulgence. The after effect was similar to that which follows a momentary doze at the wheel of a car. Never again!

Punctuality and Pride

MUCH HAS already been written about the period I spent at both Woodford Halse and Ipswich depots as Shedmaster, a period so full of incident and experience that it seems unbelievable it extended only from September 1949 to August 1952. But Woodford Halse, on the Great Central section, was my first command and, young as I was, I learned so much in those first few months and may well have left a certain reputation behind me.

At Woodford the lives of many of our staff were bound closely to the countryside, whether through village life, the Church, farming, small holdings, cricket, football or an interest in field sports. My own love of cricket enabled me to play for my village XI and it is fair to say that my present zest for hunting was born out of the interest and enthusiasm of our railwaymen for what, in those days, was not the controversial sport it has become.

The Woodford staff, many of whom had additional and sometimes profitable ways of earning a living, were nevertheless dedicated railwaymen, faced with many problems which had to be solved, indeed fought by us all. We were desperately short of firemen and young men had to be imported from northern cities to lodge in Woodford, perhaps at the railway barracks in Percy Road that gave me so much trouble. In seven months we recruited precisely one engine cleaner against the 35 we needed to fill the roster and clean a few engines. But, although we never realised it, the writing was on the wall for steam: young men were no longer prepared to join – or stay – with a railway where shiftwork, difficult hours, cold, heat, dirt and dust would be rewarded by very average rates of pay and conditions. One had to have 'steam in the blood'!

The artisan staff were Great Central shopmen, with conditions very different to those employed elsewhere on the Region. Their duties were spread round the clock so that there rarely seemed to be the depth of talent at any one time to tackle the heavy maintenance requirements. On the other hand, to a man, the fitting staff were excellent and the boilermakers waged continuous war on leaking tubes and dirt in the barrels, for Woodford water was terrible.

Now Ipswich was very different: a first impression was that, compared with Woodford, there was a lack of respect for authority and yet this proved to be a complete illusion. I found that the temperament of the Suffolker is such that he will not be driven an inch but that he is prepared to accept the lead of a management interested and dedicated to the cause.

The depot at Woodford Halse was relatively modern and well set out and although there was only limited equipment for repairs, there was plenty of room and an excellent 'cenotaph'. On the other hand, Ipswich was desperately cramped, there were insufficient pits for the examinations of locomotives to be carried out in an orthodox manner, there was next to no covered accommodation and the coaling of 91 locomotives was either by hand straight from trucks or by digger picking up coal thrown by hand on to the ground. However, there was a very reasonable machine shop and wheeldrop, far better than at Woodford.

At the latter depot, the artisan staff were constantly up against it and fought a just perceptibly winning battle against the whimsicalities of the Gresley 'V2s' and the diabolical riding tendencies of 'B17s' and WD 'Austerity' 2-8-0s. On the other hand, the Ipswich maintenance staff were well on top of the job. Consequently, one could anticipate the requirements of enginemen who were allocated regular engines, and the co-operation between fitters, boilermakers, footplate staff and foremen was outstanding. But one could not sit back and relish this teamwork because all parties expected the Shedmaster to take the lead, to pay the closest attention to what was going on. This meant that I had to maintain a firm hold on the running of the depot, the maintenance of the best possible standards of punctuality, cleanliness and smartness, the need to acknowledge, in person, a job well done, and equally important, to take personal action against slackness, laziness and poor workmanship, on the road or in the depot.

The men at both Woodford and Ipswich welcomed a Shedmaster who was prepared to try and carry out this philosophy and to take a close interest in footplate or depot work out of normal hours. But now and again this interest could be taken the wrong way. In 1950 the 'B12' 4-6-0 No 61570 was run by Drivers V. Trenter and F. Gibbs. This particular locomotive was in first rate condition but for some reason I cannot remember went off the boil, not badly but instead of standing over mile after mile at 175lb/sq in she hung around 150. She would time her trains with this pressure but there was something wrong. The booking had been made on a Friday morning by Driver Trenter and I accompanied Fred Gibbs in the 17.05 Liverpool Street as far as Colchester to satisfy myself before we carried out any work. The journey confirmed the shyness of steam but it was impossible to diagnose the cause on the road as everything appeared to be in good order.

It was not possible to stop the engine for repairs until the Monday and I thought no more about it until just after 16.00 when there was a knock on my office door. A little man, in serge jacket, waistcoat and black trousers, shirt devoid of collar but with stud in position, stepped inside. He wore a non-uniform cap, was smoking a newly lighted cigarette, and one glance at his face told me that he was very angry indeed. The first thing to be settled was the removal of the cap and the disposal of the cigarette. This made him even more angry as he was denied the opportunity of flouting convention and the normal code of manners. But the coast was now clear for Driver Vic Trenter to tell me in no uncertain manner that he knew his engine and that, when he booked a defect or specific symptoms, he did not take at all kindly to a young Shedmaster riding with his opposite mate to check whether he, Trenter, knew what he was talking about. He stormed out of the office, banging the door behind him and worked the 17.05 to Liverpool Street with the spare 1500, No 61577. A few days later, his own engine back at work, Vic was a happy man, particularly as he had given valuable information leading to a total cure.

In those early days I doubt if I could have expounded a philosophy of leadership or management but there is no doubt that the lessons learned at South Lynn, Woodford and Ipswich sharpened my ability in these fields. But one worked for what now seems a totally different set of values. The stringencies of budgetary control at local level were unknown and, as a Shedmaster in my twenties, my objective was to play my part in the running of the railway. If special trains were required, one moved heaven and earth to find engines and men, achieving the impossible at times. Part of one's life centred around the engine availability figures, the efficient deployment of locomotives and their reliability in service. One did not think consciously of costs but of achieving the highest standards of maintenance, performance and punctuality. One took a great pride, sometimes almost fanatical, in the cleanliness of all Ipswich locomotives – passenger or freight and extended this to the tidiness of the shedyard and its comparative freedom from ashes and loose coal. This was achieved at a cost, of which I was unaware. Yard and engine cleaning gangs worked on Sundays at overtime rates and the crack engines were cleaned seven days a week. On the other hand, every man worked hard, down to the cheekiest cleaner boy. It is easy to say that engine cleaning is unproductive, that yard cleanliness is only a detail and that costs could have been drastically cut. That would have to be the argument today, but more than 30 years ago a young man who passionately loved his work and all that it stood for thought differently.

The words Industrial Relations had never been used in 1950. We were certainly aware of the strength of the two large trade unions and, at disciplinary hearings, a Shedmaster, especially a young one, had to be on his toes if he were not to be outwitted by that professional advocate, the trade union organiser. On the other hand, almost without exception, the organisers wanted to see a show well run and had little or no respect for a management that was weak or ineffective, although not necessarily saying so!

At both Woodford and Ipswich, there was a standard of representation, particularly for the footplate grades, that made our joint work of raising depot standards and performance all the more rewarding. I am not likely to forget the warmth of the co-operation, help and guidance that I received from the accredited representatives of the footplate staff at these two depots. Do not think for one moment that we always agreed or that harsh words were not exchanged for one was dealing with experienced, mature men with a lifelong interest in their work and their conditions of service, but there was rarely any pettiness in their make-up. One can work with such men and the values we set ourselves included the need to improve the standards and performance of the staff, to improve their morale, their working conditions, to bear in mind and act on, where possible, the sensible suggestions of practical men and never to shrink from the difficult decisions that may affect the lives of the men employed at the depot. They would far rather hear the truth and with this goes the crucial need for local management to do what it has promised to do – nor to promise to undertake what it cannot be sure of achieving.

Two letters that I received over 30 years ago will show exactly what I mean, letters which ring true because of their simplicity. Just before Christmas 1949, I posted a notice to all the Woodford staff thanking them for their co-operation over the 31 months I had been in charge. I received the following reply from George Wootton, Driver, and Acting Secretary of the Woodford LDC and dated 24 December 1949:

Mr Hardy
Dear Sir,

May I, on behalf of the footplate staff, thank you for the good wishes you extend to us per your Notice of yesterday. It is my pleasurable duty to tell you that the good wishes are reciprocated. We wish you a very Happy Xmas and a very Happy and Prosperous New Year. We hope you will spend many more of them in your new home. I feel I may say that the staff appreciate the efforts you have made on their behalf and will give you the assistance you ask, in any direction that will improve conditions at our depot or on British Railways generally.

Thanking you again,
Yours sincerely,
Geo. W. Wootton

The promise made by George Wootton was faithfully upheld in the few remaining months I was to stay at Woodford.

It must have been February 1949 when C. H. D. Read, my Chief from Colwick with very high standards and strong views, visited me and was hurriedly recalled to deal with a strike at his principal depot. The only way to get him back quickly was by a light engine which was quickly prepared by the Driver, Tom Floyd who was given our solitary 'N2' 0-6-2 Tank engine No 69560 – a deaf and dumb contraption if ever there was one but just the engine to run light, downhill, on a cold day. When coasting, most GC men tended to lengthen the valve travel on a piston valve engine to a drifting position of about 45%. However, the instruction of the Motive

Power Superintendent, L. P. Parker, was that cut-off should be fixed at 15%. He took a very firm line on this and implemented his instructions through his District Officers, of which Charles Read was one.

I walked across to put this point to Tom Floyd who replied very directly that he was the driver, that he would work the engine his way and that if C. H. D. Read interfered, he would get a dusty reply. In fact, the journey was completed without any acrimony and Charles Read quickly got back to Colwick to deal with his difficulties. I saw little of Tom in my last few weeks but, during the summer months and before we moved to Ipswich, I was on a short course in London and travelled up with the 6.20 from Woodford each day on the locomotive. One week, I had Tom Floyd. He immediately delegated the driving to me, became a firm 15% man and was extremely interested in the course I was attending. As a result I tried to get him to a National Tribunal hearing but without success. By this time, I had moved my home to Ipswich and received this reply to my letter.

4/11/50

Mr Hardy,

Thanks very much for your letter. I am sorry nothing much can be done about attending National Tribunal but I am very grateful to you for trying to get me in. I fully understand the position. I am glad you have settled in at Ipswich and hope you will be O.K. I don't mean by that that I am glad you have left Woodford. In fact, I will sincerely say that you brought a breath of fresh air to Woodford Loco and I can assure you that I and the majority of the men at Woodford are extremely sorry that you are not still with us.

Yours sincerely,
T. Floyd

Here was a letter to be treasured, because Tom Floyd meant exactly what he said and because it was a confirmation that one's efforts to achieve the values in which I believed had not been in vain.

Now a few words on punctuality. It is not an understatement to say that the importance of punctuality in train running was bred into one, on the LNER and on the Eastern Region, although I was soon to realise that there was still much to be learned when I transferred to the Southern Region. That is not to say that timekeeping was always good but passenger train punctuality and a certain proportion of the freight running was subject to very close scrutiny. One incident stands out clearly from my Ipswich days, to illustrate how personally one took one's responsibilities.

During the summer of 1951 one of the afternoon expresses from Liverpool Street to Norwich was doing badly, the engine a 'Britannia', the driver Burton of Norwich. We received a message to say that he wanted assistance, Ipswich to Norwich. There was something of a crisis at the depot and the outlet was temporarily blocked by a derailed locomotive. I told the Running Foreman that I would go round to the station and have the station pilot on the front. We never put our best engines on this job and No 61668, a 'B17' ready for Shops and by no means a good one, was shunting carriages when I arrived. I told the Station Inspector we were going to take his engine but he demurred to the extent that I told him we would make a straight change so that he could keep the 'Britannia' to do his carriage shunting until we could get it replaced. What a fool I was! Burton ran down with the 'Britannia', I asked him to change footplates which he did with a not unnaturally bad grace. The change of engines took the best part of 10 minutes, making the train 25 minutes late from Ipswich. Once 61668 was on the train, Driver Burton created his vacuum quite correctly, with the small ejector only, whilst I fumed on the platform in silence at this self-imposed delay. It only remains to say that No 61668 lost 10 minutes to Norwich. A fool I most certainly was but the point is that the delay haunted me for months. One did not shrug it off as 'one of those things'. It never need have happened and I took it very personally to heart, yet another lesson learned.

When one has been involved to such a great extent with footplate staff, it is inevitable that much of what one writes touches on their often charismatic lives and experiences. My own knowledge of what went on the road was perhaps deeper than my knowledge of the work of the highly skilled running shed artisan, with his understanding of what to do to make an engine roadworthy in the shortest possible time, or how to cure an elusive fault on a locomotive, standing mute and cold, devoid of life.

At Ipswich, we were blessed with dedicated craftsmen. Wilfred Brown, Chargeman Fitter and once a Lentz valve specialist, still writes to me 32 years after we worked together. Jack Percy, the Leading Fitter ultimately became Mechanical Foreman. What he achieved with steam has long been a memory but after they had all survived the trauma of the reconstruction of the depot, Jack was promoted in 1956. Soon afterwards I heard from him, and knowing that I was now at Stratford, he added a PS: 'Engine 61669 (Easterling fame 1955) now in Shops, 96,000 miles. She will carry the Headboard for 1957 – warning to all Stratford Foremen, not to be borrowed, loaned or stolen, keep your hands off her'. I doubt if any other depot in the country could have got 96,000 main line miles out of a 'B17'. That was Ipswich for you.

IV

'Dahn the Lane'

Nor long ago, I was asked which of the positions that I had held had given me the greatest satisfaction in terms of total involvement. I had no hesitation in saying that my time at Stewarts Lane from August 1952 to January 1955, as Shedmaster, was the toughest, hardest, seven-day-a-week assignment that could have been offered to a 28-year old railwayman. Whether in locomotive running and maintenance, in real front line railway work, or in leading and sometimes infuriating some 730 individualists, the utter involvement required – indeed demanded – by those 2+ years' unremitting labour had to be experienced to be understood. In my time there were about 126 locomotives and not only did one have the whole responsibility of the running of the depot, and punctual running of trains by Stewarts Lane engines and men, but also there was a responsibility for some 130 motormen on electric train work. For the first time in my life I had this responsibility, and little knowledge of the electric train drivers' work which was far less simple than I had always imagined. My early inclination was to concentrate on the steam side of affairs but as time went on I realised this was a mistake and that the motormen had their problems which had to be handled as effectively as those who still had steam in the blood.

Much of Stewarts Lane's great work had been achieved by paying men more than the hours that they worked. This had got out of hand so that, by 1952, the simplest tasks were negotiable. As the majority regarded it as their right to be paid overtime whether worked or not, and as some of those who could not be so rewarded received compensation indirectly and even more illegally, and as even those responsible for the compilation of the pay sheets booked themselves overtime, it can be realised that those of us who were determined to put a stop to such practices were in for a difficult time. But when I reported to my new Chief, Mr G. L. Nicholson, early in August, I was blissfully unaware of what lay ahead. I had applied for the position on impulse. I had been appointed, after thorough research and probing interview because I was young but experienced. Gordon Nicholson had put several personal questions to me at the preliminary interviews, including 'What are your faults?' I knew them only too well but an honest reply did me no harm at all. On the other hand, he knew that few older, perhaps more conventional men, would have been prepared to stand up to the rigours that he knew I must face if I were to carry through his instructions to clean up the depot in every way but, at the same time, to improve its performance in every aspect of the business. Nevertheless, I have to say that my opposite mate Charlie Boarer at the Bricklayers Arms in the Old Kent Road was around 50 years of age, had every bit as rough

It was in the early 1930s that the author's family moved to Amersham in Buckinghamshire and this brought him into contact with the locomotives operated by the Metropolitan Railway; these included the four 0-6-4Ts – Nos 94-97 – that were supplied by the Yorkshire Engine Co during 1915 and 1916. Designated Class G by the Metropolitan and Class M2 by the LNER, No 96 *Charles Jones* was one of two that survived to be withdrawn by BR in 1948 (as LNER No 9077). *J. Joyce Collection/Online Transport Archive*

The erecting shop at Doncaster Works in the early BR days; it was at Doncaster Works that the author commenced his railway career in January 1941. *John Meredith Collection/Online Transport Archive*

John G. Robinson's 'D11' 4-4-0 No 5505 *Ypres*.
John Meredith Collection/Online Transport Archive

The workhorse of the LNER in the West Riding – a Class N1 by H. A. Ivatt.
No 69472 is seen at Drighlington & Adwalton with an Ardsley-bound train on
31 July 1951. *Tony Wickens/Online Transport Archive*

John G. Robinson's four-cylinder mixed-traffic 4-6-0 Class B7, No 1390 recorded in 1947. *John McCann/Online Transport Archive*

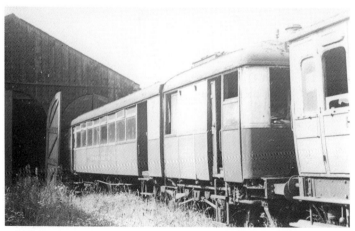

It hasn't proved possible to locate a view of a Sentinel railcar in the West Riding so this view of No 273 *Trafalgar* – new in May 1928 and disposed of in March 1946 – must suffice. In cold weather the boiler and chimney were good friends, but then a bad rail on a heavy gradient would reduce speed to a crawl and the water level in the boiler. Nevertheless, the LNER made good use of them, particularly the North Eastern area and the author's first journey in one was from Harrogate to Ripon in the mid-1930s. *John Meredith Collection/Online Transport Archive*

Class B3 4-6-0 6168 *Lord Stuart of Wortley*; sister locomotives Nos 6164 *Earl Beatty* and 6165 *Valour* were reallocated to Copley Hill shed for a few months in 1942. The latter deputised for 'K3s' and 'V2s' on a the heavy Colchester to Leeds turn and was regarded as a thoroughly good engine. It was named in memory of all those Great Central employees who gave their lives for the country in 1914-18. *Les Collings Collection/Online Transport Archive*

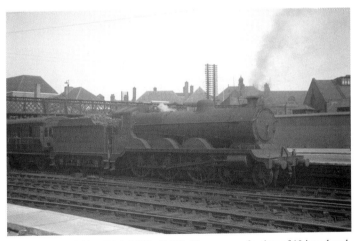

Ex-Great Central Class B4 4-6-0 No 61488; this was one of a class of 10 introduced in 1906 to a design of John Robinson. The locomotives spent much of their lives in the West Riding where they were transferred after the Grouping. Fourteen bogies to London from Leeds on a Sunday excursion was a fair proposition for a relatively small engine. All had been withdrawn by the end of 1950. *John McCann/Online Transport Archive*

Class C14 No 67444 stands outside Ardsley shed on 5 October 1952. These locomotives were widely used on the ex-GNR lines in the West Riding; even with relatively light trains, no chances could be taken with any locomotive on the heavy gradients that were such a feature of these routes.
Tony Wickens/Online Transport Archive

Robinson Class J11 'Pom Pom' 0-6-0 No 64364 at Woodford on 14 July 1951.
Compact, hand and economical, the 'Pom Pom' took some beating for all but the
heaviest and fastest work. *Tony Wickens/Online Transport Archive*

Class B17 No 61667 *Bradford* seen at Walton-on-the-Naze some time after the
locomotive had been transferred away from Woodford shed. The author records that
the 'Sandringham' was one of two that were transferred to Woodford Halse from
Lincoln in early 1949; Nos 61647 and 61667 were, in his view, 'two of the hardest
and most uncomfortable cabrankers it was ever my misfortune to come across.'
J. Joyce/Online Transport Archive

On 31 March 1955 Class C12 4-4-2T 657386 with E82350E and E86141E is
pictured departing from King's Lynn heading towards South Lynn.
John Meredith/Online Transport Archive

South Lynn station in 1950. *Real Photographs*

Opposite: In 1951 the author greeted three Doctors to Ipswich shed including the
well-known photographer Dr Ian C. Allen. The visit was recorded for posterity on
Class D16/2 No 62590, pictured here on another occasion. The locomotive was that
'Super Claud' in service and was withdrawn in January 1952. Right up until the
end the locomotive did good work on express passenger and stopping trains to
London, Norwich and Cambridge. Driver Arthur Enfield took No 62590 to London
towards the end of 1951 in about 80 minutes which is not surprising if you knew
Arthur. In his firing days he is reputed to have joined as astounded driver after a
consultation with the guard via the roofs of the carriages of a lengthy train.
John McCann/Online Transport Archive

On 12 June 1957 Class B1 No 61004 *Oryx* is seen shunting in the carriage sidings on the south side of Ipswich station; there was a small locomotive servicing facility at the station, visible on the extreme left of this view, with a 45ft turntable and coaling stage. By 1957 it was approaching the end of its life as closure came during 1959. *John Meredith/Online Transport Archive*

On 19 July 1952 Class J15 0-6-0 No 9 65459 stands at Saxmundham with the train for Aldeburgh. *John Meredith/Online Transport Archive*

One of the outstations under the author's control when he was shedmaster at
Ipswich was the shed that served the ex-GER branch terminus at Framlingham.
Pictured at Framlingham station on 19 July 1952 is Class F6 2-4-2T No 67230.
The coaches comprise six-wheel brake No E509451, composite No E63612E and
brake third No E62386E. *John Meredith/Online Transport Archive*

Pictured on 24 September 1949 at Laxfeld, the terminus of the Mid-Suffolk Light
Railway and another of the branches that had a shed under the control of the
shedmaster at Ipswich, is Class J15 0-6-0 No 65459. The train comprises six-
wheel brake No E62331, six-wheel composites Nos E63404 and E53405 and a
four-wheel goods brake. *John Meredith/Online Transport Archive*

Again on 24 September 1949, a second Class J15, No 65457, is seen approaching Haughley from the north a train from Bury St Edmunds to Ipswich. The track of the Mid-Suffolk Light Railway line to Laxfield can be seen on the right.
John Meredith/Online Transport Archive

The 'B12' 4-6-0 was a feature, during the author's time in East Anglia, of services on the ex-GER main lines; here No 61533 is pictured at Ipswich on 14 October 1956 having arrived with the 10.6am service from Cambridge.
Marcus Eavis/Online Transport Archive

On 26 June 1948 'Battle of Britain' No 21C163 *229 Squadron* pictured at Stewarts lane shed on 26 June 1948; the locomotive was renumbered 34063 in early 1949. *John Meredith/Online Transport Archive*

Class T9 Mo 30718 recorded at Synsthurst Crossing in September 1955. The author relates how a Battersea man would tell you that a Drummond 'T9' 4-4-0 was a great engine, almost to be compared to a Rebuild – a compliment to the South Western design to be sure. *Alexander McBlain/Online Transport Archive*

On 3 April 1951 Class D1 4-4-0 No 31145 is seen at Canterbury East with a
portion of the 'Thanet Belle'; the author recalls a trip in September 1956 when
Peter Handford travelled to Kent to record Class E1 No 31019. The return journey
was behind No 31145 with the author firing the 'D1' – an eventful trip that the
author will 'live with for ever with every exhaust beat sharp and clear above the
deafening racket within the cab.' *Derek Cross*

Class N15 4-6-0 No 30763 *Sir Bors de Ganis* – the first of the class to be
completed – pictured at Durnsford Road with a Down service in 1957. The author
comments: 'One did not think of the "King Arthurs" as fliers, nor indeed were
they. Speed had to be won gradually and then maintained whereas with a Rebuild
and a suitable train, or on a "Schools" no appreciable effort was needed to get into
speed.' *Fred Ivey/Online Transport Archive*

Class H 0-4-4T No 31005 in use as the works shunter recorded at Ashford station in June 1958; the author records an incident one winter evening when the locomotive had a close encounter with a 'King Arthur' class 4-6-0. As the author notes 'My task next day was to find the culprit. I might as well have tried to time the "Golden Arrow" with a "P" class tank.' *Derek Cross*

On 19 September 1954 Class C 0-6-0 No 31576 was employed to haul the 'Palace Centenarian' which was the last passenger train to use the old LCDR station at Crystal Place at the end of the branch from Nunhead.
Fred Ivey/Online Transport Archive

Britannia class No 70004 *William Shakespeare* on the Down 'Golden Arrow'
emerges from Martello Tunnel en route for Dover. In '*Railways in the Blood* the
author records his displeasure at the condition in which the locomotive was
returned to Stewarts Lane after a brief sojourn to the LMR; he wrote: 'I am sorry
to say that outwardly nothing has been done to this engine which has returned in a
disgusting condition to this Depot.' *D. Kelk/Online Transport Archive*

A virtually brand-new 'Britannia' – No 70001 *Lord Hurcomb* – looks in superb
external condition as it prepares to depart from Liverpool Street with the Down
'Norfolkman' on 18 April 1952. *Neil Davenport/Online Transport Archive*

The new order on the Eastern Region: No D200 – the first of the future Class 40 –
departs from Liverpool Street on 22 June 1959. The old order – in the guise of
'Britannia' No 70006 *Robert Burns* – can be seen in the background.
Neil Davenport/Online Transport Archive

The last locomotive that was prepared for use on the royal train whilst the author
was in charge at Stewarts Lane was No 34088 *213 Squadron*, which was used to
haul a special from Portsmouth to convey the emperor of Abyssinia, Haile Selassie,
to London. On a slightly less august occasion, the locomotive is seen at Shortlands
on 14 June 1958. *Peter N. Williams/Online Transport Archive*

A total of 14 Class 'N7s' – Nos 69600-11/26/27 – were allocated to Wood Street shed; these were not fitted with vacuum brakes and so were restricted to the Jazz services. As is the case with No 69608 pictured with a Chingford service at Liverpool Street on 9 June 1957, the crews kept the locomotives in remarkably good order. *Julian Thompson/Online Transport Archive*

Enfield shed was located alongside the station at Enfield Town; this was one of
the small sheds that was closed – on 30 November 1960 – with the electrification
of the GE suburban lines. Pictured outside the shed in its final months is Class N7
0-6-2T 69656. *J. Joyce/Online Transport Archive*

Class F6 2-4-2T 67219 about to leave Stratford (Low Level) for North Woolwich.
This was as easy place to start, being on a curve and rising gradient.
J. Joyce/Online Transport Archive

Opposite: Ex-LT&S Class 3 4-4-2T No 41949; this was the locomotive that was
'modified' with Great Eastern type under footplate injectors.
Harry Luff/Online Transport Archive

Class Y4 0-4-0T No 61825; when the Devonshire Street 'Pots' came to Stratford shed for a wash out from Devonshire Street, they would sometimes damage their front cylinder cocks when bouncing along faster than they should have done.
Harry Luff/Online Transport Archive

Class J39 0-6-0 No 64871 heads an Up goods at Little Benton South on 13 June 1960. The author encountered the class in his time at Stratford and Ipswich. The former used the class quite frequently on express passenger turns. While they held together and despite 5dt 2in driving wheels, they were very effective, but their motion was suspect after about 20,000 miles and an edict was issued in about 1953 on the Eastern section prohibiting their use on passenger trains of any sort. They were strong, capable and had an excellent boiler but were a fitter's nightmare. We had several at Ipswich and rostered the best to the 19.45 Ipswich-Whitemoor, a fast freight train. But after two evenings spent disentangling the motion somewhere between Haughley and Bury, we booked our spare 'B1s', Nos 61054 or 61056, to the job and kept the 'J39s' for less onerous duties.
Alexander McBlain/Online Transport Archive

Opposite: Class B1 No 61054 stands at Liverpool Street awaiting departure with the Down 'Easterling'; this named service, which operated between 1950 and 1958, ran through to Lowestoft and Great Yarmouth. No 61054, as an Ipswich-based locomotive, was one of two 'B1s' regularly employed on the 19.45 freight from Ipswich to Whitemoor in place of the Class J39s.
John McCann/Online Transport Archive

One of the specially prepared station pilots at Liverpool Street – Class N7 0-6-2T No 69614 – stands in the station awaiting its next duty in July 1960. *LRTA (London Area) Collection/Online Transport Archive*

Class J69 No 68619, pictured here on an RCTS special, in black livery was another of the regular Liverpool Street pilots. The early turn crews earned their cleaning bonus particularly on busy days when the pilots were required several times on the shift. *J. Joyce/Online Transport Archive*

Class N7 No 69663 with a Down suburban service formed of a quintuple articulated set at Bethnal Green on 3 March 1957. *Julian Thompson/Online Transport Archive*

In 1963 the author had his first experience of the Brush Sulzer Type 4 locomotives Nos D1500-19; by the time he had been transferred to Lincoln in November 1963 he had come to regret that BR had selected this type over No DP2 as the standard Type 4 diesel-electric. *Harry Luff/Online Transport Archive*

No DP2 was outwardly similar to the contemporary 'Deltic' class then being built by English Electric but internally was very different. Designed as a prototype for a class of Type 4 locomotives, No DP2 was destined to be a one-off. However, to the author, this was a mistake as 'It was reliable and it was fast and had it become the standard Type 4 locomotive in the higher power ranges, we should have been on a winner.' *Harry Luff/Online Transport Archive*

A two-car DMU formed of Nos E56444 and E51285 stands at Barton on Humber on 17 May 1969. The author recalls his first visit to the terminus shortly after his appointment to the Lincoln District a few years earlier 'on a cold and windy December night'. *John Meredith/Online Transport Archive*

A two-car DMU departs from Liverpool Lime Street with a service towards Warrington in 1969. *Geoffrey Tribe/Online Transport Archive*

When recorded here in 1969, the high level section of Liverpool Central station was a pale shadow of its former self with the surviving services operating to and from Gateacre; when these were withdrawn in April 1972 the high level station was finally closed. It was largely demolished the following year.
Geoffrey Tribe/Online Transport Archive

By 1976, when this scene of three EMUs was photographed, Liverpool Exchange station was living on borrowed time. The number of platforms at the station was reduced in the early 1970s to permit the construction of the Merseyrail underground link; the completion of this led to Exchange being closed on 30 April 1977. The bulk of the station was subsequently demolished although the façade was retained and incorporated into a new office development. *Geoffrey Tribe/Online Transport Archive*

Very much the mainstay of the West Coast Main Line when the author was in Liverpool, two Class 86s stand at the north end of Crewe station in 1986. *Geoffrey Tribe/Online Transport Archive*

No 63762 – seen here with a freight at Radcliffe – has no official name. To railwaymen, he is a, 'ROD', a 'Superheater', a 'Tiny' or an 'O4'. Call him what you will, he has won two wars, the simplest, strongest, most reliable heavy freight locomotive ever built in this country. He has worked in Australia, Egypt, France and Persia. He was the backbone of the LNER Southern Area coal traffic. He also worked for years on the GWR virtually unaltered, and that is the great accolade.
D. Kelk/Online Transport Archive

In October 1968 Class 231G No 231G61 is pictured at Boulogne Ville; sister locomotive No 231G42 was fired by well-known engineman Sam Gingell on an eventful trip to France in 1969 when he was 77 years old.
Harry Luff/Online Transport Archive

SNCF Class 231E No 231E17 pictured in 1961 at Calais about to depart for Paris with the *Fleche D'Or*. Sister locomotive No 231E16 had fascinated the author on his visit to France in April 1958 when he had had the opportunity of firing the legendary Chapelon Pacific between Amiens and the French capital. The author describes the '231E' as 'the most remarkable, brilliant locomotive on which I have ever worked ... Its brilliance became apparent when there was real work to do.' *John McCann/Online Transport Archive*

An SCNF Class BB67000, No 67348, hauls a passenger train at Les Sables on 18 August 1988; towards the end of his railway career, the author was on a train hauled by one of the class when a French railwayman described him as 'une figure legendaire'. *Geoffrey Tribe/Online Transport Archive*

The dramatic lines of an SNCF Class 232S 4-6-4: in May 1971, when the author was invited to travel on the last steam-hauled service from Paris to Calais, one of the class, No 232S002, hauled the departing train from Paris to Lille, where it was replaced by Class 231E No 231E36. *John McCann/Online Transport Archive*

SNCF Pacific No 231K82 stands at Albertville after arrival from Calais on
29 September 1968; three years later, on 26 May 1971, this was the last French
Pacific on which the author travelled when he had a footplate ride from Amiens
to Boulogne – despite concerns as to the damage that might arise to his suit!
Alexander McBlain/Online Transport Archive

Romney, Hythe & Dymchurch 4-6-2 No 1 *Green Goddess* at Hythe in June 1956.
Phil Tatt/Online Transport Archive

a job, and gave every bit as much to it as I did. Our methods were different but the results were the same.

My greatest asset was G. L. Nicholson himself. He was a young man of 38 which was rare indeed for a Southern Region District Motive Power Superintendent. He was Southern through and through and yet quite different from any of his colleagues with whom I dealt, on that Region. To see him stalking down the signing on corridor, through our outer office to his own sanctum – tall, military bowler, white stiff collar and furled umbrella – was a revelation even though his efforts to get me into a white stiff collar did not meet with the success they deserved. To know that he was steadfastly behind one was great, to experience the strength of his leadership and inspiration was even better for he was no extrovert leader who claimed the limelight but one who would delegate but expect his men to emulate his principles to the full. Furthermore, he had both a sense of humour and a sense of the ridiculous so necessary in running a Motive Power Department in which the actors were so often hilariously funny, sometimes quite unconsciously.

My right-hand man on the running side was the Chief Foreman, Fred Pankhurst, reared by the South Eastern & Chatham Railway. He had the repartee answer to most people and situations. Dead these 20 years, he will never be forgotten whether for his greatest assets of loyalty and incredible memory for engine working, for his gentleness and kindness on occasions, or for his bigotry, his temper, his extreme pungency of expression and his love of barter. He was as acrid on anything to do with the LB&SCR, its engines and many of its men, as he was on the subject of some of the less well-endowed SE&CR depots – Gillingham, Faversham and certainly the long-closed Slade Green. It was said that, in the past, his temper had involved him in a punch up, followed as his assailant fell flat on his back, by the quoted remark, 'Next time, I'll bleedin' hit you, you Slade Green bastard'. Mercifully I did not have that sort of situation on my hands but soon after I had started at Stewarts Lane I witnessed the exhibition of the sort of barter that he loved but to which I would shortly have to put a stop.

In the corridor outside my office, at 14.00, a conversation was taking place between Fred, who was adopting a wheedling tone, and a comical hunchback character with a strawberry nose, by the name of Bailey. So it may have been, but to everybody at Stewarts Lane he was Snozzle or Snoz. He looked after lavatories, wash places and messrooms, was civil, smiling and a master of craft. Snozzle's roster was 08.00 to 16.00 and Fred had just said in the nicest possible way:

''Ere, Snoz, where the bleedin' ell you think you're going.'

'Home, Mr Pankhurst, I'm off at four.'

'Come off it, Snoz, you'll do a couple of hours in the stores for me?'

'OK, Mr Pankhurst, I'll do that for you but book me off at six.'

'Alright, you greedy old so and so, get in there quick.'

Thus the emergency was covered and at a cost. This was a very simple example of the very complex problem that I had to stop, bit by bit, but how did one start when so many were involved, and those that were not feared that a firm hand would weaken their ability to get the job done.

You may well say that Fred Pankhurst had no right to be a foreman. He had no education, he could barely write, he was heartily disliked by some and, no doubt, today he would never have been considered as suitable – too rough, too uncouth, too likely to rock the boat. I would simply say that his dedication and his memory made him the sort of man who could not be ignored and his loyalty to me to the shape of total support was unforgettable, and this despite the fact that he had been brought up in a world of barter to achieve the impossible. My job was to improve the impossible without the barter.

A fortnight after I started, I walked out into the corridor with Charlie Bayliss, my Chief Clerk who introduced me to Driver Percy Tutt. Percy was the true South London Cockney, a humorist but a railwayman of the highest class. How I wish that the like of Percy Tutt had lived to come to our retirement party for he epitomised all that was great in the Londoner's character. Furthermore, no more conscientious engineman nor better mate ever took duty. In 1954 O. S. Nock rode to Dover and back on engine No 30769, *Sir Balan*, which was driven by Percy and fired to perfection by Reg Wilkes. Their work was described in glowing terms but their company as the most congenial that could be encountered on the footplate.

However, Percy twinkled at me and asked: 'Do you swear, Guv?' I had to reply: 'Not all that much, I suppose'. Said he: 'You'll soon learn, Guv, at Stew Lane'. A month later, after we had embarked on a hard driven plan to tidy the depot and clean the engines, I came across a fireman – none other than Reg Wilkes, emptying a smokebox inside the Shed.

Already there was a large cone of hot clinker opposite the engine cab, which practice I had expressly forbidden in the interests of cleanliness and safety. Reg must have realised how angry I was and suspending operations he jumped down to the ground. I said more than enough to a splendid man who was, in any case, under orders from the Running Foreman, turned on my heel with a final threat and walked straight into Percy Tutt. 'Cor Guv, didn't take you long to learn, did it!' he said, with that impish smile that calmed me at once.

Not a word have I yet said about the locomotives based on Stewarts Lane. It is the men that one remembers as one gets older, but then, where many of them are no longer with us, some of the locomotives are alive and well thanks to the efforts of those who have rebuilt and run them.

But there is no 'Rebuilt Coppertop', Class D1 or E1 and it must be more than 24 years since I heard the unforgettable bark of one of these glorious engines. At Stewarts Lane we housed six of these little machines, Class D1 Nos 31743 and 31749, Class E 1 Nos 31019, 31067, 31504 and 31506. To us, they were all 'Rebuilt Coppertops' or 'Superheaters', and they will for ever be a locomotive that had something extra, the thoroughbred and, for my money, these six and others of their breed had that spark of genius that made them fascinating to control, unpredictable, sometimes wayward, but exciting and powerful as they tore into their work. The Rebuild's whole conception was remarkable; the transformation of a saturated, slide valve, truly conventional but extremely competent early Edwardian design to an up-to-date, long travel valve, powerful flyer with a long sloping firegrate. The front end took great gulps of steam and used them effectively with the exhaust bark that one would have associated with a large modern locomotive. But, by order of the Chief Civil Engineer of the day, axle loads were to be no heavier than in the original engine whilst the performance was to be revolutionised to meet the ever increasing train loads. A real quart into a 52-ton pint pot but this time without the normal penalties for such daring.

A Battersea man would tell you that a Drummond 'T9' 4-4-0 was a great engine, almost to be compared with a Rebuild – a compliment to the South Western design to be sure, for Chatham men had little room for most of the soft-beating Drummond types as compared to their own lusty, free-steaming, fire-throwing machines. But they found in the 'T9' an ability to run and pull quite out of keeping with their modest size, but an ability, more gentle, more subtle but still comparable with that which they had come to accept as normal with the Rebuild.

When I came to the Southern, the day of the Rebuild was coming to an end. On the Eastern, larger 4-4-0s were confined to very secondary work and yet, on my first Saturday at Stewarts Lane, there was a little 4-4-0 roaring over the Battersea arches, the white feather showing, and Driver Alf Murray whistling to the Shed Foreman down below that all was well with the world; 10 corridor coaches bound for Ramsgate, to be restarted at Bromley on 1 in 95, on a curve and 1 in 100 at Whitstable, Herne Bay and Birchington, never without noise but always without distress.

The boat trains of the 1950s were beyond their power, of course, and their promotion to a 'King Arthur' job was not always appreciated. I recall the dignified Driver Harry Wing, then in No 2 Link, with a face of thunder and some precision of speech, asking me what he was supposed to do with 'this thing' (31743) on a heavy Ramsgate turn. But we knew our engine and our man – who was making a token protest – and the storm having blown over all was clear for a good day's work.

The Stewarts Lane Rebuilds had their final fling on the Saturday 11.50 from Victoria and the 14.02 from Dover, both via Faversham, with 280-ton trains, little less than the weight they were designed to haul on the boat trains of 1920 and the ideal load for a little engine over a heavy road in the twilight of its career.

People began to hear about the 11.50 and one or two visitors came across to see for themselves. William Hoole of King's Cross, no less, and old Sammy Gingell made his hair stand on end. And not only Sam but the others in No 2 Link and the spare men from No 4 when they covered the turn.

In September 1956, after I had left Stewarts Lane, Peter Handford, then of Transacord, travelled in the front van of the 11.50 to record for posterity the voice of No 31019. Our driver was Sammy Gingell and he had with him his regular mate, Jim Williams. I did the firing to and from Chatham, and coming home from Faversham we had the 'D l' No 31145, a Dover engine. If the outward journey was exciting, the return was an extraordinary exhibition – for that is what it was as the timetable demanded no such running and we set out to see, for the first time, what could really be done.

Sam started away from Victoria, helped by the banker, like a shot from a gun. I put the left-hand injector on as we struck the 1 in 64 of Grosvenor Road bank when the safety valves began to hum and never shut it off until we stopped at Chatham, eight minutes early. From time to time, I used the other feed to keep the boiler level constant and to check any blowing off while we roared up Nunhead bank, and lurched to the right for Catford and then again when we checked at Bickley and St Mary Cray. By the time we passed through Swanley after the checks, we were up to 50mph before we began our headlong dive for Farningham Road, the engine vibrating with life and energy as it tore up to 80 down the gradient. Jim Williams and I stood on our respective sides of the cab, whilst Sam with his head outside was oblivious to everything except his task of doing what had never been done before. After Farningham Road the gradient changes abruptly to 1 in 100 up and Sam reappeared to juggle with the steam reverser before the speed fell off. Unfortunately, at 80mph and full regulator, she grabbed the lot and for a couple of seconds our world went mad. In those seconds my fire, built up high at the back end of the long sloping grate, filling the firehole, disappeared with the long uphill stretch ahead of us. There is only one solution when a big fire goes down the slope. Lumps, the bigger the better, under the door in no time at all. However, these had to be fetched from the top of the tender but, working hard with hands, shovel and boot, the job was done and the fire rebuilt. Sam paid not the remotest attention but the old engine did: the 175lb/sq in became a rock-like 140 but still she roared through Meopham and then Sole Street at an incredible 60 up 1 in 132. Then it was all over, we could relax and pause to wonder if Sam would ever get down to 35 round Cuxton Road. He always did!

When we left Faversham with No 31145, we knew nothing about her at all. The Dover men had little to say, the coal was the usual soft Kent mixture of Chislet or Snowdown, sometimes good, sometimes not so good. We left Chatham five minutes late. We had plenty of steam as we attacked the long 1 in 100 of Sole Street bank, starting, so I am told, at 39mph. The harder Sam worked this engine, the better she steamed, not using a great deal of coal, whilst the water stayed constant near the top of the glass and the pressure rose to blowing-off point as the speed gradually increased to 42mph. We couldn't believe it – not with the doubtful coal – and we stood there, with the firehole door well open, fearing some terrible priming was about to start. But no; we went through Sole Street in tremendous form so Sam, waiting for just such a situation, wound our old engine and its 300-ton train up to 87mph, not on a long down grade where it could be done gradually but an 87 which I will live with for ever with every exhaust beat sharp and clear above the deafening racket within the cab.

Swanley stopped us as it was entitled to do but Sam's blood was up and a photograph of us taken by Dick Riley must be the final compliment to the Maunsell Rebuild. It was not a warm day, we were going uphill and working hard and yet there was no smoke, not a breath of steam showing at the chimney top, at safety valve, at gland or injector. There was nobody in sight on the footplate and we were working-parts of our little machine. By doing our job, by thinking and planning ahead, we had helped her to do what; in theory, would have been impossible.

Old Sammy Gingell's name has come into this story more than once and it will reappear again before long. And as all engines were 'marvels' to Sam, we should never get the truly comparative story if we quoted him. But if you had had the good fortune to come across the lovable Percy Tutt, and had questioned him on the worth of these little engines, he would have replied, as he did to me, 'Cor, Guv, they were wonderful jobs. We got 3d a day extra for firing them on the boat trains when we were knocking about spare, but we never got *mogadored* with them!'

One did not think of the 'King Arthurs' as fliers, nor indeed were they. Speed had to be won gradually and then maintained whereas with a Rebuild and a suitable train, or on a 'Schools' no appreciable effort was needed to get into speed. This may have been an illusion with the 'Arthurs', for their heavy construction and solidity, and their hard riding, made one think of them as ponderous old things. Maintenance costs were pretty light and a hot axlebox was a rarity.

One more thought about the 'King Arthurs'. The scene is Stewarts Lane depot at 16.00 on a blazing hot August Sunday afternoon in 1954. The summer service is at its height, the yard is full of locomotives, and despite their complete clearance by lunchtime, the disposal pits already show signs of the number of engines that have had fires cleaned and smokeboxes emptied. These pits can soon become a danger if the ashes are allowed to accumulate so we have to keep a daily eye on crafty old Smithy from Deptford and his two strong and splendid Poles, Frank Butowski and Con whose name I could neither spell nor pronounce. Frank, in particular, has wide, powerful shoulders and, when not allowed to work overtime, is capable of abusing Fred Pankhurst and especially me, in a splendid mixture of Polish and English, the latter bearing some resemblance to South London Cockney.

The 4pm Driver is a light work man, confined to Shed duties, the fireman has had three firing turns and is just 16 years old. The Running Foreman is waiting to tell them that their first job is to put No 30764 away, throw the fire out, empty the smokebox and put it in the Shed for washout. Easily said but what faces that little lad, who, if he had not turned up for work that lovely Sunday afternoon, would have been disciplined? No 30764 has been round Kent and the fire has not been skilfully handled. It is level with the firehole door, there is 6in of clinker beneath the fire, the smokebox is full of ash to the crossbar and, being a 'King Arthur', there is no drop grate. The little lad tackles the job, maybe he is sent some help and maybe not; his driver is over 60. Think of drawing every bit of that fire up through the firehole door and throwing it between the cab doors, using a long clinker shovel, heavy for a start but almost impossible for the unskilled youngster. This sort of thing is no exaggeration, it would be their first job of a day's work and although in theory an hour's work, more likely to take two. And then pause to realise why we could not always keep our youngsters, why they went to less demanding work, why they were absent and had to be done, as a result. The steam locomotive could never have come back again as a regular proposition. We know there are no longer any facilities; but never would it have been possible again to attract the youngsters, or indeed the adults, to clean fires, empty smokeboxes, clean tubes, wash out boilers, the dirty jobs that were once taken for granted.

V

South London Personalities

IN LONDON, particularly in the South and East, laughter is never far away however difficult the circumstances, and it is as impossible to imagine a Londoner devoid of humour as it is a Liverpudlian.

My predecessor at Stewarts Lane, a certain Alex Bennett who had once been at Edge Hill, Liverpool, sat in his office and answered a knock on the door which, when opened, admitted Driver F. Brown. 'Country' Brown was very much a Londoner although he originated from Colchester. A few days before, he had been on the carpet for losing time with one of the Boat Trains. 'Brahny' had a very loud voice, raised at the slightest provocation: he was tallish, bald and heavy featured and he clomped into the office with a bucket of small coal which he, far from gently and without any ceremony, up-ended on the Shedmaster's table. 'You did me for losing time last week, Guv, what am I supposed to do with this bleedin' stuff?' History no longer relates what happened next, but never before or since have I heard of a bucket of coal finishing up on the boss's desk.

It has always been a marvel to me that so many men with whom I have worked have a great gift of repartee and turn of phrase. The written word could be laboured but the spoken word had the spice of humour. Nicknames abound and every depot, of course, has its private army of Smiths. At Stratford there were Bookstall, Chopper and Hammer, to name but three, and at Stewarts Lane we had three F. H. Smiths – Dabtoe, 2 Gun and F. H. 3 – we had our Guardsman and more prosaically our Big Bill Smith. 2 Gun used to own an air pistol and there was some story of him taking potshots through open windows near to the line on the North Pole Bank, between Kensington and Willesden.

Needless to say, the foremen, very much in the limelight, did not escape. George Kerr, habitually looking at his watch, was naturally, 'Tick-Tock': Syd Bailey, with heavy jowl, growling and slurred speech, and trilby hat pulled well down, was the 'Clapham Junction Yank' and, of course, there was 'Von Kluck', Algy Harman of the shaven head.

One winter evening, Engine No 31005, Class H, stood in the running shed. It had been there for some time, was well cleaned, had been lit up a few hours previously and was, quite legitimately, being left to its own devices as it was not booked out until the early hours. No doubt when placed in the Shed it had been left with the lever in mid-gear, regulator shut, cylinder cocks open, hand brake hard on. Men would be sent to prepare it, by and by. Driver Charlie Stewart and his mate had brought a 'King Arthur' into the depot, cleaned the fire and smokebox, coaled the tender and then turned ready for an early morning duty. On their way into the Shed, they paused by the Running Foreman's Office for a perfectly friendly verbal conflict with Algy Harman. The conversation was conducted through the office window and from the height of the footplate in such a way that it could have been heard a hundred yards away. Now Charlie Stewart was one of the great characters of Stewarts Lane. He had been a fighter in his youth, had a bull neck, and was a very tough old man. His shirt was open to the waist, winter and summer, his neck and chest red raw and his nickname, appropriately, 'Rasher'. He had an even greater command of the shouted word than his antagonist of the moment.

Back in the Shed, the 'H' class tank sighed, stirred, made a tentative movement, cleared itself of water and silently, oh so silently and gently, made its way down the Yard. Its speed had reached 5mph and it was on a collision course. The argument was at its height when, with a colossal bump, the 'H' class hit the stationary locomotive on the blind side. Charlie picked himself up from his corner of the cab and dispute forgotten, and shouting blue murder, joined Algy Harman to vent their fury on the crew of No 31005 only to find that their combined profanity was vented – much to the amusement of the gathering crowd – on an empty cab!

My task next day was to find the culprit. I might as well have tried to time the 'Golden Arrow' with a 'P' class tank. The bland faces, the civility and smiling and innocent composure of those who might have been involved, told their own story but there was not a shred of evidence to say who had unscrewed the handbrake, put the lever in foregear and opened the regulator. Steamraisers, cleaner boys, gland packer, fitter and fitter's mate, clams to a man, artists in the cover up!

That same Charlie Stewart, contrary to appearance and demeanour, was a very conscientious man: one morning he was working a peak hour train from Faversham to Cannon Street and his mate was taken ill at Chatham. Rather than cause a delay, Charlie took a porter from the platform, sat him on the fireman's seat and told him to watch where he was going. He then did both jobs to London Bridge, where a Bricklayers Arms fireman was waiting. His engine was a 'West Country' and he was not far off 60 years of age. Ask him to do this sort of thing in cold blood and he would have very forcibly refused. Indeed, he would never have been asked. And yet he instructed the station staff at Chatham to give him that porter, in order to save a delay and that was that. A challenge, self-imposed, quite irregular. He saved a serious delay and lived on the story for years.

And yet there was the time when he refused point blank to carry out an instruction in Stewarts Lane depot. He was sent to the messroom and the foreman came to see me to report the matter and no doubt to get things rectified. How to deal with Charlie Stewart? First of all, there must be no third party present, no gallery, nobody to chip in. The direct instruction to get on with his work would have brought forth a highly embellished negative: the raising of my voice, the thumping of my fist would have been fatal, for Charles could out-shout and out-thump most of us. No, L. P. Parker had taught me a thing or two and the quietish voice, a touch of steel here, sarcasm there, and an indulgent smile checked the bluster before it had chance to surface and our friend, civilly thanking me for my time, departed to do as he was bidden.

My own personal bodyguard, the Chief Clerk, the Chief Mechanical Foreman, Bert Wood and the Chief Running Foreman, were all men old enough to be my father and none of them is alive today. But very much alive is 'young' Johnny Greenfield, very much alive despite extremely severe illness in recent years. John was a Class 4 Clerk, just off the bottom rung in the clerical grades. He had a passionate interest in railway work and in the staff of the depot. He had not an idle bone in his body, he had courage and a splendid temper, very fiery, not the typecast clerk by any means. He could obtain a reliable report on loss of time on the road from a driver who had no intention of being cross-examined by a clerk and, in view of this sort of achievement, it is no wonder that he eventually made his mark and rose to the post of Motive Power Officer in the Central Division of the Southern Region, before ill-health forced him to retire. We had many similarities – I was also fiery, but we were both passionately interested in playing our part in running an efficient and punctual railway. Naturally, our friendship grew and in later years I was able to take John behind the scenes on both Eastern and London Midland Regions, although he remained a true and unrepentant Southern man until he retired.

Of all our journeys the one I shall remember most clearly was to Leeds Central from King's Cross with none other than William Hoole and engine No 60007 *Sir Nigel Gresley*, the Class A4. William and I had only known each other for a few years. He was a Scouser and that accounted for a lot, although I did not realise what until I had spent some wonderful years in Liverpool. He was a Great Central man and he had a certain reputation which is well known. It is as well that we never actually worked together – suffice it to say that he had a wonderful charm that had come to his rescue on many occasions, that his staff history was far from being without blemish, and that his enginemanship was, shall we say, uniquely vigorous. Having said that, he was a splendid mate to work with and gave his fireman opportunities to gain main line driving experience.

Our journey was made in 1956 at which time I was the Assistant District Motive Power Superintendent at Stratford. Nothing out of the way took place between King's Cross and Leeds. Nevertheless, John and I were a bit shop-soiled when we entered the Great Northern Hotel and immediately got swept up with a large number of people, very smartly dressed, about to attend a Masonic gathering. Having got that sorted out, we were examined closely and with no favour by the receptionist and packed off to our rooms to wash, eat and relax. Next day was a Sunday and before we left Leeds, we knew, because of engineering work at Nostell, that we were to be routed by Pontefract and so Bill needed a conductor from Wakefield Kirkgate, who turned out to be a very stout gentleman from the LMS Shed at Wakefield. With difficulty, he hoisted himself through the side doors of the 'A4' and took stock of the situation. Bill, charming as ever, offered him the driver's seat which he accepted without comment. Having settled himself, he forced back the rolls of fat at the back of his neck and addressed us critically, 'Which of you – lot's the fireman', to which I replied at once, I am, mister'. Having already taken a look at the fire, he replied to the effect that I would do well to put some more coal on the fire as the road via Pontefract was rough; and even a 'Black 5' would have all on with the heavy train. More coal was quite unnecessary standing in the station, nor would I do as I was bidden but any argument was curtailed as we got the rightaway. It was very evident the gentleman had never handled a Gresley engine before. We slipped and slipped, and all the while he compared our engine very unfavourably and very vocally with everything LMS and told me to get more coal on – which was rapidly becoming necessary due to his antics. Having turned to our driver and said critically, 'Your mate's not much bloody use', Bill took the law into his own hands, opened the regulator wide, pulled the lever up from 55% to about 25, told the old boy, 'For Christ's sake, get a move on' and we were on our way. In fact, once he settled down, he did the job quite well on a strange engine and it was no problem to give him steam over a road that if it presented problems to LMS engines, certainly did not do so to us. Once we got to Shaftholme Junction and back on the main line, he hoisted himself out of the seat and took up a warm and strategic position with his back to the fire. His parting shot, as he lumbered off down the platform, 'Don't think much of your engine and that's all the so-and-sos are going to get out me, today'! I never had the heart to tell him who I was!

Ever since the days when Benny Faux taught me that rules can be bent or manipulated, I have learned many things to my advantage through keeping my eyes and ears open and my mouth shut, listening to this story, noting that comment, putting two and two together and now and again making five. Some irregularities one could watch until the appropriate time for action arrived, some could be allowed to rest, but others needed immediate and effective attention.

The importance of men signing on at the proper place, at the proper time had been drilled into me by my own training and experience. At a Motive Power Depot, not only is time vital but because there are certain notices that drivers must read and for which they may have to sign when they take duty, it is very important indeed that they are aware of any newly imposed speed restrictions or changes of route before they leave the Shed.

When the terrible Eltham accident occurred in 1972 I was Divisional Manager in Liverpool. It was clear to me from reports that the driver involved could never have signed on duty at the proper time and place. My own experience had taught me how easily this type of action could be condoned and I immediately instructed the area Managers in the Liverpool Division to examine the signing-on arrangements, at all points and assure me, without a shadow of doubt, that men were signing on correctly. I remember extremely well the first time – and the last time – that I came across the irregularity at Stewarts Lane.

The Brighton Goods Link worked the famous 18.10 Victoria-Uckfield – 'the terrible 6.10', which was a very heavy train usually hauled by a London Midland Class 4 Fairbairn 2-6-4 tank engine, with a moderate degree of success. The engine was taken out from Stewarts Lane early in the afternoon to Eardley Sidings, Streatham, and ultimately brought up a set of cars for what was probably the 17.50 Tunbridge Wells West. The afternoon men were relieved by the main line crew who had signed on at Stewarts Lane at about 16.45, read their notices and travelled over to Victoria via Battersea Park.

In April 1953 I stood talking to the List Clerk in the Timekeeper's Office at 16.40. A telephone rang nearby and instead of leaving it for the Timekeeper, I picked it up and said, 'Stew Lane'. A voice at the other end said: 'Harry Pearce, on my way to Victoria, book me on for the 6.10 Uckfield, will you, mate!' I said quietly, 'What do you mean, book you on'. The conversation then went something on these lines:

H.P.: 'Bleedin' book me on and anyhow who the hell's that?'

R.H.: 'Hardy'.

H.P.: 'Christ'.

R.H.: 'Yes, Harry, and you can come down here and book on properly and read your notices'.

H.P.: 'But I'll miss my turn, Guv'.

R.H.: 'I can't help that, you come here and sign on. It's me who takes the can if you land yourself in trouble and I'm not having that'.

H.P. (relishing the thought of eight hours' P&D work): – !

An isolated case perhaps? But I could not leave it to chance. Harry lived at Earlsfield and was in fact ringing on the internal phone from Earlsfield station.

A thousand to one nothing would have come of it and the temptations on Supervisors to condone this sort of arrangement were great, particularly in the early hours when the train service was sparse and it might make all the difference to a man if he went straight to his work rather than spend two hours in the depot messroom overnight before booking on.

Not so long before I retired, I was asked why certain plans which had been carefully made had not been executed successfully. I felt bound to point out that, whereas plans can be shaped to cover every contingency, their introduction needs attention to detail, and inspiration from those responsible in order to ensure success.

The thought of the success of the 'Lock & Key' system at Stewarts Lane has always been tremendously warming and if I am to be remembered by the staff of that depot, I think it is because of that success against all odds. It so happens that I have kept copies of some of the original notices to staff and some of the notes I sent to colleagues and to my own staff.

It was great to be 29 and fired with passionate enthusiasm: one wrote with a sincerity that experience and advancing years might possibly have muted. Readers of *Steam in the Blood* will remember how and why the system was introduced. The Southern Region required all tools to be left in the tender cupboards and the 10-minute allowance to the fireman, when booking off in order to take tools to the Stores, had been abolished. The result, bearing in mind our national and well-loved habit of taking short cuts, was chaos in that no engine ever had what it should have had and every time a locomotive was prepared, a game of hide and seek for tools took place. Once a set had been mustered, it only needed the momentary relaxation of the crew, for a felony to be committed on their engine. And this was no way to run a railway: quite apart from lack of efficiency and consequent loss of interest, delays were caused by engines leaving the Shed behind time and the Southern tradition would not allow this. On the other hand, the Motive Power Headquarters, having issued instructions in the first place, did little to see that they were carried out. There was only one person who could do this, the local guv'nor, and there was a mansize job to be tackled when and if he had time.

We decided to make the effort. We had to muster with the help of Headquarters the correct tool equipment for all of our 126 locomotives. Each tool box was to have its own key with the engine number attached. No cupboard could be unlocked without the key being booked out by the Timekeeper to a driver; keys had to be booked in when the locomotive was stabled, once more the name of the driver returning the keys to be recorded. Bill Allen, the Toolman, was responsible for frequent spot checks, and to say that he was an ally is putting it mildly. How simple it all sounds! But you try to get 600 men who have been used to helping themselves to toe the line. But we did and there was no man, in any grade, who did not acknowledge success and the effect it had on his work. No longer did a crew have to ransack the place for equipment; they picked up the keys, unlocked the cupboards and found all the tools they needed, and, more often than not, enough oil to do the preparation. Of course there were still robbers but we caught them and did what was necessary or frightened them. Dover and Ramsgate no longer could help themselves, as they had done in the past, to our equipment from our engines. Our men going to a foreign depot to pick up a Stewarts Lane engine, would refuse to leave the Shed without a full set of tools, safe in the knowledge that their engine must have had its full complement when it arrived!

So much for the enginemen. What about the others, who never had to take a locomotive from the Shed. Surely they would not be bothered. Ah, but they were, for every man at the depot seemed to need a paraffin flare lamp – fitters, mates, boilermakers, cleaner boys – and where had they come from? Not the stores where the cupboard was usually bare. We had found a broken lock and a flare lamp missing from the kit and no way of finding out who had done the damage to the lock. But why just a flare lamp? Prevention is better than cure and we realised what had happened and why. We had many priceless assets at Stewarts Lane but, in this context, there was no greater one than our tinsmith – a rare grade in a running shed and probably a throwback to Longhedge Works. He designed and produced at an incredible speed an improved and distinctive flare lamp to be issued on an individual and personal basis to all staff not strictly speaking entitled to such equipment. We also recalled the BR flare lamps to augment the stores supply and from then on, if one saw an artisan with a standard flare lamp, one knew a felony had been committed and could act accordingly.

You may think perhaps that there was much ado about nothing. It meant very hard work by a number of us; with a few minutes to spare, I would range round the Shed, casting an eye on to every footplate to check the lock, climbing on to an engine being prepared to check all was well and there was nothing missing.

We knew that drivers occasionally and absentmindedly took the keys home which meant the next crew had to use a master key. There was a cure for this, used from time to time, if the driver lived near to the depot, which was to wait until one hoped he was asleep and then send round for the keys. Driver Frank Box, given, I think, to telling the tale, informed me one day that he had taken the keys home and had not realised the fact until he emptied his pockets on going to bed. He dressed again and brought the keys back to Stewarts Lane, returning home with a lily-white conscience. True or not, this was the spirit our scheme created but it might not have lasted 24 hours but for our dedication.

And yet, but for a chance meeting, it could have failed on a psychological issue. I had set out the Timekeeper's record book to include the time that keys for every engine were issued, and correspondingly the time that they were returned. In each case the driver's name and engine number were recorded.

To understand the problem, one has to realise that there was an agreed time for the disposal of each class of locomotive. If the fire was not particularly dirty or the smokebox comparatively empty and the pits and coal road clear, the work would be completed under the scheduled time. Men involved with disposal could therefore get away home when their contract work for the day was complete.

A couple of days before we were due to start our scheme, Charlie Sands, Driver and member of Sectional Council No 2, sidled up to me near the disposal roads. Charles could adopt a lordly attitude, climbing on to his poop deck and looking at one with a certain superiority. He asked me whether I was expecting my scheme to work and before I could reply, he announced magisterially that it would do no such thing. I thanked him very much and after a few minutes during which he enjoyed himself at my expense, he advised me not to have a record kept of the time the keys were returned. It was not necessary to show this time which had been included for the sake of tidiness. Once I was satisfied that Charlie Sands was right, I made the alteration. Men had already said (but not to me) there was a catch in it somewhere and that I would be checking on the actual time against their contract time. Had I not listened to advice and acted at once, the outcome could well have been failure.

One is tempted to say more for, as soon as one starts to write, one relives the passion and excitement of those days over 30 years ago; but I will content myself with saying that a simple scheme, with everybody playing a part, revolutionised an important aspect of work of the depot. It is impossible to calculate what we saved in oil and tool equipment. Thousands of pounds, no doubt. It was absolutely clear that the despatch and efficiency with which preparation and disposal was carried out was improved out of recognition. It was clear that our staff had achieved something at a rough London Shed that had not been achieved elsewhere. They knew it and they were proud of it. Was it all 'cost effective'? Who knows and, quite frankly, who cares. We had achieved the impossible.

From the notices that I posted and letters that I wrote, I have selected six which are of general interest. They gave some idea of our enthusiasm, and in the case of poor old Dave Humphrey he must have loved me when he received the note. Nor can I resist the inclusion of that written about engine No 70004. We were fiercely independent Southern men with a healthy contempt for that railway across the water!

NOTICE TO ENGINEMEN ENGINE TOOL EQUIPMENT

THE LOCKS AND KEYS HAVE BEEN GOING A YEAR ON SUNDAY, 17TH, 1954. I MUST THANK YOU ALL FOR THE PART YOU HAVE PLAYED.

I MUST ADMIT TO YOU NOW THAT I NEVER BELIEVED IT WOULD LAST MORE THAN TWO MONTHS, BUT I BELIEVE THAT NOW IT WILL NOT FAIL US AND **WE SHALL NOT RETURN TO THE APPALLING CONDITIONS WHICH PREVAILED AT ONE TIME.**

ONE OR TWO POINTS I MUST MAKE AGAIN:-

(1) FLARE LAMPS ARE ALWAYS A TROUBLE AT THIS TIME OF THE YEAR, 2 IS THE NUMBER FOR EACH ENGINE. SHOULD THERE BE ONLY ONE, PLEASE REPORT IT AS A MISSING TOOL. I CAN THEN TAKE UP WITH THE FOREIGN DEPOT IF THAT IS WHERE THE LOSS IS FROM.

(2) DO NOT, PLEASE, RELAX IN YOUR EFFORTS TO KEEP THINGS RIGHT: IT IS ONLY BY REPORTING MISSING ARTICLES THAT WE CAN KEEP CONSTANTLY ON TOP.

AS YOU SAW, THE ARRANGEMENT MUST HAVE SAVED MANY HOURS AND MANY TEMPERS DURING THE SUMMER MONTHS. I THANK YOU FOR THE HELP YOU HAVE GIVEN AND FOR THE EXAMPLE YOU HAVE GIVEN TO THE MEN OF OTHER DEPOTS.

R. H. N. Hardy
SHEDMASTER
15th January, 1954

Toolman W. Allen,
STEWARTS LANE

ENGINE TOOL EQUIPMENT

The tool equipment scheme has been going exactly a year on Saturday next. It has been a great success and this is very largely due to the endeavour that you have put into the job and for which I must express my most sincere thanks.

I hope you will be able to play your part for many years to come.

R. H. N. Hardy
SHEDMASTER
15th January, 1954

Driver C. Higgins,
STEWARTS LANE

ENGINE TOOL EQUIPMENT

Unfortunately, I cannot write this note to you as Secretary of the L.D.C., and so I must combine my thanks for the way you carried out those duties, with my thanks for your help and example in making the Engine Tool Equipment Scheme the great success it has been.

I owe you and Jack May a lot for the propaganda and example you both set when the scheme started and for spreading the gospel at other Depots so that, ultimately, they had to follow the example of much scorned Stewarts Lane.

Although you now do not hold official office, perhaps you will still continue to help us in the way you have done in the past and "educate" some of the lesser mortals in the art of locking up boxes and checking tools, especially at foreign Depots.

R. H. N. Hardy
SHEDMASTER
15th January, 1954

NOTICE TO ENGINEMEN
IMPORTANT

THERE ARE SOME KIND PERSONS IN THIS YARD WHO ARE PRESUMABLY IN POSSESSION OF A MASTER KEY AND WHO UNLOCK ENGINE TOOL BOXES, TAKING EQUIPMENT AND, OCCASIONALLY, THE LOCK.

I SHOULD BE GLAD IF ANY MAN WHO SEES THIS SORT OF THING GOING ON WILL REPORT THE FACT TO ME AT ONCE. I DO NOT WANT TO SEE OUR TOOL SCHEME DAMAGED IN ANY WAY, AND I AM SORRY TO SAY THAT SUCH ACTS AS THOSE MENTIONED ABOVE ARE TO BE DEPLORED.

PLEASE CONTINUE TO REPORT ANY MISSING ITEMS FROM TOOL BOXES ON ARRIVAL

R. H. N. Hardy
SHEDMASTER
15th January, 1954

Driver D. Humphrey,
STEWARTS LANE

ENGINE NO 1904

The above engine was brought in by you last night off Sp1.103: will you please note the following tools were found missing this morning by my toolman:-

Lock and key	$\frac{7}{8}$ SE	$\frac{3}{4}$ Ring.
$\frac{5}{8}$ Ring × $\frac{1}{2}$	$\frac{5}{8}$" Open x $\frac{3}{8}$ Ring.	
Flare Lamp.	Hand hammer.	

Will you please inform me (a) why you did not report the missing articles (b) if they were present on the engine when you left it, why you did not lock the boxes.

It appears at the moment that there has been negligence. Please note that I cannot tolerate this sort of thing, and in future cases shall be obliged to take further action.

R. H. N. Hardy
SHEDMASTER
21st June, 1954

District Motive Power Superintendent,
STEWARTS LANE

ENGINE NO 70004

The above engine has returned from the L.M. Region today.

I am sorry to say that outwardly nothing has been done to this engine which has returned in a disgusting condition to this Depot. The engine was sent to Crewe in a state of cleanliness as if it were going to work the "Golden Arrow", in the hope that the L.M. people might be inspired to rise from their perennial state of filth and lethargy.

I would add that when my Driver arrived at Camden Depot to take charge of the engine, there were no tools of any description on the engine, and although there is presumably no hope of getting these tools replaced, I should be grateful if you would take the matter up strongly.

The following tools are missing:-

3 head Boards	4 Head Lamps
1 Hand Brush	2 Torch Lamps
1 x 2-gall. Oil Can	1 Bucket
1 x 1-gall. Oil Can	1 Coal Pick
1 x ½-gall Oil Can	1 Tallow Pot
1 Feeder	$\frac{3}{8}$ x $\frac{5}{8}$" Open Spanner
1 Dart	½ x $\frac{5}{8}$" Ring
1 Pricker	$\frac{3}{4}$" open x $\frac{3}{4}$" ring
1 Clinker Shovel	1" open x $\frac{7}{8}$" ring
1 Hand hammer	$\frac{7}{8}$" Single End Spanner
1 Gauge Lamp	1$\frac{1}{8}$" Single End (Gauge Column Spanner)

R. H. N. Hardy
SHEDMASTER
6th May 1954

VI

This Happy Breed

WHAT MEMORIES! What splendid conflagrations one had, rarely with any malice born by either side. Sometimes I won my point, sometimes I did not. Sometimes I was beside myself with barely controlled rage, sometimes it was my contestant.

I remember the Saturday morning that a certain fireman was intent on having me for defamation of character: another stoker was going to sort me out after I had enquired into a mythical Saturday afternoon hospital appointment. Teddy Champion, an excellent fireman recently passed for driving with several main line turns to his credit, stood in front of my desk, white with passion when booked as a *fireman* on a special to Hoo Junction on a 'C' class. He went on the job but he had had his say, so we were both satisfied. Reg Coote, passed fireman, driving for the day, swept some coal on to the ground as I was passing his engine. I told him to pick it up. He replied, 'If you want it picked up, pick it up yourself, Guv'. Victory to Reg Coote. I had it picked up and my angry retort amused him sufficiently to remind me of the incident a few years ago.

It can be said that enginemen are a unique breed and therefore our differences of opinion – fierce, angry, humorous – had a certain savour. But what of the Shed staff, especially the skilled craftsmen, fitters, boilermakers, and the carpenter, Harry Milton, a gloriously contrary genius. Rarely were there differences of opinion on workmanship but often on issues of timekeeping, disappearances from the job and old Spanish customs! It was surprising how people melted away about 8 o'clock in the evening or on a Sunday afternoon. If I was feeling really nasty, I would ring up from home and ask to speak on the phone to a particular individual and would listen, to my secret delight, to the fabrications eventually given me by the Timekeeper to justify the temporary absence of the man I sought.

Jim McTague, Irishman, had served an apprenticeship at Inchicore, Dublin and had joined the Southern Railway soon after the war. Usually employed in dealing with whatever were the current eccentricities of the Bulleid Pacifics, he needed to be one of our best fitters. But Jim either could not or would not get to work on time so, after many warnings, I was obliged to take disciplinary action. Had I then been a disciple of Somerville & Ross I would have realised I was wasting my time, but the zealot in me pursued the matter to the suspension stage. Only then did Jim lose his Irish charm but he did manage to arrive at 8.00 for a few weeks. In a short time, he went back to his old ways, leaving Altenburg Gardens just before 8.00 and bicycling slowly and resignedly over Lavender Hill and down to the depot. I gave up the struggle, and Jim stayed with the railway until the end of steam when he took a job as a Shift Engineer at the pumping station on the embankment by the Grosvenor Bridge. Until his retirement, years after, every Christmas Eve he would ring me, the man who had done his best to regiment the Irish personality, to wish me all the best.

The Longhedge fitting shop, once part of Longhedge Works, was subject to the Factory Acts and therefore visits by the Inspector: but the running shed was not (and a good job too). Many artisans in 1953 still came to work in overalls but change was in the air. I had been brought up on the principle of allowing no time to wash up at the end of the day and making an immediate start on arrival. At Doncaster I had had no locker and my washing facility was a bucket of paraffin. Ipswich was unbelievably primitive but, here in Longhedge shop, we had modern washplaces, clothes cupboards and even a changing and drying room. All very nice but foreign to me. I recall my righteous indignation, when very seriously taken to task with the dogmatism of a Thatcher, by a lady Factory Inspector for not having a formal agreement with our staff on time allowances to use these splendid facilities. To agree to such a proposition seemed impossible in those far off days and furthermore, exploiter of labour that I was, I would not do so! I also have a feeling that I bridled because the proposition was put to me by a lady for I believed that Locomotive Depots were no place at all for the fair sex. It took me until 1977 to change my mind on that one!

Despite the constant uproar, the wrangling and jangling with one's staff, our depot ran remarkably well. I suppose things were simpler in those days, leadership could be more positive and, provided one was prepared to fight, one was rarely thwarted in what one set out to achieve. As I have previously indicated, budgetary control at depot level did not exist and yet, many times I have felt that the costs of our various achievements, had they been weighed, would have been justified by favourable results in terms of performance and efficiency. However, there were cases where simplicity was carried to an extreme in order to achieve a result.

When I arrived at Stewarts Lane, I found that the cleanliness of the Shed and Yard, which left a great deal to be desired, was left to Smithy, his two Poles and his ash loading crane and also to a PW ganger and his men. These gentlemen were the property of the District Civil Engineer at Purley and were responsible for track maintenance in the area. A number of labourers, on our pay roll, were supposed to carry out the work but were employed elsewhere by Fred Pankhurst on a variety of unorthodox and unauthorised duties.

One of my first tasks was to sort this out and give each labourer a patch of the Shed and Yard which was *his own* and where, incidentally, he could be found if required. Smithy, despite his devious protestations, was galvanised into regular action and the whole Shed and Yard was cleaned and clear of ashes by 15.00 each weekday. Conditions were better, for no longer were there heaps of ashes in the darkness of the Shed, a danger to life and limb, and the men with the broom and barrow could, and did, take a pride in their work and, on the whole, people were happier and certainly safer at their work.

However, Mr Murphy, the ganger, buttonholed me and told me, with gentle but sinister Irish charm, that 'Governor, you're taking the bread out of our mouths'. I could not see why I should be blamed for this difficult act but it transpired that Mr Murphy's arrangement with Fred Pankhurst had been to the tune of two hours overtime, per day, Mondays to Fridays inclusive, and the correct deployment of our staff had the effect of putting Mr Murphy and his men on to what passed for a bare week in the PW Department. There was no question of my striking a bargain with the gentleman but he did not allow the matter to rest until, months after, somebody had a brainwave. The Shed buildings needed internal painting, so did the messrooms, the carpenter's shop, the offices, a never-ending list.

However, it would have taken years to get this work done through the proper channels but we did gather from Mr Murphy that he and his gang would, given brushes, paint and ladders, repaint the entire depot, messrooms, everything. He took care that this job became permanent, rather after the style of the painting of the Forth Bridge, and my own concern was not with the splendid results but whether there was a catch somewhere. On asking Mr Murphy how he squared his Inspectors, he replied that he simply charged the overtime to 'Motive Power'.

All very unorthodox, indeed irregular. Permanent Way staff painting offices! Who was financially accountable? Goodness only knows. We got our depot completely done over, the gloomy, crude, rough messrooms made tolerably presentable. We wasted no days, weeks, months, trying to make a case to a department overburdened with demands, we saved time and no doubt we saved money. Above all, working conditions were improved, again to the safety and advantage of our staff.

Victoria Eastern, high summer 1953: Saturday afternoon, a derailment outside the station. The Stewarts Lane steam crane, with Bert Wood in charge, cleared the road by 18.00 but traffic had been heavily delayed and disorganised. What more is there to say? Sunday follows Saturday, a tremendous booked service. Engines could be sorted out during the night but men can claim the 12 hours rest agreement and many crews who had been delayed several hours were booked on duty on the Sunday morning. And what of the 'squeezers'? Now the 'squeezer' was part of the Southern scene. It had to be in order to cover the work. Stan Reynolds, our bland and armour-plated list clerk, knew who would work any and every Sunday, who would require 12 hours' rest and who, booked late on a Saturday, would work an extra Sunday and an early morning Monday as rostered with perhaps nine hours' rest between duties. To book a 'squeezer' today – with diesel and electric traction – would be downright folly but, paradoxically perhaps, it was different with steam.

When I went home late on the Saturday, I was concerned about the Sunday working: would we be responsible for delays and, worse still and unforgivably, for cancellations if our men claimed the 12 hours rest to which they were entitled. I came back to the depot at 7.30 next morning. I never need have worried for I saw, with my own eyes, a Southern miracle. Every man to work and on time. Every man on his booked job and right time from the Shed. Marvellous!

We were lucky to have so many main line men of quality. They stretched through the links down to many of the passed firemen keen to tackle the fast train work as soon as they were passed for driving. More than a few could aspire to great heights of enginemanship but I shall remember old Sammy Gingell to my dying day for he was quite unique. We thought he would live past the 100, which he was always trying to do when engaged in what he called 'a little sprint' but he did, in fact, die on his 85th birthday. I have written much about this remarkable man who had his first 'traction' experience in the Abergorkey pit in South Wales, in charge of a pit pony named Cardiff who knew how to derail tubs of coal when he needed a rest, by putting a hoof in the points as he passed along the railway, far down in the bowels of the earth. But Sam, the strongest of the strong, knew how to get the tubs back on the road, and his faithful Cardiff back in business.

He loved his work in a unique way. He was not a unique engine man, nor had he outstanding technical knowledge. He loved to go fast and hard and he knew exactly what he was about. He was a remarkable mate who never got excited, upset or unnerved but this, in itself, is by no means unique. Sam put the same dedication into his work, whether he was going round Kent or on the coal road 'P' class shunt. He had an enormous capacity for work and would never jib at anything; he would take any engine on any train and get there, if his engine was grossly overloaded, he would get there just the same, not specifically because of his or his mate's skill, but because they were both infused with his indomitable spirit to achieve the impossible. 'She's a marvel' or 'She's a good machine', he would say about a rough engine as he came, black as a sweep, to the foreman to see if there was anything more he could do to help before going home. I doubt if he was ever mean or petty in his life, and he never let anybody down. He was generous to a fault. He loved his glass of 'Sherbert' at Ramsgate or Dover and always looked after his mate on these occasions. He was shy and reserved and yet made many friends for he was always courteous, helpful and humorous, and shy though he was, he treated all men in all walks of life in exactly the same way, for he was a gentleman, the greatest gift of all.

His record over 44 years of main line work was clear – if one excepts a little matter of making smoke at Holborn in 1919 – and it is rare for even the best of men to go through a footplate career with an unblemished record, as any Motive Power man knows full well. He had one day off sick in 44 years – with German 'flu in 1918, going to work next day to the despair of the doctor and coming home none the worse. He was never late, never absent, never missed his turn.

In 1970 there were signs of cancer and by May 1972, when I saw him at his home on a quick visit from Liverpool, he was in acute pain with only a few weeks to live. Mrs Gingell, six years older and crippled with arthritis, was by no means active but miracles can happen. The Gingell family ties were strong; for example, morning and evening for six years Les Gingell called at the little house in Earlsfield to get his father and mother up and to put them to bed. In the summer of 1972 the apple cheeks began to return, the boxing glove hands regained their strength and apart from the loss of the use of his legs, Sam was in good form to move around in his 'machine' as he called it. We thought he had once again done the impossible, but on 28 July 1977 a great railwayman and gentleman died, never to be forgotten by the men with whom he worked or by those who were honoured to know him.

Sam did many extraordinary things in his life but nothing was as bizarre as his trip to France in 1969, at the age of 77. For the first time since 1918, he had been ill; with shingles. When the day arrived for his journey across to Calais, he was still far from well but he refused to listen to the advice of his family and set off on his great adventure. I had made plans for him and the fact that he could not speak a word of French mattered not one bit – the railwayman's language is international.

Sam arrived at Dover, and found on the *Invicta* Stanley Gardiner, a cabin steward of glorious memory, who no doubt said, 'Now, sir, my time is your time, can I get you anything'. A bottle of whisky arrived and the two of them got down to serious drinking in the cabin. On arrival at Calais, Sam was beginning to glow and the reception committee introduced him to Mécanicien André Delrue, who was working Train 34 to Amiens with an American '141R' class locomotive. He should have ridden with Delrue to Amiens where his old friend Henri Dutertre would be waiting with food, wine and No 231G42, to bring him back on Train 39 to Calais. However, when Train 34 arrived at Rang-du-Fliers, André Delrue told Sam that he must get off and return by Train 27, this being *my* normal procedure before an evening meal with the Calaisians. Sam couldn't argue and got off after shaking hands. Rang-du-Fliers is in the Northern plain, a few miles south of Etaples and the railwayman in him told him to cross the tracks, where he saw the magic word 'Buvette'. A good cognac and he emerged to see another '141R' coming up over the horizon, which shortly came to a stand in the station. Old Sam padded up to the engine, showed his pass, and was welcomed by the driver as though it was the most normal thing in the world to find a 77-year old Englishman climbing aboard at a wayside station on a Saturday afternoon.

On arrival at Calais, he walked up to the depot, found his room in the dormitory, washed and decided to go out for a meal. He had just made a start when he bumped into Lucien, his driver, homeward bound. 'Come home with me, Sammy', he said and in a few minutes, they were drinking Ricard in Lucien Ducrocq's front room. The old man was no great talker and yet his genial nature always carried him through so that Lucien soon found out what should have happened at Amiens. At 22.00 Sam was back at the depot, then the wicked youngsters took the old man out drinking. What they filled him with, they later refused to divulge but, probably for the first time in his life, Sam, the man with hollow legs, had to be put to bed, 'avec les oreilles rouges'.

However, he slept well to get up at 06.00 as fresh as paint. He shaved and dressed and, having no doubt given the dormitory attendant a generous 'pourboire', joined the crew of the 07.57 to Amiens where he was received, welcomed, dined and generally given the treatment by Rene Gauchet, his driver on Train 19, the 'Fleche d'Or'. Thus inspired, Sam picked up the shovel and fired No 231G42 as far as Abbeville as he would a West Country.

Having handed over a huge fire and a full boiler and head of steam, he retired with honours to be promoted driver to Etaples. No doubt, the speed limit of 120kph seriously concerned René Gauchet as Sam had his head outside and was blazing away as if he were going up the straight to Ashford. That problem was solved by gentle use of the brake behind his back! 'That'll do for you, Sammy', they said and he became the third man for the rest of the day. Black as Newgate's knocker, he shook hands all round, gave René and his mate a cigar apiece, and boarded the *Invicta* where Stanley quickly had him away into a cabin to wash and relax. Relaxation! They finished the whisky and the 77-year old S. Gingell went home, cured of shingles. And that is the truth!

The time came to leave the Lane in January 1955. My last morning's work was 2 January and with my three splendid henchmen, F. Pankhurst, B. Wood and C. Bayliss, I sat drinking crème de menthe, of all things at 11 o'clock on a Sunday morning. We reflected on our successes and our failures and I knew that I had been through a great experience which I would remember for the rest of my days. Let others speak for me once more – I received some lovely presents when I left, some kind things were said, and the letters that I received will never be lost or even mislaid. Yet again, some of these came from men to whom writing must have been a burden but I quote them without alteration. Two were written in January 1955 and that from Tom Banton two years later when I was at Stratford.

Sir,

Please do not be offend by this small gift I am giving to you but I feel I must repay you somehow for all you have done for me.

Yours

Driver S. Gingell

Dear Sir,

It is with regret that I learned that you were leaving us so soon. Without any fear of contradiction I can honestly say that Stewarts Lane has benefitted by your short period of Shedmaster.

An official who can demonstrate with his own hands how a job should be done gains respect if not admiration from all thinking footplatemen.

For what you have done and tried to do at our Depot during your stay the majority of us I am sure are grateful.

Yours respectfully,

Driver A. Murray

Driver Tom Banton and I had had a dust up in 1954 and I had said more than I ought to have done on the footplate of No 70004 on the matter of topping up the tender with coal after the cleaners had finished their work. Tom had refused to speak to me for some weeks but recovered his normal good temper after I had fired to him on a West Country and he had flayed the living daylights out of me. The letter from Tom Banton was written in 1957 from Brompton Hospital. He had not long to live but his letter was well and firmly written. I have only quoted a small portion near the end:

'Well, Sir, I see you have to live up to the times and have a go with the Diesel, well I have only one thing to say and that is you will master anything. I have always said a man that can master Stewarts Lane can master all.'

Do not think that I am vain: always remember the rough and tumble that I had to endure, and the fact that one had to wait until the day was won before receiving the words of appreciation that mean so much. Remember too that men rarely wrote if they had anything uncomplimentary to offer. They called to see me, stopped me in the Shed, sometimes in the street, and they let me have it!

Stratford for Ever

By 1955 the Stratford District had come down to some 500 locomotives, and apart from the large central depot, the District Motive Power Superintendent was responsible for the sizeable depots of Colchester, Parkeston and Southend, sub-depots at Clacton, Walton, Braintree, Maldon and Southminster, small depots at Wood Street, Enfield, Bishops Stortford, Hertford and Epping, and any number of places housing one or two engines only, such as Chelmsford, Buntingford, Devonshire Street, Goodmayes and Canning Town. There were the electric train drivers' signing on points at Ilford, Gidea Park and Shenfield and the terminal point at Liverpool Street. There were 350hp shunting engines at Temple Mills, Drewry 204hp locomotives at Spitalfields and two petrol locomotives, one at Brentwood and one at Ware, both of a type used in World War 1. Brentwood had a gradient on which to start, whilst Ware, especially when Driver Perry was relieving, often had to rely on a push start. Steam was one thing, a big Dorman four-cylinder engine very much another.

The old Stratford has vanished into the past; the new depots, visited by Her Majesty the Queen in 1962, are now well up the seniority list. In 1955 the 'New Shed', built in the 1870s, had no ventilation and so, apart from diesel shunters, no live engines were repaired there. The Jubilee Shed as befitted its name was newer, bigger, draughtier, dirtier and intolerably smoky. The offices lay between the two Sheds and not only the offices but the New Shed are still there.

The old Stratford was a shanty town. Everybody had a messroom or an office or a sentry box or an air raid shelter and only the most skilled foreman knew who lived where. In 1954 a certain Mr Adkinson from BR HQ was deputed to carry out a purge on principal London depots and arrived unannounced. My predecessor, Bill Dixon, was instructed to take Mr Adkinson to the fitters messroom but landed quite unintentionally in a temporarily empty shack. Full marks to Bill!

Thus, in a few words, the incredible Stratford which must never be confused with Stratford Works.

Nothing was conventional, labour relations were an extraordinary mixture of the co-operative, the volatile and the restrictive, steam locomotives were generally in pretty poor condition and yet the spirit of the place was unquenchable, as it is today.

Nevertheless, we were gradually sinking, and dieselisation, with electrification of the North East London services, came none too soon. The technical revolution that we had to endure over four short years was eclipsed only by the human revolution where the lives and conditions of 3,500 men were completely changed, with all the training, planning and scheming that this entailed. We were in the van, we had the joy of making our own decisions on what we were going to do and how to set about it. We converted fitters into electricians and paid them 10/- (50p) a week for the privilege of practising two crafts. And then, in 1960, the new depot was opened, the old restrictive practices became a thing of the past and standards of cleanliness were achieved and maintained that have left their mark on Stratford locomotives to this day. A tremendous spirit of cooperation and understanding existed between staff and management; and the old order was finished, the new had arrived. Stratford, slimmer and now a shadow of its former self, is forever Stratford, a remarkable place, where people have thrived on adversity as well as on success, where the unorthodox has always been normal and where, in the high noon of steam, a Chief Mechanical Foreman frequently came to work in a Palm Beach suit and dancing pumps!

Appointed Assistant District Motive Power Superintendent at Stratford in January 1955 at a salary of £1,000 a year, I served under T. C. B. Miller for nine months. He was promoted Assistant to the Motive Power Superintendent at Liverpool Street, Mr E. D. Trask, and after W. D. Dixon of Lincoln had held the fort for a time A. R. Ewer was appointed in February 1956. The old Motive Power Department ceased to exist on 31 December 1957 and Mr Ewer and I moved our offices to Hamilton House, Liverpool Street where, for a time, we felt lost and out of touch in the Traffic Organisation. A year later on 26 January 1959, A. R. Ewer moved on and I was placed in command on a temporary basis. I was appointed in August 1959 and my title became District Running & Maintenance Engineer, Liverpool Street. My salary was £1,815 a year and I assumed responsibility not only for locomotive running and maintenance but all carriage and wagon, plant and machinery and road motor maintenance repair in the District. My Assistant was the wonderful Bert Webster who was sent to complement my youth. I was 35, he was 57, and between us and a super management team, we reckoned we could move mountains with some 3,500 staff to help us! We had a Traffic Manager, H. W. Few, of outstanding qualities, and grand colleagues in the Operating and Commercial departments and on the Civil & Signalling Engineering sides. We needed them, for between us we had to achieve that human and technical revolution and run the railway to the highest possible standards while we were doing it.

VII

Terry Miller's Stratford

A s A SHEDMASTER, working seven days a week at very high pressure, running my own show with the minimum of interference, I had begun to have a fair idea of my own ability. I knew something of Stratford but I certainly had not analysed the changes in work about to be wrought by my transfer to the District level. I had much to learn.

At Stratford the District Motive Power Superintendent, his Assistant and the Shedmaster worked very closely together. We shared the same clerical and secretarial staff and our offices were cheek by jowl with those of the Running Foreman, the Mechanical and Boiler Foreman and the Chief Locomotive Inspector. It needed a spirit of understanding and co-operation not to tread on each other's toes and to ensure that our efforts were truly corporate. On the other hand, there was enough to do to keep us busy 24 hours a day and the Stratford of 1955 was a treadmill.

In the first few weeks I found it difficult to adopt my new stance. I was too aware of the difficulties of depot life, I saw too many things I did not like. In my enthusiasm, I forgot how G. L. Nicholson had stood back to allow me to be my own boss. In time, and thanks to the influence of my Chief, Terry Miller, I began to settle down. The third member of the group, the Shedmaster, was Dick Robson who had acted for 12 months as the Assistant and was then passed over when I was appointed. Despite his disappointment and my early interference, he became and remained a wonderful and humorous colleague until his untimely death in 1962. He went to King's Cross in a similar position to mine in 1956 so his loyalty was soon rewarded.

Our relationships were friendly but formal. That was the way in those days. To my Chief, I was 'Hardy'. Everybody was on the same footing, all addressed by their surnames by Mr Miller. Mary White the Secretary, a legend at Stratford, was 'Miss White' to all of us but such formality was absolutely natural, and begat mutual respect for our Chief. Terry Miller and I shared our problems daily across the desk: I saw everything that he saw, except the strictly private, in order that I could gradually become a true deputy. I learned much from him in the art of handling difficult and complex correspondence and he taught me that an effective reply to a rude or inconsidered letter from a senior colleague was best dealt with by returning it to the sender, asking him whether he had read it before signing it! No longer did I work on Sundays, and I went home at lunch time on a Saturday. More accurately, I tried to do this but no man worth his salt can walk out of a crisis and we spent a fair amount of time dealing with the causes of potential strikes which would flare up for obscure reasons amongst the maintenance staff on a Saturday.

Terry was an extremely able engineer and administrator and had borne the heat and burden of the District since 1947. I was by no means as good an engineer but nevertheless, I had to tackle engineering problems because the Assistant DMPS was expected to concentrate his efforts on the maintenance of locomotives throughout the District and to take a large share of the responsibility for the satisfactory running of the extraordinary conglomerate of out-depots.

Used to the stringent punctuality practices of the Southern Region, I could not bring myself to accept the enormity of delays to freight trains, seemingly condoned by the Running Foremen. Whilst one of the Chief Shift Foremen would be here, there and everywhere, driving his men on, another would be comfortably and placidly leaning back in his armchair, safe in his assumption that a 320-minute late start from Temple Mills to Hither Green was not the end of the world. After all, he said, on the West London Railway, what passed for a timetable was based on the calendar. A punctual start on a Monday would often be converted to a very late arrival or indeed no arrival back at the Shed. Multiply this many times and you have a picture of the chaotic cross London and local freight services. Add a touch of bad weather, shortage of staff, specials, cancellations, absenteeism and bad maintenance, and, by Thursday, a Running Foreman would take duty to find few of his freight services covered and a large number still waiting power on the previous shift. Small wonder that the Stratford Supervisors learned the arts of deception and robbery; small wonder that one or two of the weaker vessels would give up the struggle, leaving it to those with great spirit and resource to tackle an insoluble problem, day after day, seven days a week. No other depot escaped their felonious intentions and one could overhear a country depot plaintively asking for the return of one of its best passenger engines, to hear the Cockney reply: 'Sorry, mate, she's on the Sarfends'. An Ipswich 'B12' on the 'Sarfends' was manna from heaven, release from the bondage of No 61613 – most villainous of all the Stratford 'B17s' – or from the cotter throwing '1500s' that Stratford kept up its sleeve against a shortage of the best Southend engines

On the other hand, the 'Britannias' based on Stratford were regular men's engines shared by three sets of men and booked out personally by the Shedmaster. No foreign depot was permitted to pinch these engines, however serious the emergency: the footplates were scoured and polished daily and the immensely proud majority of 'Britannia' drivers in the Norwich link would never have allowed their engine to be taken away from them. The names of the men who ran the 'Brits' in 1955 are always fresh in my mind. Few were much over 50 and what fine enginemen most of them were. When they moved up the rosters into the L1 suburban link or on a voluntary basis, to shunting duties at Temple Mills, their places were taken by others, equally competent, although now and again there was trouble, for example, when W. H. Lee was promoted from 61233 to 70037. No '37, under Griffin, Rolstone, and Sampher, had been a crack engine but Walter Lee thought otherwise. Walter knew what was what but he had special methods of expression to which we shall come later.

Maintenance of locomotives, or lack of it, was the basic problem at Stratford. It always had been, even back in the 1920s under the direction of L. P. Parker, although it was often the place for the outstanding or the unusual. Those who had never worked at Stratford had the solution to the problem: the steel hand in the steel glove which was necessary to restore law and order, but such a solution would not have commended itself to the artisan staff represented by J. J. Groom, Boilermaker. A man of ability, flexibility of mind, and many talents, Jim Groom had his methods in dealing with the apparently hard men. He knew, as I was to find, that the rough, tough superintendent is frequently a broken reed who dismayed him not one bit. In the late 1940s he had led the famous boilermakers' strike when 313 locomotives were finally stopped for repairs, and the steam suburban service, not unnaturally, disintegrated. On the other hand, he was not only dedicated to the improvement of conditions of artisan staff at Stratford and indeed throughout the country: he expected management to be of uniformly high standard, direct, intelligent, well-informed and a match for him: and it was in management that we had our problems.

The mechanical staff were in charge of a man of brilliance, he of the Palm Beach suit and the dancing pumps, Arthur Aubrey Day. He was assisted by Syd Casselton, who became the greatest Stratford character of my time and the finest breakdown foreman whom it has been my honour to see in action. The District Boiler Foreman worked closely with them and that was that – no more salaried men to look after the welfare of nearly 500 locomotives. Chargehands covered the shifts, men of reputation and ability in most cases, artists in making do and mending.

Arthur Day was a very clever man, in some ways a genius. He was also a fixer and hardly a week went by without a threatened stoppage of work brought about by his ingenuity in trying to circumvent the restrictive practices which had gradually been forced on management in an endeavour to thwart his sleight of hand. Each case had to be thoroughly investigated, but whilst J. J. Groom and his Committee almost invariably stuck to the facts, Arthur Day's actions would be supported by his vivid imagination, inventive genius and remarkable gifts of elaboration. Eventually this had to stop but while Arthur was with us at Stratford, we were frequently in the centre of a whirlpool of make believe. His brilliance lay in the unusual, the unorthodox, the inventive. He loathed the daily rough and tumble with its inevitable treadmill: he was quite capable of releasing an engine from the 'stops' for service, without a blastpipe or a piston; he had favourites and he played into the hands of the Workshop Committee. But give him a special engine to prepare, a valve setting problem, or the arrival of a new class of engine and he was totally absorbed, the man for the job. Like Douglas Pope, Running Foreman, he could rise to the great occasion and organise the impossible, but with great changes to be made in organisation and method, it was better that he should move on with the onset of dieselisation, to become a Headquarters Technical Inspector.

One of my predecessors, George Weeden, knew both Day and Groom extremely well. In December 1949, dismayed that he would ever gain the co-operation of the Workshop Staff, he had started a series of informal fortnightly meetings with Groom. It must be realised that people like George and indeed myself were of a different generation and outlook and we could see the need for consultation, whereas many older men, management and staff alike, were far from convinced. Although successful in gaining a genuine measure of trust and understanding, the meetings lapsed when Weeden left Stratford; but, acting on his advice, I reinstated them three weeks after my arrival when the Shedmaster and I met Jim Groom and one of his three Committee members. Our activities were based on simplicity. Each request was discussed and noted; the action to be taken was noted; once the result had been achieved, the item was crossed off by red pencil. Groom and I kept identical books and I still have the book that George Weeden started and which was used until my appointment as District Running & Maintenance Engineer in 1959. There were no minutes, no time wasted in fruitless argument and deception, and if we could not agree, the item went to the formal Workshop Committee for discussion with T. C. B. Miller.

The working conditions and messing facilities at Stratford were primitive, even by the standards of the 1950s. Many of the points raised, if ignored, would have festered and lead to stoppage of work with a solution at pistol point. When management demonstrated that it would say yes or no and that it would keep its word, our troubles began to diminish although, whilst we had Arthur Day with us, life rarely became easier!

On the other hand, our close understanding was rewarded when, with dieselisation in full flow and the new maintenance depot on the verge of opening, we had the closest co-operation from the staff and their representatives in creating a new Stratford, cleared of the restrictive practices of the past.

Here is a typical afternoon's work of the informal committee:

Meeting 14.00. Hardy, Robson, Groom, Oddy.
    Items discussed informally:
    White metallers positions.
    Tom Field.
    Men on shift work.
    Making up shifts in Jubilee Shed.

Covering of mates in the Arc.

Item

214	Drainage of Back End of Lifting Shed.	D.E. agreed to go ahead.
215	Crossings Jubilee Shed. End and Centre.	Running Foreman & Shunters to keep clear.
216	Clogs. Issue to individuals on Diesel Fuelling. Oil-skin individual issue.	Arranged
217	Storage of Brake Blocks. 14 Road.	Blocks moved to Backend, Jubilee Shed.
218	White metallers window to replace.	Advice to Crane Shop.
219	New Examiner's Roster.	Agreed to start Nov. 9th.
220	No. 1 Rd. New Shed. State of floor.	No. 1 scrape and sand regularly.
221	Fitters Asst. L. Maund (Breakdown Gang) Telephone.	To D.M.P.S. Work-Shop Committee.
222	Compressor. Machine Shop.	Silencer fixed outside building.
223	Door No. 6. New Shed. Damaged.	D.E. Stratford will renew.
224	Washing Room. 16 Road Messroom. Hot Water Boiler	New boiler ordered: plugged temporarily.
225	Rats & Mice in Fitters Messroom.	Rodent Officer will deal with quickly. Depot cats have not done the job properly.

How trivial some of these items may seem to the reader. They were not: behind each item lurked potential trouble. Nipped in the bud, dealt with, their danger disappeared to the advantage of us all.

I have mentioned Syd Casselton, who died in 1981 aged 62. He was Stratford through and through, his father having been a driver on steam before transferring in 1949 to Ilford. I had known him since 1946 when he came out of the Army and was selected by L. P. Parker for special training. He followed the legendary breakdown foreman, W. Hunting, on his retirement, and even those who marvelled at Hunting's ability acknowledged that Syd's was greater and that his personality was more endearing.

Breakdown work has changed since those days: it is not so demanding of men's time; there are fewer locomotives, trains, wagons and fewer derailments. Men joined the gang to make money. If they were out all night on a breakdown, and they were booked on duty at 8.00, there was no nonsense about going home for a rest. The money was good and the spirit was willing.

There were few changes needed in the gang, but anybody who did not measure up or turn out was unceremoniously bundled off and his place taken by another carefully vetted man from the pool of those waiting to make their fortune. And if this point is still not understood, in the winter months prior to electrification the crane would have some 50 calls in a month, plus bridging jobs for the District Engineer at weekends. Those men were strong, independent, resourceful and disciplined. Syd Casselton gave the orders, nobody else. He brooked no interference from anybody once started on a re-railing job. He was perfectly capable of telling the District Operating Superintendent exactly what to do with himself when being pressed for possession to allow a train to pass. His men worshipped him. He was large, florid, loved his beer, and reckoned he was an engineer and not a railwayman. When work was done, he encouraged relaxation in the riding van – good food and drink where appropriate – and perhaps of all the Stratford humorists he had the sharper wit. He also presided over the fortunes of the Crane Shop. Yet again, this was a unique Stratford institution which covered all the sidelines, so to say – water column and water tank maintenance, carpentering, and the construction of just about anything under the sun.

Syd Casselton, underneath his flamboyance, was a very sensitive man, quite easily hurt. He was a close and loyal friend. He saw our family grow up when now and again the Stratford lorry or van driver with Syd and maybe Charlie Wilcox, the carpenter, found its way, on some errand, to our home for the fashioning of the small jobs which were in those days an acceptable thing rarely if ever abused. Times have changed and the personal attention and service can no longer be applied readily. When I finally left Liverpool Street in 1963, I asked Syd if he would call at Hamilton House in Bishopsgate, remove my railway pictures – of which there were many from the wall and then take them to my home in Surbiton. He duly did this and on his return he was laughing and obviously had something to say that he relished. I said: 'Did you take the pictures home Syd'. 'Yes, I did and asked the Guv'nor (Gwenda) what to do with them.' She told me, 'Put them in the outside lavatory Syd, that's the best place for things like that'!

The District stores was naturally based on Stratford Depot. A stores; how dull you may say! There was a clerk in charge, an assistant, and Mr Charlie Lock. And wherever Charlie happened to be, there was always action for he *was* the stores, and the clerks did as they were told by the little man in overalls, humped back and smouldering Park Drive, who was quite formidable when he wanted his way.

Whereas the clerk-in-charge, quite rightly, attempted to curb the amount of stock on hand, Charlie's idea of a stores was to supply any item for any locomotive several times over, immediately and *triumphantly.*

The stores was situated under the main water tank within the tall brick building on which the tank was constructed. It consisted of a labyrinth of passages, nooks and crannies which had grown upwards, floor by floor, until by the earlier days of the diesels, it was so cramped at the top that even Charlie had to stoop!

He loved to nurse a particular grievance on grading and, after years of argument and fulmination, he was suddenly promoted, almost to his disappointment, to Storekeeper, Special Class. But he was at his best when called to the counter from the heights of his empire, by the call of 'Chawley', in order to clarify what it was that a particular *craftsman* required. Charlie's post was 'semi-skilled' and so the investigation would open with a colourful description of the appellant's character and general incompetence. The item could not be fetched, until Charlie had been allowed, in an outraged voice, to determine what was needed. He would then disappear, go straight to the pocket concerned and, sparks flying from his Park Drive, return to bang the part *triumphantly* on the counter.

The old stores closed in 1962 with the completion of dieselisation on the Eastern section. Charlie was moved to the new Diesel Depot Stores but he was no longer the same power in the land. Had Doctor Beeching arrived a couple of years earlier, it is certain that Charlie would have met Philip Shirley, Vice-Chairman and trouble-shooter. His methods and Mr Shirley's dictates were diametrically opposed and both believed passionately that they were right. Both were violently outspoken when roused and neither would ever give in. But it was not to be, and when and if a visit did take place, the old man was in the twilight of his service and the old all-embracing steam locomotive stores was a thing of the past.

Colchester and Parkeston depots were largely self-supporting. They had Shedmasters, John Bellwood at Colchester, Fred Hulme at Parkeston. The former depot was a very cramped affair and at weekends, locomotives had to be serviced and then stored away from the depot. Parkeston, by contrast, had plenty of room and, in some ways, was a model depot. John Bellwood had 'B2s' at Colchester and the 'B17s' at Clacton outbased for the London services. With this lot, one could only wish him the best of British! Until 1958 the four Clacton 'B17s' were worked by the same eight drivers and never shall I forget them.

61650: A. Smith and G. Brown
61651: G. Amos and H. Hudson
61662: H. Alston and W. Weyman
61666: H. Sparke and S. Pittuck

In the down direction, the Clacton buffet car trains were slackly timed. Consequently, the running was even slacker. It only wanted a spot of bad coal and some of the drivers would lose time on a 32min schedule to Shenfield. In 1958, with the 'Britannias' about to be introduced on vastly improved timings, I dreaded what would happen should a 'B1', 'B2' or 'B17' be used in place of a 'Britannia'. In fact, the elders were trained for the Colchester-Clacton electrics, the remainder took marvellously to the new timings – a revelation – and the vacancies were filled by young drivers and passed firemen in whom the fires of youth had not yet died.

Herman Sparke was a gloriously erratic old gentleman. He was frequently on the carpet either for losing time or for refusing to do as he was bidden. I was hearing a discipline appeal when he informed me with relish that 'he prided himself on his obstinacy'. And, occasionally, the pendulum would swing and he would achieve something worthy of commendation. But he was never dull! On the opposite shift was Stan Pittuck, tall, thin, charming, answering variously to the names of 'Chocolate' or 'Split Pin'. In Stan's hands No 61666 rolled back the years. She strode over the ground as she had at Neasden before the war on the 3.20pm Marylebone-Manchester, the 4.55pm Leicester, or the 'Newspaper'. 'Chocolate' was self-effacing, the most obliging of men, a marvellous engineman who loved his work. When determined to prove to the Clacton men what could be done with a 'B17' on the coming 'Britannia' timings, I chose Stan with No 61666. And did we go! Our net time to Colchester was 52½ minutes and the job was done to perfection. Perhaps the differences between Herman Sparke and Stan Pittuck reflect our very English way of doing things. Until the advent of electric and diesel traction, we brought up enginemen purely by experience, we gave no formal training so we had created a race of fiercely individualistic men. One man could know comparatively little and yet have a natural flair as an engineman, another could be an expert on rules and his knowledge of the locomotive and yet have no flair whatsover, and little confidence. The French trained their enginemen to achieve a remarkable degree of uniformity of performance, so rarely, if ever, would one meet a Herman Sparke in France – erratic, outspoken and sometimes brilliant – never would you meet a mécanicien who 'prided himself on his obstinacy', but rarely did one meet the untrained, completely natural, genius of the Pittuck's, the Gingell's or the Hailstone's and a thousand other self-taught British enginemen. Parkeston men had extensive route knowledge and the younger men, bent on promotion and driving turns, would go anywhere and do anything: fast freight, passenger work, ferry traffic, the Parkeston foreman could always find engines and men.

But when it came to the small depots in the 1950s, life could be hard! Southend, prior to electrification, was a very rough shop indeed. The Shedmaster had the minimum of assistance and his artisan staff was insufficient to maintain either the Southend 'Bongos' or the ailing fleet of 'B12s' in first class order. Punctuality drives would be mounted, Inspectors imported to man every peak service up and down for a week, but to little permanent avail.

Southend, though small, was a main line depot but, when it came to really hard work, Wood Street, Walthamstow was the place. Enfield Town, a very similar depot, was tucked away at the side of the station. Fourteen 'N7s' were allocated there and 12 of these were treble-manned. They were kept in remarkably good order, many of them polished in the cab, one or two cleaned externally by the crews. Enfield seemed to run itself and it was rare for time to be lost by men from that depot. But Wood Street was different and hard graft for the man in charge. The allocation was 14 'N7s', Nos 69600-69611, all Westinghouse engines, with no vacuum brake and therefore of no use away from the Jazz services, with Nos 69626, and 69627 as spare engines. The Chingford road was certainly heavier than the Enfield line but there was a certain unruliness about the place. The footplate staff varied – I seemed to have more discipline cases at Wood Street than the rest of the out-depots put together – whilst the resident fitter, very competent indeed, was nevertheless on occasions a man of dark mood and stubborn demeanour.

The man in charge of the depot could be a young man on the first rung of an engineering career, or a practical man drawn from the fitting grades at Stratford on his way to Mechanical Foreman or Shedmaster status. In 1957 the Chargeman at Wood Street was not on the salaried staff, and therefore not in the salaried staff pension fund; he was expected to work every Sunday for which he was given an allowance but he was not permitted to book overtime. He had 14 locomotives, as previously mentioned, was assisted by a temperamental genius of a fitter, worked alternate early and middle turns opposite a Shed Chargeman (non-clerical) drawn from the 'Conciliation' grades, and his group of helpers was completed by a fitter's assistant and two boiler washers. He was in charge of some 78 footplate staff and today he would most certainly be salaried and well paid. His engines had all gone into service by 07.30 and returned between 18.45 and 02.00 next morning so the work to be tackled on the two engines stopped each day was considerable.

Jack Barker, the Chargeman, would get to Wood Street on the early turn at 05.20. He would immediately start work on running repairs on the locomotives about to depart: brakes adjusting, big end cotters adjusting, Westinghouse pump difficulties. By 07.00 he should be seeing the wood for the trees, so he would move into his office – a primitive affair – to start on a routine. He had to check and sign all drivers' timesheets, make out next day's engine list, write out the availability sheets, and book up all repair, mileage and washout cards. He would check on the coal used in the previous 24 hours, balance the coal and oil books and wire Stratford with this and other daily details required. Without interruption, this would take him until about 9.15. He had then to prepare the daily supplementary enginemen's rosters and, where necessary, the weekly and Sunday lists and he shared this work with the Shed Chargeman who appeared at 11.00. At about 10.00, he started work on the Shed engines, working with the Fitter who had arrived at 08.00. He was liable to be distracted by telephone calls, distribution of cloths or soap or a visit from the Assistant District Motive Power Superintendent. About noon, he washed up and then started on correspondence. On Sundays, he was full-time fitting on the Shed engine from 08.00 to 12.00 for which he was not paid! How on earth did we find men to do this sort of work? The answer is quite simple. There was a great challenge; master Wood Street and a man had achieved something really worthwhile.

In *Steam in the Blood* I told of my journey to Wood Street with Driver Harry Hibbert on the 14.20 Liverpool Street to Chingford, fast to Hackney Downs and all stations to Chingford. We had a longvalve-travel Stratford 'N7', the Westinghouse pump struck work and we lost our air by Clapton, and no amount of hitting the top head with a coal pick, and associated opening of the steam cock above the firebox, would start that pump. The 14.24 Cambridge and Norwich was on our tail and we had to get out of the way, so with a good guard and a good strong arm on the engine handbrake, we reached Wood Street a couple of minutes behind. My intention was to leave the engine at this point and visit Jack Barker in the depot but he was waiting on the platform with his tools and a spare ball lubricator and reversing spindle. My first reaction was that this was nonsense and a delay in Wood Street station would possibly cause cancellation of the return working from Chingford. However, Jack jumped up on the gangway behind the pump on the smokebox side and started to strip down the top head. Meanwhile, we tackled Highams Park bank and arrived at Chingford three minutes late. At Chingford, he fitted a new reversing spindle to the top head, checked the lubricator and, in no time at all, our engine was serviceable and attached to its return working which left to time with a Westinghouse pump in excellent order. The job done, we travelled back to Wood Street on the cushions, satisfied that we had avoided a delay to a main line service and that not only a failure but a cancellation had been averted. There were no medals for any of us, Driver Hibbert, Jack Barker and myself, no report would appear in the Central log, nobody else would know what had taken place but the job was done, the challenge met and that was what mattered.

VIII

A Hard Year

A T THE END of May 1955 the Associated Society of Locomotive Engineers & Firemen called a strike on pay which was to last for 17 days. The National Union of Railwaymen did not support the stoppage and instructed their driver members to report for duty at their rostered times and to undertake the class of work for which they would normally have been booked. And so the old sores of 1924 were reopened.

T. C. B. Miller, still at Stratford, took the day shift and I spent 12 hours on nights in charge of all Motive Power activities within the District. Sadly, I have lost the log book that we kept but we enlivened a miserable period with much humour and laughter and lived on Rob Roy biscuits and tea. Needless to say, there was competition between the two shifts and we were delighted to score off the day shift, on which were the more powerful members of the District team!

One might think that, with few trains running, here was a heaven-sent opportunity to get overdue repairs carried out and to some extent this was done, but when engines had to be moved and no driver was available, they had to stay put. I have very firm views on the handling of locomotives, which are the product of the generous training that I was given in my early days. The driver is the captain of his ship and I have never, in my life, asked to have hold of a locomotive. (On the other hand, when given the opportunity, I have rarely, if ever, refused, remembering that I must make no mistake and that my performance must be exemplary.) Nor have I ever instructed an unauthorised person to move a locomotive and this principle held good during the strike.

When it sets comrade against comrade a strike is a particularly detestable business and I had hoped never to see another in my railway career. At 02.30, on the Sunday morning, the first NUR driver entered the depot. He had had a jangle with the pickets on the way in and went white when he realised that he was truly the first to report, perhaps the only man. But several more had arrived by the time I went home at 08.30 and, within 24 hours, a skeleton service had been organised, largely steam but with one or two electric trains in the London area.

By no means all the NUR drivers and firemen came to work. Some turned away and went home after discussion with the pickets, some others joined ASLEF there and then and went on strike. Some men came to work and were never the same again, others came in silently and shattered, others believed in the rightness of the NUR policy and were quite unshakeable, and a few signed on as though everything was normal, with not a care in the world. Some insisted on carrying out the precise work for which they were normally rostered whilst others, such as the driver who went twice a day to Southend with a 'Britannia' (and came back tender first) with 10 coaches, covered work far removed from his normal duties but did so voluntarily.

Old Charlie Lavender, well over 60 and who had done it all, reported for duty regularly for suburban work. He was normally in No 17, the Woolwich link, and was very outspoken. He recounted what he had said to a picket who had once been his fireman and for whom he had had no use whatsoever! 'And do you know, Guv'nor' he said, 'he had the bleedin' cheek to lecture me on matters of principle, that bleeder who couldn't fire to a kettle'!

Nevertheless, the picketing was fairly conducted by men who had had years of experience and membership of their Trade Union and the resident policeman was never required to intercede. My own entry and exit was regarded as perfectly normal and acceptable. And then, suddenly, it was all over, from midnight men returned to work, first a trickle, then an unstoppable wave. To this day I shall never know how Doug Pope, Running Foreman, got them sorted out. He was a true Stratford fixer of the old school and engines were sent out one after the other, simply as passenger, suburban and freight to be sorted out by an equally overwhelmed Traffic Department. The strike was over but the undercurrents lasted for many years. There was little or no anti-management feeling, we carried on where we had left off but men did and said very harsh things to one another and, as I have said, a few who worked were never again the same men. But before long, the good sense and humour of the East Londoner returned and most of the feuds were forgotten in the mêlée of Stratford life, with its trials and tribulations, its rackety engines and its great comradeship which, as ever, thrived on adversity.

So far I have dealt largely with events in 1955-57 but we were shortly to endure formidable changes. By the end of 1957 we were no longer answerable to our Motive Power Superintendent, E. D. Trask, and we felt the finger of doom was pointing straight at us for we had a great loyalty to our Regional Chief. He was a splendid boss, a fine engineer and administrator, very direct and straight, and scrupulously fair. He wanted deeds, not words. Today he loves to talk of his experiences – but nearly 30 years ago he was gloriously abrupt. I had been Terry Miller's assistant a few months when he took some leave and I was temporarily in charge. During this period, E.D.T. rang me to ask for an explanation of some incident. I knew all the answers, launched into a detailed and accurate account, to be interrupted by my Chief: 'Now, look here, Hardy, come to the point, will you'. Splendid treatment!

But now E.D.T. was an assistant to the General Manager and here were A. R. Ewer and I in splendid isolation in Hamilton House, Bishopsgate. We were losing our grip at the very time we needed to be in the centre of activity with diesels arriving each week.

Steam locomotive performance started to slip seriously. I had been Assistant District Motive Power Superintendent in the centre of one of the most difficult Districts in the country. Now I was Motive Power Assistant, my Class 1 salary £1,235/year. A Class 2 Assistant Traffic or Commercial man was rated at £1,285. We were the poor relations and, for the first time, I became bitter and discontented and aware of what others were earning. I was indeed sorry for myself, jealous of the success of others, but I was soon to learn that to be sorry for oneself is a ridiculous, though natural, attitude of mind.

How little did I realise that, within a year, I should be placed in charge of affairs, a prelude to four wonderful years of responsibility and joint achievement.

Nevertheless there was much to be done, and within a year a proper Technical Section had been set up. But in the meanwhile much of the compilation of reports, shopping proposals and casualty reports fell on my shoulders to be carried out at home and at weekends. I was never so happy being miserable and for a time I must have been a trial to Gwenda although she never allowed the ups and downs of railway life to interfere with our happiness. Even this was a quickly passing phase and, as ever, she was a wonderful support to me. Lucky is the man who has such a wife, who encourages, aids and advises unobtrusively and wisely, who is the guv'nor at home but who never tries to run his professional affairs!

Ron Ewer and I had brought with us to Liverpool Street a posse of Locomotive Inspectors under the direction of Percy Howard. They were initially lost in their new surroundings, on the top floor of Hamilton House. Gone were the days when a driver could bang on the door of the District Loco Inspector's Office at Stratford and get a problem solved at once. Percy Howard had completed his express train work years before. He had thrived on the West Countries in 1952 when they deputised for the 'Britannias'. He would have been on suburban or shunting work had he not become an Inspector but now he had the great advantage of a man who had done it all. The other Inspectors, Elmer and Lockwood, Shelley and Warren, had the same advantages and this little team, aided by Wally Mason, a driver temporarily upgraded, performed wonders at a very difficult time. Yet again, we were all partners in adversity and we thrived on the challenge.

Each morning, the log would be examined and a plan of campaign drawn up on the deployment of the Inspectors. Any driver who had lost time on the previous day would either be tackled personally or ridden with, no matter the hour of the day or night. Enginemen would certainly not come voluntarily to alien Hamilton House, and the Inspectors were far better employed in their rightful place, on the footplate of a locomotive, dealing with difficulties on the spot.

Percy's contribution was not his knowledge of rules and regulations, nor of steam or diesel locomotive theory and practice but in his ability to cooperate and support me in 1958 and the better years that lay ahead; he got things done. His anticipation, and the practical knowledge of his Inspectors, saved us thousands of minutes of potential delay. They minimised the need for paper and maximised their influence on enginemen, when they mounted the footplate, exuding the confidence born of experience. They were the men for the moment and they knew that when time had been lost the previous day, it would not be lost when they were present. 'Right Time' was our policy and although we were to lose ground in 1958, triumphantly we had won back our reputation by the middle of 1959.

Whilst on the subject of punctuality, let us now consider my last port of call on an evening once we had transferred to Liverpool Street. The Motive Power Department had an office at the country end of No 11 platform which commanded a view not only of the station but of the turntable, coal plant and disposal pits. Here was stationed the Running Foreman who covered the early and late turns, assisted by a clerk on days, a turntable man who was the Foreman's Assistant, and a coalman who manned the small coaling plant, necessary for turn-round purposes. The Foreman's principal responsibility was the punctual departure of all trains from the terminal.

Whilst the electrics caused him few worries (except in the early days of 25kV when they drove everybody insane), both steam and diesel locomotives had their difficulties. And here again, all concerned fell back on the services of an unsung hero who must have saved hours and hours of potential delay. This was the Liverpool Street fitter, Sam Chapman, to be called to deal with the sudden failure of a component, or one of those mysterious half faults that would beset a steam locomotive, or perhaps a vacuum ejector that would not maintain the correct degree of vacuum. Sam would have three choices; to fail the locomotive and cause a delay, to effect a permanent or temporary repair, or to take a chance – for which he would be held responsible based on a profound knowledge of what could or could not be left to chance on a steam locomotive. Despite a lifetime of experience on steam but not on diesel traction we expected exactly the same service from Sam and it was a marvel to me how this man, all but 60 years of age, grasped the new technology and fulfilled the same requirement of making diesel locomotives roadworthy at the shortest possible notice.

Ted Carron, Running Foreman, had a very loud voice and a gammy leg. My half hour with him or with his opposite mate was always an enjoyable prelude to the journey home. Over a cup of tea, I could keep an eye on the locomotive position, the standard of punctuality, the occasional refusal of somebody to clean the pilot, and it did no harm for people to know that I was around particularly after I had become District Running & Maintenance Engineer in 1959.

And so, 1958 dragged wearily on. The loss of departmental impetus weighed heavily on us, I was seriously (and stupidly) considering leaving the railway, and our performance sank lower and lower, to be unfavourably criticised by the Great Eastern Line Manager, W. G. Thorpe. Early in 1959 some changes were made. My Chief, Ron Ewer, was to some extent held responsible for the bad state of affairs and was moved to Line Headquarters. Some people in high places felt that I should also take my share of responsibility but there were others, not least Terry Miller and my Traffic Manager, H. W. Few, who felt that I should, as a young man, be given the chance to run the District. Had they not won the day I should have remained an Assistant and when I analyse myself I realise that, except in very favourable circumstances and answerable to such as T. C. B. Miller, I was not a good Assistant nor was it my forte to be No 2; indeed, from 1949 to my retirement in 1982, I was lucky enough to be No 1 except for the four years at Stratford and Liverpool Street.

A Human Revolution

1959 to 1962 were great years: complete dieselisation on the non-electrified Great Eastern Line, a major suburban electrification scheme introduced, conversion from 1,500 dc to 25kV ac on the Shenfield, Southend and Chelmsford services, electrification to Clacton, a revolution in traction, and a human revolution. And all the time, the standards of performance on the Great Eastern Line were rising despite the early difficulties with the high voltage electrification, and despite the immense amount of bridge building, track relaying, weekend work, resignalling and station rebuilding that involved many people in hours of planning and careful thought and gave no respite at weekends to operating men who were driven hard enough during the week.

It was the most unforgettable experience. We were up against time, against problems we had never faced before, managerial and human problems. It was essential to think ahead, to plan and scheme, and to keep the staff in good heart, for I was determined to keep on top. It may seem ironical to my readers when I say that no one tried harder than I did to complete the dieselisation programme in our Division, but equally, no one made greater efforts to get the best from the remaining steam locomotives before the last were moved away in November 1962.

In *Steam in the Blood* I wrote about the life that we led through those four years and gave some idea of what was achieved. In the next chapters I shall enlarge on the same theme. No day passed without incidents of some sort; it was a tragedy that I kept no records, no diary and it never occurred to me that we should be making history. However, a few letters that I wrote between 1959 and 1962 are still in existence, together with a hilarious file on DMU brakes which somehow found its way into my possession. And so what I have to say should be read against a background of continuous change where men, many of them grumbling and swearing as is the British way, were making do in the most marvellous manner and giving us our new railway without hindrance and obstruction.

Our own leadership was outstanding. W. G. Thorpe was Line Manager (Great Eastern) and he was an inspiration to us all. Furthermore, he had picked the right team at Line level. From my point of view, they let me have my head and nothing could be better in such critically changing times.

IX

Each Day a Fresh Challenge

M EMORANDA ARE INTERESTING, if brief and to the point, so here are a few which reflect the spirit of the times; some I wrote between May and September 1959, others in 1962.

To

H. Brinklow, Passenger Assistant (Operating).

I travelled on the footplate of the 18.07 Liverpool Street to Bishops Stortford on June 12th. On arrival at B. Stortford, the guard complimented Driver E. Whitehead of Stratford on his performance for the week. I should be glad if you could have the guard thanked for the part he played. I have never seen smarter, better work carried out at stations than by this particular man.

To

Loco Shedmaster, Stratford. Personal

Absenteeism

I am only too well aware of your difficulties and shortages clerically but I am extremely concerned with the development of absenteeism and I want you to take immediate steps as follows:

1. Confer with your Senior Running Foreman and obtain details (check them with your records) of, say, a dozen really bad timekeepers, drivers or firemen.

2. Issue charge sheets (Form 1) quickly.

3. Pass the details on receipt direct to me with your recommendation.

4. You can rely fully on my support and I intend to stop this trend.

Loco Shedmaster, Stratford Personal

14.30 LV. St-Norwich 19.6.59

Engine 70010 arrived in London with a Norwich driver who asked for a fresh engine for the 14.30 Down. The Running Foreman LV. St. 'oaked' the Shed at 12.50 and apparently the Shed could not or would not provide. As a result of what seems total lack of appreciation of facts, 70010 worked the 14.30 and lost 80 minutes to Ipswich with tubes leaking.

Let me know by Wednesday morning the facts, who you hold responsible and what you are going to do with him.

Inspector P. Howard   5th July 1962

I understand there is an individual with a beard riding in the cab of the 7.14 electric Clacton to Liverpool Street. He may be an Electric Traction man but on the other hand he may not. Please have him apprehended without delay.

C. L. Rowbury, Movements Assistant   3rd July 1962

I believe you had a Guard Cock of Ilford who refused to change ends at Chelmsford recently. You might be interested to know this same Guard was riding in the cab with the motorman of the 16.42 arrival (E.C.S) at Chelmsford today.

H. Conway, Traffic Costing Officer   5th July 1962

Have we any means at our disposal of determining the cost of carrying out a 6 monthly examination on a diesel locomotive, other than reducing it to its simplest form which, from my point of view, would be to find out the number of men employed and the number of hours they were on the job. I would think this is far too simple and would give me the wrong answer?

Just half a dozen memoranda which do not begin to cover the variety of our daily task. One can note our difficulties: the need for vigilance, the need to acknowledge a good job well done, and lastly, my very elementary knowledge of costing which, to most of us, was new, alien and slightly ludicrous. We were to learn very quickly that it was nothing of the sort. The chap with a beard disappeared without trace. In 1962, a beard was a nine days' wonder and the bush telegraph no doubt told him to keep out of the way or have a shave!

Most drivers who have gone right through their main line work have slipped up at some time or another. This could have been the disregard of a signal at danger or perhaps misjudgement resulting in it being passed in the 'On' position. One might say this should never happen but to do so not only ignores the human factor but presupposes that every man in charge of a locomotive will observe every signal, every incident, under all conditions of working, in all weathers, and in steam days whatever the state of the engine, however inexperienced the fireman. Whilst my readers can without reprimand jump the traffic lights or merely overshoot them, it is a very serious business for a driver to pass a signal by a few yards. Only too well do I understand the responsibility that rests on a driver's shoulders and how a momentary loss of concentration, or association of ideas, may lead him into trouble. Here is a case that shows exactly what I mean.

A highly experienced driver by the name of Harry Wooltorton was working the 08.57 Chingford to Liverpool Street with an 'L1' 2-6-4T locomotive No 67727. He was working bunker first, the engine had left-hand drive and so he was on the right-hand side of the locomotive in the direction they were travelling. He approached Signal L91 on the up suburban line at Liverpool Street. The signal was at danger and he was about to come to a stand. This was the last signal before entering the station; he had left Bethnal Green at caution, stopped at L87, three parts of the way down the bank, received one yellow as he approached L89, and clearly was on the tail of the man in front.

Wooltorton had almost stopped at L91 when the platform indicator blind, which had been blank, exhibited the figure 3 to indicate Platform No 3 into which he was booked to run. He was a quick-thinking man and he immediately opened the regulator, releasing the Westinghouse brake at the same time. But having seen the indicator at 3, he omitted to check the signal, a colour light which was still at danger. The road, despite the platform indication, lay for Platform 4 with No 99 disc signal off, an error in setting up the route in the Liverpool Street power signalbox, but it was impossible under these circumstances for the signalman to clear L91 signal.

Nevertheless, and quite correctly, I had to hold Harry Wooltorton responsible for disregarding a signal at danger, caused by an association of ideas. A thousand times he had seen the blind change and the signal immediately clear from danger but this was the odd time out. He had an excellent record over many years driving and he freely, if picturesquely admitted his error. Punishment there had to be but, in deciding on the level of punishment, one was careful to consider *all* points of view before arriving at a conclusion. Indeed, the railway disciplinary machinery, properly used, is both effective and fair and, in my experience, justice is almost invariably done and this leads us to examine another case which is of more than usual interest, because of its further implications.

I shall never forget the case of the three Stratford drivers who had been learning the road to March. They had recently been promoted to the Main Line goods link, none of them would see 50 again and all were excellent railwaymen who had spent up to 25 years firing before becoming drivers. The incident took place in the summer of 1959, and the three men had reached the last day of road learning which was a Friday. Their job done, they called on some friends in March who entertained them rather too well and they were eventually placed under the close observation of the guard working the passenger train from March to Cambridge which connected with the 19.05 to Liverpool Street. By the time they reached London, they were, to quote Somerville & Ross, in the condition best described as 'not drunk but having the appearance of drink taken'. They were met by the Running Foreman and told to report to my office at 10.30 on the Monday morning, being suspended from duty until then.

We were able to contact both ASLEF and NUR branch secretaries, men of ability and honour, explain the circumstances and discuss the matter from all angles including the severity of the punishment before the interview took place. Had they been in charge of locomotives, there would have been no question of anything less stringent than dismissal. Clearly the matter had to be dealt with severely and the punishment was to be suspension from duty with a final warning that if anything similar ever happened again, that would be that. It was far better this way, a week at home on top of the immediate suspension from the Friday. At least one of the three was said to have got hell from his wife which was exactly what was needed: she would do it better than me! I took the unorthodox but effective course of dealing with the charge, settling the punishment and inviting an appeal all at the same time. As arranged with the branch secretaries, the invitation was declined, and the three men with their advocates were out of the office inside 10 minutes. Nobody spoke except me and to this day I can see the faces of the three drivers and their look of relief when I said what was going to be done, after I had given them some effective home truths.

From time to time, one reads of the incidence of drinking amongst trainmen. The Eltham accident of 1972 concentrated people's minds on the evils of drink and when incidents have arisen since then, there has been critical comment in the press. This is perfectly right but there is another side, as ever, to the coin.

In the very hard old days, it is fair to say that there was more strong drink taken than today. On the other hand, men worked uncommonly hard and their loyalty to one another was undisputed; they saw each other through. And I have to say that, between 1947 and 1973 when I was almost continuously in charge of large numbers of men, I had to deal with only four such cases, including the one already related.

Nevertheless, each individual incident has to be correctly handled and, of course, experience cannot be gained easily. One thing is very certain: the steam locomotive at its roughest was very effective as a pick-me-up, whereas a man in charge of a diesel or electric locomotive must be at his clearest from start to finish; and, furthermore, no more Sunday squeezers for he must have his laid down rest period of 12 hours between duties to maintain that clearheadedness.

During my time as District Running & Maintenance Engineer, I had to deal with an unusual set of circumstances involving a certain very highly strung driver who had reached what was still called the B1 link, although much of the work was diesel hauled. In a matter of a year, he would be on the Norwich jobs. I was far from happy about this but there seemed to be no grounds for refusing to allow him to take his seniority when the time came because he was basically a competent and conscientious engine driver.

My concern was compounded by a number of incidents, some of which I can still remember. The first would have been about 1959, when he was on the verge of the passenger links. He was working a Great Eastern Class J17

0-6-0 on a freight train between Ipswich and Goodmayes and coming up the bank between Chelmsford and Ingatestone he was banging on a bit. The fireman, perhaps unwisely, walked across and somewhat ironically examined the cut-off indicator whereupon the driver closed the regulator, came to a stand at Ingatestone and instructed his fireman to get off the footplate. Eventually order had been restored and the incident was dealt with, bearing in mind that a driver's authority should be upheld wherever possible.

A few months afterwards, the same driver lost a few minutes with a 'B1' without an adequate explanation on the 17.57 Cambridge, a heavy and tightly-timed semi-fast train. The following night when sent to ride with him, the Inspector, Bill Shelley, who had been an outstanding driver with No 70039, had a very difficult time. The driver quite unnecessarily insisted in taking water at Bishops Stortford: this caused a delay of five minutes and I was beginning to feel that a form of inferiority complex existed as there were often ructions between the driver and the firemen who had been booked with him.

Let me make clear a principle. The driver is the boss and it is the fireman's responsibility to measure up to his requirement. A driver could, after complaint, be given a new fireman but it was totally against my principles to have a fireman moved from his rostered driver because he had made a complaint about his driver.

Some months later the same driver was involved in an incident with D202 near Bethnal Green. There was disagreement in the cab and I was obliged to discipline the driver for failing to carry out the laid down tests and causing an unnecessary delay. While the case was pending, I became more than ever convinced that he was having difficulty in fulfilling his responsibilities and I asked the Regional Medical Officer to carry out an examination to see whether there were medical reasons for his erratic behaviour. The examination took place but the doctor was not prepared to go further than to pronounce him fit for main line work. The discipline case was then settled by a punishment which he took very hard.

I was still unhappy and steeped as I was in railway work, I knew that something really serious could happen. Soon, a further delay took place, only a couple of minutes, but as it involved both driver and fireman I decided that was to be the end. My wise old Assistant, Bert Webster, and I had talked it through on many occasions: Bert said that the driver would be perfectly happy when he had nobody breathing down his neck. I sent for the driver and told him I was taking him off locomotive work and placing him on the railcars, or DMUs as they are now known. He was very upset, and appealed to the Local Departmental Committee which, quite naturally, was annoyed at my apparently high-handed attitude.

I explained the full circumstances, the whole background and my fears. I made it very clear that although I could not expect the Regional Medical

Officer to take the driver out of his normal link position, I knew there was something seriously wrong. Furthermore, the Norwich link loomed ahead, and a disagreement could so easily lead to a mishap; without doubt, I should be held responsible if the driver's state of mind had contributed to a mishap. I was not taking that risk and nobody was going to make me do so. Our driver went on to the railcars and, so far as I know lived happily ever after.

One must not forget that we were going through a human revolution. A thousand day-to-day things were happening; some were the effect directly or indirectly of the revolution, others were incidents that would probably occur in any case and, to add to our problems, we were taking on responsibility for carriage and wagon, plant and road motor maintenance. When a reorganisation takes place, there is nearly always somebody who has a crazy notion about what should be done but, mercifully and again thanks to my own Traffic Manager, we were able to reorganise with care and with due consideration to the men who had given years of service to a particular department and were now being swallowed up in alien surroundings.

This was a sign of those times: had we been in any way restricted by bureaucratic influences, we could never have changed the way of life of hundreds of drivers, firemen and artisans in so short a time. And I say we because, although I may have taken the lead, the support that was given by so many people selfless and genuine support – was out of this world. Cast your mind back to the variety of diesel main line and shunting locomotives that was loaded on to us, some for short periods; there was a need to train drivers for 25kV electrification on our short piece of 25kV railway between Colchester and Clacton and the urgent requirement to train the maximum in the minimum time, whoever they were and wherever they were based.

In *Steam in the Blood* I have described how we had to train our artisans, virtually all steam men: how we converted fitters and boilermakers to electricians in the homespun school tucked away in the upper offices. Craftsmen are intelligent men and such a change was therefore not so revolutionary as our critics would have it. Of course we had our critics, but what was the alternative? Diesel locomotives were arriving every week, we were responsible for maintenance, we had to get on with the job. The critics were as wise as ever after the event and continued their theme long after I had left Liverpool Street but the fact remains that these men who had added to their experience had become the backbone of the Stratford artisan staff. Electricians, members of the ETU, joined the railway or transferred from Scotland after the new depot was in operation and the basic staff had settled down. These new men did not at first like what they found but, as time went on, the liberal attitude of the East End of London sorted out any ill feelings and the various groups worked well together under common supervision: even this was revolutionary. In the past a boilermaker chargehand gave instructions to a boilermaker, a fitter chargehand to a fitter, and now we had one supervisor for the three crafts. Perfectly normal today but it was our pleasure to be in the lead and to decide these things.

The training of drivers and firemen in their new duties was undertaken by experienced and competent drivers, paid as instructors. These were very carefully selected by the Chief Locomotive Inspector and the Shedmaster, cool headed men with the ability to impart knowledge with authority and understanding. They were fascinated by the challenge that diesel traction offered, they knew we had to change and that their steam experience was a marvellous platform on which to base a mastery of the new machine. These men had no technical training, no engineering education but they listened, they learned from manufacturers and artisans alike and they were enginemen with the skill and experience to instruct with confidence.

The list of instructors was complete but it then became necessary to select another to cover the additional work with which we were constantly loaded. And this is where the famous Mr W. H. Lee comes in the picture.

It can be said without a doubt that Walter Lee was the biggest menace to foremen and artisans that I have ever met in my railway career. It can also be said that his knowledge of locomotives and the rule book was profound, and if he were alive today he would be proud to read these two sentences about himself.

He was tall, commanding, angular, and had a voice that penetrated, leaving the listener in no doubt that Walter was right in everything he said and did. Rarely was anybody cute enough to turn the tables on him. I did it once: I walked into the Running Foreman's office one morning at 08.45; Walter was booked Clacton with 61089; he had already failed it and had been given another 'B1'. The Foreman was gloomily talking of a late start. I asked him whether he had a pilot and crew: he had and he told them to slip off quietly to Liverpool Street. Quietly, because Walter had eyes in the back of his head and he also liked the small amount of mileage that Clacton would bring. Suddenly, he realised that everything had gone quiet, that he had been left to his own devices and that he could not get the 09.36 from the terminus. He stormed into the foreman who blandly said 'you'd better go and see Mr Hardy'. We met in the corridor outside my office, his wrath was very real and he was going to claim mileage because we had sent another man in his place. I was joyfully able to tell him he hadn't a snowball in hell's chance.

One last Walterism before we come to the point. Before my day, he wrote to his District Locomotive Superintendent by registered post telling him that he had instructed his solicitors, in the event of his death when in charge of a Stratford engine, to hold the railway responsible in view of the condition of that depot's locomotives! He was perfectly serious in this and everything he did, for he was very short of that Cockney gift, a sense of humour!

And now we needed another instructor. By this time, Lee had finished his main line work and was crashing about on the 'L1s' on main line suburban work, failing one engine after another and then losing time with the replacement 'N7s' which he was forced to take. He had previously made it clear that his ability entitled him to consideration as an instructor: we all felt that he had been such a nuisance over the years that he could never fit into our plans. To cut a long story short, I eventually bowed not only to his personal pressure but that of the staff representatives whom he had massed behind him, and appointed him as a driver instructor. I was disgusted with myself but I need not have worried. Walter was a poacher turned magnificent gamekeeper, his victims went through hell but they knew their stuff, and for the first time in years there was a smile of genuine pleasure on his face.

X

A Royal Visit

IN MARCH 1983 agreement was reached on one-man operation so that training of drivers could begin between St Pancras and Bedford, but in 1959 we were experimenting with single manning under the 1957 agreement. We started quite quietly – the best way – with the Southend parcels. Even in steam days the guard had travelled in a vacuum fitted goods brake, sitting in warm solitude therein. The vans required no heat and we provided our Type 1 NBL or GEC/Paxman locomotives. These machines were poor affairs and all stations to Southend and back in eight hours was about their mark. In the light of later experience, it is interesting that our drivers did not object to driving on their own, with the large bonnet ahead. If they could not see, they crossed quickly to the other side of the cab, before the dead man control came into operation. The Southend parcels service ran itself and I, for one, was amazed.

But the test was yet to come. Once steam heat was no longer required in the summer months, certain jobs could be single manned and we should then see how our men would measure up on their own after years with a mate. In most cases they measured up remarkably well. Only 18 months before had the first diesel main liner come over the horizon and yet here were the men in their 50s and 60s tearing about on their own. Certainly a fault could result in a dead stand until rectified by a cool head and hands; certainly old fashioned remedies still persisted to torture the manufacturer's service engineers. It was hair raising to know that fire-bell clappers were stuffed with newspaper to stop them ringing unnecessarily and thereby causing a delay to put out a fire that had not taken place.

The early problems were not on the road, indeed some men liked a few weeks away from their firemen. But the 1957 agreement insisted on two men on a light engine. Within station limits, the second man could be a driver, a fireman, a shunter or a guard. Outside the limits, the second man had to be in the line of promotion to driver or fireman. The agreement made it quite clear that drivers would not be required to couple and uncouple locomotives from their trains, and of course, smartly uniformed passenger guards did not normally tackle this dirty and fairly strenuous job.

What had been simple in the past now became complicated. On the GE section firemen had always hooked up and now there were no firemen. Nor were there surplus traffic department shunters in Liverpool Street station. The HQ Diagram Office said that it was not going to get involved with rostering secondmen at the terminals and that it was up to us: in other words, a national agreement had been made and the local lads, as ever, would make it work.

We started by using adult engine cleaners from Stratford, paid as shunters. But hooking up is not as easy as it looks and none of these men had had railway experience so they were no use as lookouts. Later on and not without reservations, we rostered firemen working under the direction of the Liverpool Street Running Foreman. These were standby men without any booked duties and they were an expense that, no doubt, had been written out under the agreement. These men met incoming trains which were single manned, unhooked, and rode with the locomotive to the station sidings, the depot or straight on to another train. Likewise they would pair up with a driver to take an engine on to a train and hook up. All very complicated, probably unforeseen, but time and experience have changed the agreement that, 25 years ago, we worked to the letter.

Anything that cuts at the established practice or at the employment of one's staff is bound to be critically received but we helped ourselves by anticipating the problems in order to find the solutions before the difficulties had arisen. So single manning came into being without either national or local trauma which pleased us, for we were deep in our human revolution.

At the same time, high voltage electrification at 25kV was on its way. The power had been switched on in 1959 in North-East London, and despite warning notices posted at depots and personal booklets of instructions, men had still hopped up, by force of habit without thinking, on to bunkers and tank tops to do their work, touching the wires with terrible results. This was heart-breaking. I had seen what the French had done to save lives by changing the habits and practices of a lifetime and had campaigned unceasingly for the introduction of warning notices and signs on locomotives and grilles on the cab roofs to prevent men going on to the tender.

The Clacton branch was the first to be electrified at 25kV in our part of the world. In August 1959 we were on holiday not far from Clacton and it was a gloriously hot summer. One evening Gwenda and I slipped down to the beach for a twilight swim – the North Sea warm and still as a millpond. When we returned, the phone was ringing: Bert Webster to tell me that Fireman Reg Rowe lay badly burned in Clacton Hospital. Next day I went to see him and I shall never forget his smiling face and his courage, nor his amazing recovery; but he was a lucky lad. He and Driver Bob Nichols had turned their 'B17' after coming down from London. It was fitted with a Great Eastern type of tender which could be scaled readily from the cab. They were coming out of the depot to pick up their homeward bound train when young Reg, without thinking, sprang up on to the coal to make some last minute adjustment and touched the overhead wires. A blinding flash, a bang, he was knocked unconscious but mercifully not to the ground: he was young and wiry and he lived.

Next month it was the turn of poor Hugh Reed of Colchester. This most conscientious of men had done much for his comrades as a union representative, and as an Improvement Class Instructor. He was also the maker of the miniature gnomes which adorned the Colchester offices and messrooms, and he died on 29 September 1959.

He and his fireman were taking an 'N7' Tank engine light engine to Walton-on-Naze. They had helped each other prepare the locomotive at Colchester and had dropped down towards Hythe when they were stopped at a signal. The fireman left the footplate to telephone the signalman and could neither see nor hear what Hugh was doing. He had climbed on to the right-hand side tank to clean his driver's side cab window. He completed the job, straightened up and touched the wires. This was an absolute tragedy: my views were forcibly expressed yet again and this time reached the right quarters.

In August 1960 the Diesel depot that altered the whole working environment at Stratford was opened. It was a great day, the culmination of months of endeavour and planning by management and staff alike. We had turned our backs on the filthy and primitive conditions of the past and we achieved standards of cleanliness that were the equal of those that I had seen in Switzerland and which I was determined we should emulate. Gradually, our locomotive allocation was increased; artisans were transferred to diesel work as the rundown of steam gathered momentum, leaving sufficient for the needs of the old locomotives which were based in the remaining half of the Jubilee Shed alongside the new depot.

But there were still steam locomotives at Stratford on 15 February 1962 when the Queen visited first the Works and then our place during a tour of the Great Eastern and LT&S lines. Thanks to the selflessness of my superiors, it fell to me to take Her Majesty on a tour of the new depot and it was an experience that I shall never forget. It was up to me for those 24 minutes and I was oblivious to those behind me who included Doctor Beeching and members of the Eastern Region Board.

Stratford ran true to form from start to finish. Days beforehand people were resignedly saying that the Queen ought to have something better to do than visit this dump but the scene gradually changed. Our depot, which had been clean, without apparent effort became a palace, the floors almost dangerously polished, the locomotives shining, the staff expectant, the whole ensemble a triumph to Norman Micklethwaite, the Shedmaster and to all those who had done so much to recreate a new and better Stratford.

A royal visit must be organised to perfection. None of us had had the experience but we had done our share of organisation in other fields. Everybody must know exactly what he has to do and he can then cope with the unexpected. I still have copies of the written instructions and here is the letter that I wrote to Norman on 9 February.

N. Micklethwaite, Esq
Depot Master,
STRATFORD

Dear Mr Micklethwaite,

I attach a copy of the detailed arrangements for the Queen's visit to Stratford and we will go through this book step by step.

I shall visit Stratford on the 14th February and I would like you to arrange for a brief meeting to be held in your office at 2pm at which the following people will be present:-

Running Foreman Smith
Electrical Foreman Lawson
Electrician Oddy
Clerk Lubbock
Driver Wilshaw
Driver Munday
Driver Searle
Passed Fireman Burrows
Driver Turpin
Plant Foreman Nicholson
Carriage & Wagon Examiner Oxford

I shall not keep these people for long, but I just want to be quite sure they know exactly what to do.

Will you please arrange for Mechanical Foremen Gill, Groom and Chapman to report to your office at the normal changeover time between the day and afternoon shift on the 14th February. I presume this will be 3pm.

Will you also arrange for Mr H. King to be present and for Mr D. Barrett to be present at this meeting together with any other supervisor that you think should come along.

Further points to note:-

1. Driver Turpin should be covered by another driver on the 15th February, until after the Queen has left Temple Mills when Turpin will take up his normal duties. It would be advisable for this driver not to be a man in No 1 link, but at the same time to be a competent hump shunting driver, well below Turpin's seniority.

2. Ensure that the locomotives working on the hump of this particular day are in good order and radio fitted. It may well be possible for these locomotives to be cleaned at the weekend.

3. Ensure that there are one or two steam locomotives outside the Jubilee Shed, but these must not make smoke.

4. Ensure that there are one or two diesel locomotives outside the front end of the Diesel Depot. The engines must not be running, but they will be started up once the Queen is in the train and ready for departure.

5. Ensure that there is a rope barrier between the two roads, in accordance with Mr Wiliams wishes to No 3 Shed

6. Ensure that a competent driver instructor is put in charge of D5694, and he will act in accordance with the instructions given to him at the meeting on the 14th February. I will leave you to arrange for the attendance of this individual at the second meeting.

You will, of course, arrange for a Diesel Fitter and a Carriage & Wagon Fitter to travel on the train and Mr Noden will also be in attendance.

I am preparing a roneod form which I will hand to each member of the staff to be presented, giving him exact instructions as to where he is to be and at what time.

Yours sincerely,
R. H. N. Hardy

In Paragraph 6 there is a reference to the selection of a Driver Instructor and the choice fell to George Marler, most conscientious of men, always with a smile.

The day arrived and I found myself being presented to Her Majesty. Away we went, past the big English Electric 2,000hp locomotive, the 'B1', and the little old Westinghouse goods engine paraded nearby and lined with spectators rather after the style of Promontory Point. Up the steps, through No 1 Shed and pausing for a word with two crippled ex-drivers in wheelchairs. It was marvellous and it is difficult to describe one's feelings beyond the fact that one felt respectfully uninhibited, for the Queen was showing great interest in everything and everybody.

We paused to look at the actual component that had caused a Type 2 Brush diesel to fail a year previously delaying the train on which the Queen was travelling and then moved on to D5694. I followed Her Majesty into the cab: she sat down in the driver's seat which gave the press the opportunity to take a photograph that was not only published here but on the front page of the French national dailies.

But the photograph gave no hint of my state of mind at that juncture.

My instructions were that the Queen might like to start up the engine of D5694 but that I must not put this to her direct. With hindsight and more experience, I could have done this several ways but my plan was for George Marler, who was with the locomotive but not on the presentation list, to put the controller in the 'engine run' position some time beforehand so that when we arrived a red light would be shining brightly on the desk in front of the Queen. I would then explain that if the start button was pressed, the light would go out and, with a bit of luck, the engine would burst into life. So simple and yet I was too clever by half. When I entered the cab there was no red light, no key and everything was shut down. Knowing my Stratford, I took this as a sign that something was wrong so we talked about everything except red lights and starter buttons and, perhaps mercifully, we did not shatter the peace with exhaust roar. However, as we left the cab, we were confronted by Driver George Marler standing in the engine room doorway. He was not supposed to be there but I certainly couldn't say, 'What the hell are you doing here, George?' and so I presented him to Her Majesty. He was all smiles and, standing as he was by the electric cooker, he gave her the old story of bacon frying on the shovel and now we have a modern cooker, ma'am! This improvisation temporarily shortened my life but, on reflection, it was infinitely more enjoyable and more natural for the Queen to meet her people. My nerve soon returned and we walked on, exactly to schedule, towards the six people that we had selected for presentation. I had arranged that I would come to the rescue if their conversation with the Queen had wilted but I did not then understand the extent to which she had done her homework nor did I realise that the pen portraits of the six that I had written would have been so carefully noted.

Charlie Smith, the senior Running Foreman, stood first in the line. He had started as an apprentice fitter in 1921 and he had come from a railway family. His father had been royal train driver at Stratford in charge of one of the royal 'Clauds', 'D16s' No 8783 or No 8787 before they became Cambridge property about 1930. The Queen started by asking Charlie what were his responsibilities to which he replied, 'Provision of power, ma'am'. Absorbing this as if she had spent her life at Stratford the conversation blossomed, bursting into flower when she said: 'Mr Smith, do you come from a railway family'. 'Indeed, I do, ma'am, a real Great Eastern family, my father was royal train driver at Stratford' nothing could stop the normally reticent Running Foreman – 'and what's more, ma'am, he was the driver the first time you went on a train as a baby and it was in the papers and I had the cutting for years.'

Marvellous stuff: I could afford to feel completely at ease, guardedly watching the time, moving on to the next person and finishing with Charlie Munday, the senior diesel instructor, who had driven a 'Britannia', No 70003 *John Bunyan*, and told her what he could do with the old and with the new. She was laughing and smiling delightedly when the cameras caught us again. And so, out of the side door exactly on time, back into the General Manager's Saloon, and away she went to cheers, diesel horns, and above all, the shrill, clear Great Eastern whistle of the Westinghouse goods engine. It was very moving and the memory of that day will never be lost nor shall I ever forget the eight years in that remarkable District which enabled me to preside over the end of one era and the beginning of the next. Eight years in the life of a young man in the same District. Some may say it was too long, that I ought to have widened my experience elsewhere, at BR HQ, or at Regional Headquarters, but then I could not have played my part at so critical a time and I could not have been myself other than in the thick of the rough and tumble of railway work and all for which it stands.

It did not sadden me to see the end of steam on the Great Eastern: I had a job to do to involve myself in the recreation of a new railway world and that meant getting the best from the old whilst we made the best possible start with the new technology. I was proud that we achieved both these aims: the vast army of men had contributed far more than they realised: only in retrospect does one realise what it meant to us all to have taken part in the human revolution of 1959-62.

'Woking'

I was fortunate enough to be selected to attend the British Transport Staff College at Hook Heath, Woking in August 1962. This was to be one of the great experiences of my life from which I was able to draw inspiration for years afterwards and it consisted of 16 weeks of very onerous study of far broader managerial fields than those to which I was accustomed. We were visited by many excellent speakers, mostly in the transport business but with a leavening from the outside world, and the tight timetable and inevitable late hours, spread over 16 weeks, forged friendships that were to last for many years. We were a happy course and our relaxation whether in the College or in the grounds was complete. The oldest and the youngest member of the course swam in the pool every morning. In December a very hard spell of weather was a warning of worse to come. Our two intrepid comrades broke the ice, had their swim, and joined the rest of us for breakfast none the worse for their immersion.

The College was splendidly run by Major-General Bill Williams and his staff: based on Service principles, and none the worse for that. There were few more deceptive men than the Principal. He was gruff, abrupt, slightly slurred of speech and did not court conversation. He knew how to drink and he knew how to delegate and he knew how to rise to the occasion.

Some people came through Woking without the establishment of a bond between themselves and the Principal. Some tried too hard to do so, some were able to reach the understanding of which he was always capable. I realised after some weeks that I was in the presence of a kindly, considerate man with a great ability to administrate, to lead and above all, to appraise.

There is no doubt that he had much to do with my move into General Management. Not so long ago I was fortunate to see the confidential report he had written in 1962 on my performance, personality, strength and weakness. He knew his man, he saw down the years and it was no fault of his that I ultimately failed to achieve the position in General Management of which he felt that I was capable.

As a course, we went to France for a few days to study rail and road transport. We were royally entertained by the SNCF and RATP and I arranged with Philippe Leroy, the Locomotive Superintendent of the Nord Region, for course members to travel on the locomotives between Paris and Calais. We had a very rough Channel and when I had finished washing and had changed, I was overcome by the waves.

Somewhere about Tonbridge on the 'Golden Arrow', I made my way shakily to the Trianon bar, to find the General and his faithful aide Claude Lincoln propping up the bar. Bill fixed me with a basilisk eye and said a trifle thickly and apropos of nothing in particular, 'We're going to have you out of the "Bloody Loco" one of these days'.

X

Little did I realise that time was not far hence. When I was enjoying the halcyon years ahead, too wonderful perhaps for my own good, I would reflect with infinite gratitude on the man who had set me on a new and wider path, and whose brevity and occasionally convivial habits masked a first class intellect and great sagacity.

XI
Fresh Ground

MY MAJOR TASK at Liverpool Street was all but completed when I left for the Staff College in August 1962. I had decided that I would not interfere, during the four-month course, with those that I had left in charge of affairs as they knew exactly what was required of them and what was to be our policy in 1963.

Bert Webster was still with me but Jim Woolvett, predominantly an electrical engineer, had joined us in 1960 to handle the takeover of Carriage & Wagon, Plant and Road Motor Maintenance. He was in fact senior in grade to Bert and consequently took my place while I was at Woking.

Jim and Bert finished the job at which we had worked so hard, steam vanished quietly from the scene on a Saturday evening in November 1962, and I returned from Woking two weeks before Christmas. After 16 weeks of intense activity in a College environment, I found it difficult to settle down at once. Certain changes had been made in my absence, a new spirit of interference was abroad and I suddenly realised, despite the task that lay ahead, that I had done my job, that henceforward there would be changes in the style of management and that my method of leadership and inspiration might no longer fit this particular bill. Although I did not know it, as a result of my performance at the Staff College, moves were already afoot to move me into Divisional General Management but I was to take my first step towards this goal in a totally different direction.

At the beginning of 1963 I was asked if I would act as Locomotive Engineer to the Eastern Region, under the direction of the recently appointed CM&EE, Colin Scutt whom I knew very well. Although I had misgivings about my technical ability, I accepted without hesitation. In the first place one was schooled to accept an opportunity whether or not one liked the idea, and secondly here was the chance to move on temporarily although I felt that I was unlikely to return to Liverpool Street.

Early in January 1963, with an outward appearance of confidence, I set foot in Royal London House, Finsbury Square, and except for short spells 15 years earlier, it was the first time that I had worked at Regional level.

I followed Cecil Frederick Rose who was now concentrating his great technical ability on the problems of steam heating boilers and I was to follow him again when I went to BR HQ late in 1973. He was a very meticulous individual who had been one of the first LNER-sponsored university graduates – with a grant of £10 a year – and after a spell under L. P. Parker, had concentrated on Works Management. His methods were completely sound, not only from the technical but from the administrative viewpoint and I learned, to my advantage, to build on any framework constructed by Eric Rose.

Royal London House was quiet and thoughtful.

Gone was the rough and tumble to which I had been used, the everyday contact with men of all sorts, the constant battle with the timetable, the joys of District Management and the pleasure of joint achievement with Operating and Commercial colleagues. The change was like a douche of cold water but whereas in 1973 I was to experience the same feelings with no joy whatsoever, in January 1963 I was on the crest of the wave and any dismay I may have felt was relative indeed. Not even the tea, straight from the urn into a teapot, could put me off.

The CM&EE HQ of the Eastern Region had a small but important London office and a considerably larger Doncaster office. My opposite number, so to speak, in Doncaster was Horace Rowley, experienced in technical and production matters and we were lucky in that we complemented each other. We worked closely together and effectively, remembering the problems of communications between two head offices where one would do.

Many of the problems with which I had to deal were technical but, as the CM&EEs had only just been relieved of responsibility for running the main works, the liaison between departments was extremely close. By no means were those involved always in agreement, but there were effective means available to solve the problems that were arising constantly in the field of diesel traction.

It must be remembered that in 1963 the manufacturers and their service engineers were still deeply involved in the maintenance, modification and improvement of diesel locomotives in service on the Regions. Nor was the technical arm at BR HQ the power in the land that it was shortly to become. On the other hand, the GN and GE Line Management, to be phased out in 1964, had a strong technical responsibility, led in both cases by highly experienced and senior motive power engineers. By no means always was there accord between Regional and Line engineers.

However, we were not hampered by bureaucracy and I found that, whatever my own lack of deep technical knowledge, there were great opportunities to get things done quickly and effectively even though there were factions at Line and Divisional level that were capable of the strongest opposition to the wishes of the Regional CM&EE.

After a couple of weeks I found that my past experience was the cornerstone of any success I might have and was invaluable in getting matters moving in the technical field. One of many duties in a position that also became administratively demanding was to hold meetings with the manufacturers responsible for design and construction of the various types of diesel main line locomotives with which we were blessed.

Every third month there would be a 'Deltic' Performance Meeting. Present would be representatives from English Electric, Napiers, the Main Works, the GN Line Headquarters, and so on, some 20 engineers. The meeting was usually as controversial as were the marvellous locomotives which were the subject of debate. As Chairman I had to control a bunch of engineering specialists, and to get them to come to the point required patience, fortitude and an ability, not always given to me, to keep one's temper. It was priceless experience for me, less gifted from an engineering point of view than anybody present, to learn how to stem the flow of engineering rhetoric at the appropriate time and convert it into decision and ultimately into action. It was an education: the only preparation that I could give myself was the unorthodox visit to Finsbury Park Depot to steep myself in the practicalities of the 'Deltic' locomotive in a language that I understood to enable me to take the ultimate decisions and convert them into action.

My first meeting had been with the Brush/Mirrlees people. The 'Toffee Apples' and 1,365hp locomotives were beautifully built and relatively free of trouble. The bedplate fractures had not started to show themselves, a defect that ultimately led to replacement of the Mirrlees engine by English Electric power units. The close liaison between the three firms made the meeting a delight. Remarkably and, quite frankly, delightfully, we had nobody at these meetings from BR HQ and during the 11 months that I acted as Locomotive Engineer, I made only one visit to Marylebone. But the writing was on the wall: shortly and probably correctly, the meetings with the manufacturers began to be held under HQ chairmanship as diesel locomotives spread ever more widely across British Railways.

The English Electric men were great people with whom to work: as were, of course, their locomotives. Of 1,000, 1,750 and 2,000hp, they were well tried, rugged, splendid machines. English Electric had forgotten more than we knew in those early days and yet the company was sensitive to our requirements and realised that there was much it, the manufacturer, could learn through its own service engineers and from steam men who were learning so fast.

The Type 3s, now Class 37s, were doing splendid work on the Great Eastern line. This state of affairs had been achieved in a relatively short time and the meetings with English Electric to discuss the performance of Classes 37 and 40 were straightforward. However, the firebells of the former locomotives had a tendency to ring, a false alarm caused by what I suppose was a hot spot above the V of the diesel engine. When a second man was present, a quick excursion into the engine-room would prove that all was well, but when single manned, the driver would be obliged to stop immediately to examine the engine-room or to deal with the bell by some illegal means.

This state of affairs could not continue. The old Stratford would have removed the fire detector above the cylinders but this was the new Stratford genuinely concerned with delays, failures and the causes thereof. At the meeting, English Electric talked of trials and experiments which could delay a solution. I had to rule that the problem be immediately tackled and a solution found. We could neither tolerate delays nor firebells stuffed with paper. I think the matter was initially dealt with by English Electric in the old fashioned manner but that the detector was eventually moved to a place where it could do its work in peace.

It is a tragedy that I did not take notes of what was happening during these 11 months of great interest. Whereas my memory of people and incidents involving people is passing fair, I could only remember technical matters for as long as they were useful to me. Once we had dealt with a problem and had brought it to a successful conclusion, I dismissed it from my mind unless it re-occurred.

Although I should not have cared to remain Locomotive Engineer indefinitely – and I did not know what was to be the extent of my stay – I must say that between us all we achieved considerable progress in modifying and improving most of the classes of diesel locomotive on the Region. Doncaster Works did us remarkably well and it was interesting, once again, to set foot in the old Plant Works and to see it changing as rapidly as we had done. We had our contact, the Diesel Assistant, a certain Ron Wilburn. He had been a Doncaster craft apprentice and had come up the very hard way, as foreman, superintendent and so on. This man moved mountains for us. He died years ago but I wish he could have read these words. He and Les Thorn at Stratford Depot, both out and out steam, shop floor men, were the two who stood out amongst our own men who gave so much to the new technology. Such men understood the practicalities of railway life and railway work whilst at the same time quite dedicated to the new form of traction. Over the years it became fashionable amongst some railway managers to denigrate the men of steam. How little did these folk understand, they who had not had their whole working environment, their whole working organisation completely revolutionised in four or five short years. They had little or no conception of what we had faced and achieved.

During 1963 we were landed with the first of the Brush Sulzer Type 4 locomotives, Nos D1500-D1519. Before I left for Lincoln in November I had seen more than enough of these machines and was in a position to regret the BR HQ decision to concentrate on this Class to the exclusion of DP2. Wherever DP2 was sent it was respected: simple, straightforward, rugged, powerful, just what was wanted to back up the 'Deltics'. It was reliable and it was fast and had it become the standard Type 4 locomotive in the higher power ranges, we should have been on a winner. My readers do not need me to enlarge on the subsequent history of Types 47 and 50.

Whilst my weekly visits to Doncaster were challenging, I found that I needed to keep in touch with what was happening at the depots and, furthermore, I needed fresh air and exercise, however demanding the administrative paperwork. One afternoon, things were going well and I suggested to a colleague that we might take a trip to Colchester and back in the cab of diesel locomotives as I had not travelled on a GE footplate since the introduction of the AWS early in 1963. We crossed the tracks from No 11 to No 9 at Liverpool Street and, walking towards the Norwich men sitting on their locomotive, I saw the driver get out of his seat to pull down the AWS isolating handle. I noted this but thought no more about it until we had started our journey. Right out to Gidea Park, signals are very close indeed and the AWS bell was nothing less than an evil menace. One felt one was locked in a belfrey from which there was no escape for at that time the bell would sound for two seconds at each signal as it had done on steam locomotives largely employed outside colour light signalling areas. On the way back, we joined Stratford Driver Bob Nichols who had isolated his AWS saying quite bluntly that the noise was out of all reason. I had to insist that the AWS was put to use and was glad to get off the locomotive when we were back in London.

Nevertheless, we faced an obviously dangerous situation that needed the quickest possible solution. The way to do this was to minimise correspondence and maximise action by getting a senior S&T colleague to share the torture of threading one's way through Stratford with three signals at least within the station limits. In a very short time, the length of the ring had been drastically reduced to its present reasonable level although, even now, I regard the single 'ting' of a bell on the SNCF to be perfectly adequate to signify that all is clear.

Towards the end of September I was asked if I was prepared to go to Lincoln as Acting Traffic Manager for about seven months. Although this meant being away from home for another lengthy period, one did not turn down such an opportunity and again I was about to face a task which thrilled me and yet caused me inward misgiving.

In the first place, I was to follow Harry Graham who was retiring in October and who was a legend in Lincolnshire. He was a Commercial man with extensive Operating experience; he had been Traffic Manager since the end of 1957 and District Commercial Manager before that, whilst he knew what questions to ask of his engineers. He knew the business world inside out, and what I knew in this particular field would comfortably have gone on the back of a postage stamp. Furthermore, he was just and fair, determined and a man of great charm.

Secondly; the days of the Lincoln District were numbered and it had to be merged to form the new Division, with headquarters at Doncaster which meant, of course, the closure of the District offices at Lincoln where the staff believed in their considerable ability and regarded the merger as a very retrograde step. The staff also felt that the departure of Harry Graham marked the end of an era. Furthermore, Doctor Beeching's policies had begun to bite deep into the Lincolnshire railway scene and apart from almost inevitable closures, freight traffic was under scrutiny and the drive to concentrate coal at central points for distribution was at its height.

The General Manager of the Eastern Region appointed me to Lincoln for three reasons: firstly to gain experience in Traffic management which would ultimately enable me to compete for the position of a Divisional Manager when the rationalisation of the Traffic Managers' Areas had been completed in June 1964 on the Eastern Region. Secondly, I was required at Lincoln to take the lead, to raise morale and to keep people believing in what they were doing. And thirdly, I was to preside over the closure of the railways of East Lincolnshire, the dissolution of the fish traffic and the decimation of a proud organisation. Nevertheless, I was determined to hand over a going concern to Fred Wright, the Divisional Manager designate at Doncaster.

Some of this was to be achieved before I left in May 1964; the consultation meetings with the staff representatives had been held in March 1964, without the public lynching of R. H. N. Hardy, and I think it can be said that the Lincoln Traffic Manager's Area believed in itself and did a good job right up to the end.

XII

Lincoln a New World

WHEN I ARRIVED at Lincoln I received a warm and friendly reception. I found Harry Amos taking over Operating matters, and Ken Taylor, an old comrade in arms, in charge of Running and Maintenance, assisted by none other than the erstwhile 'Brown Bomber' of Stratford, Norman Micklethwaite. Commercial matters were in the hands of the capable Derek Burton, a young man of great energy with considerable financial experience. These were my senior colleagues and to help them with the difficult days ahead was a certain Jack Hyman, Staff Officer, on whose shoulders would rest the detail of great personnel changes with which we were to be faced. Jack Hyman came from Peterborough but he had the wit and repartee of London, where he had worked for some time. He was totally reliable, dedicated to what we had to achieve but practical, humane and humorous in his approach to staff problems.

And then, of course, the two people without whom a Traffic Manager cannot survive, a good secretary and a good chauffeur. Sheila Hazard perhaps did more than anybody to give me the confidence to tackle the commercial problems of which I had had so little experience: she was cheerful, competent and dedicated, as was my personal chauffeur, Bill Boothright. One had many miles to cover by car, not only to reach places but to get home late at night from evening functions so that a driver was essential if one was to fulfil one's many social and business obligations. Bill had been reared on rugged LNER lorries and had come back from Dunkirk in 1940. Nevertheless, my attempt to educate him in Motive Power matters was a failure. He used to do the catering on our inspection coach and one day I decided he should have a spell on the locomotive with me during the afternoon. We went forward somewhere on the East Lincs main line and our steed was a 'B1' which had seen very much better days. We were soon up to 60mph or more, my poor chauffeur very much the worse for wear holding on with whitened knuckles as the old 'Bongo' knocked and banged along – 'Never again' said Bill, 'I'll stick to the road in future'.

It is not the slightest use sitting in solitary state in one's office when one has to start to learn all over again. And so I threw myself into every possible activity both during the day and in the evenings. I lived at the 'Adam & Eve' Inn on Lindum Hill where mine hosts, the Colemans, looked after me so well. The inn was old, the clientele friendly and it was 'uphill', which carried a certain significance in the City of Lincoln.

Naturally I went home at weekends wherever possible but during the week I had my evenings free for railway work, railway business and the social side of business life which I greatly enjoyed. There was much to do even at weekends, but Gwenda and the family were the relaxation that one needed to come to terms with the exciting challenges of the week.

By 1963, I had built up a very solid base of railway knowledge from which to branch out into wider fields. On the other hand, the very fact that my knowledge of and understanding of locomotive men in all grades was considerable, highlighted my ignorance from the practical point of view, of guards, signalmen, goods checkers, motor drivers, booking clerks, station masters and many other grades. This worried me because I knew that I could not talk from experience and I knew that, amongst those with whom I grew up, little respect was shown to those without the great bond of practical understanding.

This was how I felt in my first few days as Traffic Manager in Lincoln but I was soon to realise that the previous 23 years had trained me as a practical *railwayman* which enabled me quickly to grasp some if not all of the problems that faced the operating and commercial sides of the railway business.

A very old Great Northern Inspection Saloon was at my service and in it, during the next seven months, I was to visit every corner of the Lincoln District, from Spalding and Sutton Bridge in the south to Grimsby and Cleethorpes in the north; to Colwick, Stathern and Waltham-in-the-Wold in the west, deep in the Leicestershire hunting country; and to the fascinating station of Burgh-le-Marsh, to Sleaford the station embellished with paintings of the railway scene, the work of Mr Stationmaster Rook – and to the seemingly frozen flats of Immingham, New Holland and Barton-on-Humber. Ah, Barton-on-Humber, never shall I forget my first visit to that little town on a cold and windy December night, barely a fortnight after my arrival in Lincoln. Derek Burton was and is a first class railwayman but he would perhaps admit that, in 1963, he was inexpert in the mundane task of keeping his diary straight. Derek and I were on our way by car to meet the coal traders of Spalding (to me, a terrifying ordeal) when he pleasantly announced that he was meeting the coal traders of Boston on the same evening that he was billed to debate the closure of the railways of East Lincolnshire with an NUR Organiser, in public, at the roaring metropolis of Barton.

With boyish charm, he asked me if I could help him over a difficult problem and, like a fool, I found myself opting for Barton-on-Humber when I should have told him to find his own way out of the maze. This was on a Thursday, and the debate was to be held on the following Monday. Once back in Lincoln, I set to work to brief myself on a subject about which I knew nothing.

On the day of the debate Bill Boothright drove me through rain and sleet, to arrive at Barton in what seemed total darkness. My spirits had sunk to an all-time low but my imagination was operating on full regulator.

I foresaw a hall, packed to bursting point, full of the citizens of Barton and all points east and southeast, baying for the blood of the 'cretin' who was proposing to close their railways. I could hear the jeers of derision as I made my faltering case and round after round of applause would greet the invective of the NUR orator. I longed for the assurance of a Beeching, the charisma of a Fiennes, the profound experience of a Johnson and, just at this moment, I had none of these things. But, on the other hand, I had asked Albert Bostock, Grimsby Sales Assistant, to come along as part of the audience to help me out on detail and for this and for many other things I owe a debt to Albert.

However, it was 18.00, another 1+ hours to go, when I arrived at the home of the chairman. The welcome I received steadied my nerves, and during our pre-dinner conversation my railway history came up for examination. When my host learned that I had worked with Bulleid's Pacific, a certain look, well known to me, came into his eyes and I knew I was on a winner.

A bargain was quickly struck: in exchange for a sprint from Waterloo to Bournemouth and back, I should be ensured a reasonable ride later in the evening. Whisky, wine, good food and delightful company began to take effect and when we arrived at the hall I was aglow and felt that I could take on the world.

The world proved to be not 300 infuriated citizens of Lincolnshire but maybe 20 very pleasant and enthusiastic folk, most of whom were railwaymen and who included some of the guard's representatives from Grimsby. The NUR representative had no more wish for a public debate than I had. So we each said our piece, and sat informally on the table answering questions, assisted nobly, in my case, by Albert Bostock. It was a thoroughly enjoyable railway meeting in which the audience participated civilly if effectively. And that was that. My first baptism of fire was over and never again, however severe the opposition, was I to feel so lost and forlorn although I had still to learn many lessons on the importance of preparation and the dangers of extemporising in public.

By March 1983 we were ready for the consultation meetings with our staff on the closures of the East Lincolnshire railways. The Sectional Councillors were to be involved; these were men who still held their substantive grade but who spent much of their time on what could be termed trade union business at Regional level. Whilst I knew the No 2 Motive Power Council extremely well, I had had no experience of Nos 1, 4 and 5 which represented the clerical, terminals and cartage, and PW&S&T grades respectively. No 3 Council covered operating matters and I had met it previously on many occasions. Thus representatives of all five Sectional Councils were to be present, to take the lead and, above all, to advise those who were to have their lives and employment uprooted by the closures. And they had the power and authority to make or mar a meeting.

We had heard through our staff representatives of awful gatherings on the London Midland Region where the union representatives, as a body, had walked out of the meetings which had then to be reconvened at a later date. I had heard of inadequately briefed chairmen who had lost control and I was determined that this should not happen in Lincoln. We had hired a very large hall, had taken much trouble over loudspeakers and microphones and, well in advance, we had started to do our homework as a management team, also involving other departments such as the Civil and Signal & Telecommunications Engineers based on Doncaster. We asked ourselves every question that we felt we were likely to have thrown at us and whilst we knew that we could not give answers that would necessarily give satisfaction, we aimed at simplicity and practicability together with the avoidance of that gratuitous information, that word too much, that can so often start a quite unnecessary conflagration.

We decided to allow two whole days to the staff to work on the documents and make their case. To cut down this time would have been a false economy. Whilst the various representatives were getting ready for the fray, Jack Hyman made himself available to help the Sectional Councillors over any difficult points. But while he was doing this, he was also learning what were going to be the main points of contention, a direct commonsense approach to a difficult problem appreciated and accepted by all.

6 March arrived and with some misgivings I watched the hall fill up to the tune of nearly 380 staff representatives, such was the magnitude of the closure: another 230 would be coming a week later and one wondered how on earth the railway was running without so many of its staff. Should I be able to control so big and understandably hostile a meeting; should we achieve the results that Line Management demanded of us and, in our hearts, we felt that we should do?

The Management representatives sat at the top of the hall; on the left side were ranged at least 10 Sectional Councillors whilst the local representatives themselves sat in the body of the hall. 6 March was a Friday. We started punctually at 10.00, broke for lunch more dead than alive and took stock of the situation. We had made slow and painstaking progress and there seemed no likelihood of finishing the meeting in one day.

The staff side Chairman was quietly brilliant. He missed no points, no opportunity to criticise our closure intentions but he was scrupulously fair and civil. He controlled his people with a calm strength which commanded respect. There had been one minor conflagration when the magic words 'redundant motor drivers' had been mentioned. This had caused the apparently sleeping Chairman of No 4 Council, Bert Goldfinch, to rise to his feet and in slurred but forcible terms to demand justice. I was later to know and respect this King's Cross motor driver whose eyes one had to learn to read. On this occasion the eyes were twinkling and the storm soon passed.

Lunch improved matters and there was a noticeable quickening of the pace. At 15.30 the trot became a canter and at 15.50, we were summing up. The proposals had been noted with grave misgivings but the staff side would co-operate to the full to make sure that, if closure went ahead, they could achieve the best possible deal for those they represented. Much water was to flow under the bridge before the closures were to take place but the ground had been fairly prepared.

However, there is something more to be said: it was a Friday, there was a train to Peterborough and London at 16.20 from Lincoln (change at Grantham), and the Sectional Councillors wanted to get home. They had done their work well, they saw not the slightest point in keeping a meeting going to air grievances that could well have been irrelevant, the main issues had been covered and that was that. But then, Jack Hyman was a wise old fox to pick a Friday for our meetings!

In 1956 diesel railcars had come to Lincoln and a new maintenance depot created where once had stood the Great Central engine shed. A decision was taken to employ a certain Harold Smith who had experience with maintenance of the Lincolnshire bus fleet. Harold became an institution at Lincoln. He not only brought with him a wealth of experience but he became a railwayman very quickly indeed and he gave everything he knew to his fleet of DMUs.

They had their troubles and I particularly remember the 901 Leyland Albion engine. This had wet cylinder liners which were gradually forced upwards until they caused the head gaskets to blow, giving us endless trouble. Although I was acting Traffic Manager, Lincoln, I was also Locomotive Engineer for the Eastern Region and I dealt with DMU engines as well as locomotives. Harold Smith had his solution to the problem but had not had authority to make a modification to one of the Leyland engines. Cutting far more corners than I had a right to do, I authorised, as Locomotive Engineer, the proposals that I was putting forward as Traffic Manager, Lincoln. Harold made his modification which was a resounding success, Leyland co-operated to the full, and the revised engine became the 903 and runs to this day, trouble free and reliable. Oh, that it were always possible to make improvements so quickly, based on the experience of practical men who really know the answers.

Harry Amos, who was acting as District Operating Superintendent, travelled with me when we were out with the Special. We were constantly being pressed to simplify track layouts and remove redundant assets, words which covered just about anything that would not move! A new broom wielded in the name of Philip Shirley was sweeping through the dustier corners of British Railways and, although he had many critics, he talked commonsense on the need to simplify, to count the cost, and to get rid of what we no longer needed, quickly and with profit. When he joined the British Railways Board early in the Beeching era, he was critical of railwaymen and their methods but when he left a few years later, for Cunard, he had himself become a railwayman. One learned to use one's eyes once more and it is strange that I, who had been so vehement in my demands for spotless locomotives and tidy depots, should not immediately have taken the same line about yards, goods depots and stations, until I was suddenly convinced that what Shirley was saying was sound, practical, financial commonsense.

In April, we were visited by the Eastern Region Board. Our Chairman had invited Philip Shirley and the Operating Member F. C. Margetts. We blessed him not at all and I knew that we should be on our mettle and that another L. P. Parker was abroad. Yet again, a merciful providence was to give me the necessary nudge to be prepared, to go over the ground and so we did in our faithful Special hauled by 'B1' No 61406, taking notes, getting things moving and learning the answers to the inevitable question 'why?'. To achieve this, I asked questions the like of which would never have occurred to me four months before. Wherever one looked, something could be done: warehouses and goods sheds, sidings no longer used, turntables, loose lengths of rail, old carriages as shanties – including an 1874 Midland Pullman – snowploughs, coaling plants. There were two stations in Lincoln within a stone's throw of each other and there still are but Philip Shirley gave me the grilling I knew I should get and I came through it with some credit. In a way, it was sad to see the old railway going and yet there was challenge and excitement in the air for a relatively young railwayman who had already done a fair amount to change the scene elsewhere.

Towards the middle of May I was told to my surprise and joy that I was to be appointed Divisional Manager, King's Cross on 1 June 1964. The Traffic Manager, Stuart Ward, a fine railwayman and colleague, had been promoted to the Southern Region and I was to move at once to hold meetings that would introduce the new Divisional Management.

It can be said that I had done a fair job at Lincoln. Our team had achieved much and I had learned much that I had previously taken for granted nor even remotely understood. Between us, we ensured that the Lincoln District did its job right up to the day that it ceased to exist. I have many happy memories, some fading into the past, for it was a short, hectic and concentrated stay. The 'Whitland Fish', hauled by an Immingham Class 9, left Grimsby punctually and was given priority for many miles of its journey. We knew its days were numbered, that it was one of the last survivors of a great era and that before long there would be no more railborne fish from Grimsby. But we gave it the treatment as of yore.

And there was the 'New' line, with its seemingly enormous stations miles away from anywhere with glorious names such as Tumby Woodside and New

Bolingbroke; the latter was where I was received when on a tour of inspection by a fiery, white moustachioed little stationmaster. He wore his best uniform, saluted me smartly, showed me round his manor with dignity and sent me on my way. I had visited *his* station, he was *proud* of it and he was going to do the honours to an upstart of a Loco man who had dared to break with tradition. One could *feel* his pride and his contempt for the changing scene for we both knew that the day of the stationmaster was nearly done.

The Lincoln level crossings still hold back the impatiently fuming road traffic. One was constantly bombarded with complaints and suggestions and every conceivable solution must have been discussed with the City Council; curves, closures, overhead carriageways. There are still two stations and maybe one day Philip Shirley's directive to me 'to get that station closed' will bear fruit.

The annual Lincolnshire traders' lunch 'from the Humber to the Wash' was an event in the calendar but I was spared the ordeal of public speaking to a gathering about whose business I knew so little. In time experience taught me the value of thought and preparation. These virtues, together with sincerity and humour, are priceless assets to he who aspires to speak in public. Only the most gifted can really extemporise and I was never in that category for on one occasion, long afterwards, my performance fell very short of my requirements through that most dangerous of enemy, over-confidence.

Before leaving Lincoln, I survived two visits to a great Louth character, a Mr Drinkall, who had periodically and pungently blown me up on the telephone when BR had failed him his demands for bulk grain vans. His Dickensian office, his irascible charm, his enormous glasses of sherry, I shall never forget. On my last visit he accepted with glint-eyed pleasure a miniature Hornby Dublo Bulk Grain Van – 'just in case you are ever short'.

And so my short stay came to an end but not before I had paid several visits to Apricot Hall, deep in Sutton-cum-Beckingham, home of Alan and Sheelagh Parker. Alan was the Passenger Officer at Lincoln, a phonograph expert, light hearted and humorous, who once told the Bench that he was 'a bad parker' after leaving his car on double yellow lines. Sheelagh was a noted horsewoman who inspired me to go deeper into the pleasures of riding and hunting. It was my pleasure to take Alan on the footplate in France and to send him with his highly amusing friend, Brian Elliott, to travel on Pacifics, on the 230D's and the other wonderful French locomotives that are no longer with us.

Divisional Management

And so to King's Cross in June 1964. I was a very lucky man, the first engineer to be appointed on the Eastern Region to what was to become, in July 1964, a Division: there were one or two outside of the Division of whose opposition I became aware but when one's star is in the ascendancy, one does not worry over much about that sort of thing. A long narrow Division, not large by the standards of the day and by 1968, the number of men had dwindled to about 5,000. Our northern frontier was Barkston Junction until, by a species of moonlight takeover, we lost Grantham and the High Dyke ore to Doncaster and our boundary was moved to the south end of Stoke Tunnel. We covered to Royston, the Hertford Loop, Peterborough East and Whittlesea, the ex-LNWR lines to Kingscliffe and Oundle and the Midland line from Peterborough as far as Ketton.

By 1968, the GN main line had become a high class race track relatively untouched by freight traffic. The National Sundries and National Freight Train Plans had, by 1966, reduced the size and purpose of the Ferme Park yard with its ultimate closure, and the huge organisation known as King's Cross Goods had received a nasty jolt. Nevertheless, parcels traffic was concentrated at King's Cross Goods in 1967 and a steel terminal opened, to which early Liner trains were also routed.

Change was constant: change in organisation, change in traffic flows, reductions in staff, developments in budgetary control – a never ending catalogue of ways and means to reduce costs.

We were no Sheffield or Doncaster with an industrial train load background – coal, iron and steel, limestone and ore. We did not originate on a grand scale but nevertheless, if one could now examine our services for the earlier 1960s one would find a wide variety of passenger and freight being carried: a thriving main line passenger and sleeping car service, a rather gentlemanly inner and outer suburban service serving three London terminals, transfer freights to the Southern throughout the night from Ferme Park, and main line freight trains by night and to some extent by day.

In 1964 the fish trains from Aberdeen and the East Coast ports were still running, there were fish deliveries to Billingsgate by both BR and contractor cartage, whilst off the Aberdeen meat train, there would be 50 or 60 containers delivered into Smithfield Market as soon as possible after the arrival in King's Cross at midnight.

Let it not be forgotten that, although there was no longer a procession of slower freight trains, and that Ferme Park and New England were not long for this world, there was an evening procession of fast freight trains from King's Cross Goods at 20.10, 20.26, 20.36, 20.50 and 21.05 clearing both wagon load and sundries traffic to Newcastle, Hull, Leeds and Sheffield.

Electrification was inaccessibly over the hill, any extensive modernisation was being held back and we improved our performance and tried to develop our traffic against a background of antique equipment, semaphore signals, and elderly and in some cases decrepit stations. But nevertheless, the morale was good and people enjoyed life on the railway as they so often do when a fight is the order of the day, and they could give full rein to a grumbling triumph over adversity.

XIII

Great Northern Line

IT IS IMPOSSIBLE to write comprehensively or in remotely chronological order of my years at King's Cross. I kept no notes or diaries and so, as ever, I must fall back on my memory and my heart from which I speak from time to time. As one gets older, the memories become less vivid and those that remain tend to be of exceptional happenings. But the joy of ever increasing responsibility – and occasionally the burden – make life very real, fulfilling, exciting and now and again, quite lonely.

For those who expect political revelation, I have nothing to offer. I was far too busy doing and enjoying my job to concern myself deeply or personally with senior affairs of state and this may reveal a certain lack of ambition, the absence of an appearance of personal ruthlessness regarded as necessary for ultimate success. On the other hand, I felt very strongly that although I had become a Divisional Manager, I had much to learn, much to contribute and a great deal to do. It did not enter my head to use King's Cross as a stepladder to greater responsibility. I had my mind not on the next job but on a stint at the end of which I would have mastered my job, meeting the requirements of my General Manager; not only this but I intended through our Divisional team to build up a reputation for efficiency and fair dealing and with customers and staff alike. These ideas are difficult to achieve but, in seeking a solution, I was to find pleasure and challenge.

I was to spend close on four years at Great Northern House and we built on the rock-like foundations laid by my predecessors, Geoffrey Huskisson and Stuart Ward, who had created remarkable standards of integrity with achievement over five tough years, which had seen the virtual elimination of steam, some shattering diesel experiences and the closure of the Top Shed at King's Cross, an event regarded by some as the beginning of the end of the world.

Although I had not worked on the Great Northern section since I left Doncaster in 1945, I was to find several old colleagues and many people that I had previously met. Temporarily in command was Colin Morris and those who really knew him – and they were few – realised that they were honoured to do so.

He had been a District Officer before I was appointed to Woodford Halse, had been my chief when I was at Ipswich in 1950, and here was I, 14 years later and 16 years younger than he, appointed as Divisional Manager over his head. It is right to say that his friendship enriched my life and, in return I was able to share some of the burden that he carried as Divisional Running & Maintenance Engineer. He seemed reserved to those who did not know or understand him, and slow to those who wanted an immediate answer: he worried and sometimes took personal responsibility for error where there was no need, but he was a joy to work with and the most experienced of men. Above all, he had a wonderful sense of humour which enabled him to reach solutions and find the key to men's hearts. Seemingly solemn and unsmiling, his railway stories and his assessment of colleagues past and present were hilarious, not only in content but in the droll manner of their telling.

So there was Colin, my deputy, my friend and confidant. Our railway life was always closely linked but never more closely than early on a March morning in 1968. The previous evening a train of fly-ash had left Drakelow Power Station near Burton-on-Trent for the Fletton Brick Pits around which had been created a merry-go-round railway. The train was worked by a big LM Region Sulzer and was manned by a Burton driver and secondman. The driver did not know the eastern part of the route and picked up another London Midland driver, as pilotman, at Toton.

The train eventually reached Peterborough, threaded the station at the usual crawl and passed on to the Goods Road on the south end. The night was clear and permissive working was in operation, in other words a train could move slowly along but must be prepared to stop clear of the train ahead in the section.

The secondman was a Passed Fireman from Burton, one Aubrey Dolman. He pointed out a red light ahead but the driver said it was on the main line. Once more Aubrey had his doubts and said so but seconds later they ran into the back of another fully fitted fly-ash train; both drivers were killed instantly and Aubrey was trapped and unable to move.

The New England and Finsbury Park cranes were in position and the rescue men hard at it by the time I had joined Colin on the scene. All night we wrestled with the problem of freeing Aubrey Dolman. It was impossible to use burning tackle; a fly-ash wagon was poised above what was left of the cab – poised and yet entangled, a potential menace to the life of a trapped man if we were to allow it to move. By 04.00 two alternatives were left, the amputation of a limb and the end of Aubrey's career as an engineman or a risk to be taken, a grave risk, in attempting to lift the fly-ash wagon entangled with the roof of the cab.

Both cranes would be involved, a lift of a few inches under great stress, a sudden movement which could precipitate the death of the man inside.

There would be one man in charge, Colin and real practical railwaymen on the job. I knew the quality of the crane-driver, of the foreman, and of the breakdown gangs. All I had to do was to take the decision, which I did without hesitation. The risk was taken with great delicacy and precision, a way out was created and Aubrey Dolman borne away to hospital, a brave man who had smoked his pipe and talked to his rescuers and who had the courtesy to thank us all. His reward was to be on the footplate once more within a matter of months, and my reward was not only his freedom but to have been able to take a grave decision in the knowledge that I could stand back quietly and rely on the competence, skill and wisdom of the senior Running & Maintenance Engineer of the Eastern Region.

In 1964 Finsbury Park depot was responsible for the maintenance of a fleet of main line diesel locomotives and some shunting locomotives, but only for the servicing of the Craven DMU fleet, based on Cambridge, that staggered daily and vibrantly over the Northern Heights, pausing on warmer days for cooling systems to be replenished from conveniently placed cans of water. The depot was one of the first on the Eastern Region at or near which drivers did not sign on duty and were therefore divorced from their old allies and enemies, the maintenance staff, at a time when the closest contact was necessary.

By 1964 some changes were needed and John Butt arrived from CM&EE Headquarters to help to restore the standards that had begun to slip. Finsbury Park had had an excellent reputation and it was staffed by both ex-steam men displaced from King's Cross and by artisans from outside industry. This amalgam needed a firm hand. John Butt was respected for his relentless drive and he had inherited my passionate desire for cleanliness and order when he moved on to Stratford.

Bounds Green was a remarkable conglomerate of sheds, shacks and sidings, very difficult to manage; Ferme Park yards were still in use, and Hornsey existed as a servicing point for freight locomotives.

The suburban passenger services were mainly civilised but not particularly brisk but there was an interesting and well managed interchange between main line and Hertford line trains at Finsbury Park in the evening peak. The usual rake was a Type 31 and five or six coaches, but over stations, trains, signalling, and way and works stood the spectre of electrification, which meant that there was little or no money for modernisation, simplification or improvement. If one accepts the changes wrought by the National Plans, much that we did in the Movements field was by stealth and ingenuity and we could not but long for the day when electrification would finally be authorised.

Edwin Howell was the Divisional Movements Manager. He too had the great gift of humour and turn of phrase and finished his career in Electrification Planning, a field to which he was admirably suited by ability and temperament. When he was with us, he could read with clarity and accuracy the future scene beyond the appallingly small King's Cross circulating area, with its hoardings, bothies and carriageway, and beyond the vast conglomeration of small yards and sidings. He had the sometimes dubious responsibility of running the old railway but also the pleasure of leading the planning of the new, retiring when electrification was well on the way and his task complete.

The old power box at King's Cross was the heart of the station at the busiest times which are now but a memory to those involved. Christmas Eve was invariably hectic and the operation of that station an education to a locomotive man with much to learn. I would leave the offices after lunch, visit Finsbury Park and King's Cross Goods Depot and then make my way to the main line station, where the people in charge would be very much in evidence. Whilst it is inevitable that operation occupies the centre of the scene, one must not forget that the booking and enquiry offices were constantly under siege; the old booking office on the main line station was in a class of its own for antiquity and squalor.

Trains would be standing on Holloway Bank waiting to move down, when suddenly there would be a clearance and one would then be able to witness one of the miracles of the old station, as the first train of the queue came to a stand, its frustrated passengers forcing their way through the tiny circulating area already packed with the London exodus.

The 11-coach train would be given a whisk by an army of cleaners, and examined by the C&W staff, driven on by the Assistant Stationmaster and his supervisors. Ticket collectors would be at the ready, orchestrated and conducted by a certain George Gent (yet another of the unsung, scandalously lowly-graded men who have saved the railway thousands of hours and thousands of pounds, endlessly courteous and helpful to the harassed travelling public), the passengers would be released from their waiting purgatory, herded onto the platform, and bundled into the train, which in 15 minutes from its arrival would be on its way north once more, in the charge of a big diesel. Just part of the day's work but, in fact, a miracle performed by railwaymen in a hectic, cramped, crowded and pillared Victorian relic.

Sometimes it was a miracle that a locomotive arrived on its train at such times. The Locomotive Yard had become a busy diesel fuelling and servicing point. It has gone now, and so has the subterranean No 18 platform from which trains suddenly appeared from the bowels of the earth, bisecting station from depot. All movements to and from the Loco had to be carried through by the closest co-operation between the regulator in the power box and his signalmen on the one hand and the Running Foreman and his drivers on the other. When all was going well, they were as one.

And so, on Christmas Eve, one could survey the whole scene from the power box at the end of the main line platforms. Upstairs, a regulator, always an ex-signalman, whose responsibility was to run the show together with three highly skilled signalmen whose work it was a joy to watch: downstairs, the S&T technician ready at a moment's notice to deal with faults, and failures. Although I knew perfectly well that it would have taken me time to master the work of such signalmen, I knew equally well that they and I had a close bond of understanding born out of a dedication to railway work. They were masters of their craft, the great individualists of the operating department, constantly using their initiative and taking important, irreversible decisions. About seven o'clock the worst would be over, the commuter rush now a trickle, and one could go home confident of the safe running of the railway in the hands of practical men, be they signalmen, controllers, artisans, guards, drivers or station supervisors.

If Edwin Howell intended to visit signalboxes and stations north of Hitchin and in the far distant outposts of the Division, I would sometimes go with him. We travelled by car for the sake of convenience but never would Edwin commit himself in the charge of my chauffeuse, the celebrated Miss Rolfe.

No more remarkable character sat behind the wheel of a Divisional Manager's car than Letitia Rolfe. Her driving will never be forgotten by those hardened by the experience. Always updating her knowledge, often through association with taxi drivers in their shanties, she knew her London and to a coat of paint. An avid reader, she could converse on many subjects and the three years that she drove were filled with incident, enjoyment and pleasure. Her driving was variously described as brilliant, breathtaking, powerful and diabolical. She treasured the ambition, never realised, to dice with the Paris traffic. She was by no means everybody's friend and was known to have her likes and dislikes. Mercifully, we got along wonderfully well but on one occasion, Edwin Howell, who never could attune his nerves to her judgment, told her to mind the kerb. Reprisals consisted of a 60mph dash through North London which would have done justice to a fire engine and a short cut up a one-way street in the facing direction, before my ashen-faced colleague was swung to his destination.

Her speed and judgment in London was fantastic and if she did go round corners in a series of straight lines, this added enjoyment to the proceedings, if one's nerves were suitably hardened. In the country, however, we would be dawdling along at 30mph but once we were 'up the smoke' again, Miss Rolfe would be in her element, bolt upright, the safest 'dodge em' driver of them all, who gave no quarter and expected none. But the greatest attribute of a chauffeur is to keep his or her own counsel. Often the car becomes an office, in which confidential and personal matters are discussed, but nothing must pass the lips of the chauffeurs whatever they may hear. Absorbed by concentration, Miss Rolfe did not hear what was being said and, if she did, never a word passed her lips.

It may not be known that there was a plan, well developed by the mid-1960s, to close King's Cross station. The closure was to take place with electrification. Main line traffic would have been diverted through King's Cross Goods Depot into St Pancras and electrification to Moorgate would have taken care of the inner suburban services. I attended a number of meetings, the content of which I prefer to forget, and the plan, apparently forced on the Region, caused untold alarm and despondency. No practical railwayman could see it working; no practical railwayman could understand why management should even consider so stupid a scheme, even though an undeniably valuable site would be released. Mercifully, it all came to nothing and we were able, through the medium of the communication meetings (which will be mentioned later) to keep our staff informed with the true facts; at my last meeting with over 100 London men just before I left the Division, I was able to say that the King's Cross closure was dead and buried and that electrification really was on the way. The sound of the applause still rings in my ears.

Geoffrey Wilson, Commercial Manager, was constantly searching for freight business in the London area and in the Home Counties. There was still a sales organisation at Peterborough but much of the intermediate traffic was difficult to sustain and by 1967 our Divisional freight receipts had begun to topple, the loss of sundries traffic being particularly severe. However, we were receiving freight traffic from the north and parcels traffic at the passenger station – later to be transferred to the freight terminal – where there was a thriving steel terminal, whilst Liner trains, as they were then called, appeared in 1966 or thereabouts.

On the Commercial side and in the working of freight terminals and cartage, I was still aware of my lack of detailed knowledge and practical experience. I set about making good that deficiency as best I could. I needed no encouragement to take part in the annual lunches and dinners with our customers in the meat and fish businesses, not to mention the coal trade with which Geoffrey Wilson was deeply involved.

As Divisional Manager, I was frequently in the chair at these gatherings and I learned the hard way that a good after dinner speech must be very carefully worked out, if the average speaker is to hold, educate and amuse his audience. The feeling of success, because, in my case, it was hard earned, was very real but the memory of my failures can still bring a blush to my cheeks!

We still had an agent at Billingsgate who reported to our Commercial Manager. Tom Dee was respected by the fish merchants but his life was not one of slippered ease when trains were not running punctually. Now and again he felt that the imprecations heaped on him should be shared by the Division and as a result Geoffrey and I occasionally visited the market where one would meet the fish merchants, not across the dinner table, but in their natural environment. One would receive a severe and traditionally worded castigation, followed by the equally traditional hospitality of a Billingsgate fish breakfast from which one eventually staggered up the cobbles to Lower Thames Street. King's Cross Goods Depot covered an enormous acreage, with yards, sidings, warehouses, cranes, canals, lorries, policemen and men in profusion.

Down below was the mysterious cavern known as the 'Vulnerable' which, by the mid-1960s, rarely held more than a long-lost box van with a bird's nest in the roof. Far too many people were employed there but much hard work was achieved and it was an education to watch the checking and loading of the evening freight trains, often under the watchful eye of the BR Police. There are memories of battered old Scammell 'Mechanical Horses' long since retired from the streets and used as shunting vehicles about the depot; of the ever pervading smell of rotted fish and vegetables, of the 'Bantams', the 'Sting Rays' and other larger vehicles, and of the chief foremen some of whom had spent a lifetime at the depot and who had started as vanboys in the days of horses, later to become carters with their own horse, whose welfare they had treasured.

King's Cross Goods boasted an ex-President of the NUR, Charlie Evans, a checker and a man of intelligence who knew every move in the game. The redoubtable, fiery and dogmatic Bert Goldfinch was a King's Cross motor driver and Chairman of Sectional Council No 4. Bert had not driven a vehicle for many years (even though there proved to be one allocated to him) but he had a considerable say in the running of the depot. For example, the distribution of the fish traffic was costly and uneconomic. BR HQ had involved itself not only in the traffic as a whole but in the London cartage arrangements. The plan, drawn up for consultation with the staff, excluded BR cartage to the advantage of contractors.

Albert Goldfinch, his face contorted with genuine anger, would have none of it and the fish traffic would be blacked. With BR HQ on one side and King's Cross LDC on the other, our lifestyle was becoming cramped. Protracted and heated discussion continued at Great Northern House well into the night but the final solution came with startling suddenness.

We were arguing about the retention of King's Cross motor drivers on what were known as the early morning town deliveries. We realised that a great deal hung on King's Cross men keeping, not the mass of the Billingsgate work, but the early morning calls at shops and stores to deliver fish as part of a round. BR HQ could not see the slightest objection to what was seen as a very minor concession to gain the co-operation of the staff which was promised by their Chairman. The problem was solved, BR HQ and the Eastern Region were happy, and the new arrangements went ahead on schedule. Why? Ah, that would be telling! Let us say that it is so often the minor details, so often known to the staff, if not always to the management, that can make or mar a comprehensive scheme.

The first Liner train from Aberdeen to King's Cross, conveying fish, meat and general freight in containers, was coming down the country, steadily gaining time under clear signals. It was air-braked, hauled by a Type 47 and I had left home when the train had passed through Peterborough.

Quite a few senior people had arrived to witness the making of modern East Coast freight history and there was a momentary hush in the conversation as the train ran down past the Goods, some 40 minutes early, to enable the locomotive to run round, draw forward and then propel the train under the cranes, ready to be off-loaded. The air-braked train was comparatively rare in 1966 and largely main line men had been trained in its use. However, King's Cross Top Link men, anxious to get home and on mileage payments, had brought the train from Newcastle, a splendid recipe for punctuality.

We stood smugly, waiting for movement. Nothing happened. The locomotive had run round and appeared to be ready to shunt when a foreman dashed across with the gratuitous information that the train had come from Aberdeen but that the driver was not prepared to move the last 200yd to its destination.

All eyes on the Divisional Manager who, being a man of action, set off towards the locomotive where he found a young relief driver who explained quite rationally that he had never handled an air brake in his life and if he knew how, would willingly do the job. The Divisional Manager had kept his hand in with the French Westinghouse air brake, if not with the new BR brake, but between them the job was done, the train stabled under the cranes and history was quietly made. I was glad of my Motive Power background that night.

XIV

A Western Visitor

IN THOSE FOUR years at King's Cross I was to learn much of the art of management, sometimes by conscious study, sometimes by experience, sometimes almost by default. At Liverpool Street it had never occurred to me that, as part of a management team, I was the youngest member, at times seemingly uncooperative and rebellious, passionately vocal in my views.

At King's Cross I was also the youngest member of the team but I was already more than halfway through my railway life. Youth is vital in a management team with an average age of over 50 and, by and by, a balance was created by the introduction of young managers responsible for freight terminals and cartage. Stephen Howard and, later on, Frank Markham, were both in their late 20s and were never afraid to put us old gentlemen right, for an avuncular team can soon become a mutual admiration society, whilst a few months after my appointment there burst upon us a certain Matthew Ruth, brisk and aggressive, to take our finances by the scruff of the neck and to direct our attention to the implications of the decisions at which we were arriving.

At our monthly planning group where we worked extremely hard at detail and progress, our Staff or Personnel Officer played a considerable part, for whenever changes are made they are bound to involve people, and their railway lives. Reg Clay was shrewd, tough, experienced and fair. His advice was sound for he had been reared in a hard school. Clerical in background but managerial in outlook, he was widely respected by staff and management alike. The Divisional reorganisation, with the abolition of the Line level of management, had thrown greater responsibility on the Divisional Managers than previously had been the case with the Traffic Managers.

A consultation meeting with the Clerical Sectional Council No 1 was to be held and I was to be chairman of the management side. Our meeting was the last of the seven Divisions and, apparently, some had been very difficult and had dragged on for days. But our precis had been prepared by Reg Clay to perfection. It was totally professional and I was able to do my homework equally accurately. The meeting took half a day and we were congratulated by the chairman of the staff side, a very rare occurrence indeed!

In 1959 my predecessor, on the advice of Clay, had started a series of informal gatherings with staff representatives to try to bridge the ever present communication gap; these took place every six months, there would be some hundred people present, maybe more, and Geoffrey Huskisson and his team would talk about the results that were being achieved and the plans for the future, leaving time for questions, of a general nature, from the floor. These meetings were not part of the official machinery of consultation but they had done good in keeping the staff informed of the changing railway scene.

Sometimes, Geoffrey Huskisson's patience had been sorely tried and his team had not always appreciated the hard knocks at question time. After his departure the idea was shelved, but at Reg Clay's suggestion and to the gloom of my colleagues I reinstated it – not without inward misgivings, for here was I about to face a large number of my own staff, with little practical knowledge of the King's Cross Division. There were to be three meetings one at Peterborough, two in London – and we left the meetings with the King's Cross representatives until last for it was here that I expected that I was to be put through the hoop, by the King's Cross Goods LDC, the Parcels people at the passenger station, and the guards' and the drivers' and firemen's representatives to name a few.

How I enjoyed these meetings and, even more, those that I started in Liverpool. They brought us closer to our staff, one got to know men who normally one would not have met and, above all, one was able to spell out publicly the changes that might affect men's working lives and environment in the years ahead. Rumour can do great harm and here was a means of reducing, if not eliminating, the damage for what we said was committed to paper and had to be accurate. When asked a question, if we said that we were going to do something, it had to be done. A management that does not honour its word deserves the contempt of its employees.

A certain skill was needed to control these open and informal meetings, firmness but with a light hand and a touch of humour. I wondered, when we started, if I had that touch and I knew that the sardonic and astute Charlie Evans, the explosive Bertie Goldfinch and the smiling, conversational and extremely shrewd Steve Watts, a passenger guard, were waiting for me, as was the Loco department headed by the legendary veteran of many a Top Shed battle, Driver Bob Lunnis. What did I know about King's Cross Goods or the duties of a main line passenger guard? But on the day, our meeting was tough, hard-hitting and humorous, bellicose and yet correct. I learned that one is given an extra strength on such occasions to give the right answer in the right way and never, never, never to lose control. I was to learn over the next 9+ years how to handle the pedant, the zealot, the young man out to make his name, the experienced campaigners, the inexperienced, the deep, the genuine, the light-hearted stirrers and the few whose business it was to create trouble. Once only have I known my meetings to boil over, when in later years in Liverpool a very senior person came to speak to our staff. He insisted on taking the chair, reacted abruptly to interruption and in an instant the meeting had disintegrated into a rabble from which it never recovered.

We always allowed 30 minutes for tea and general conversation before the start. I was never late, indeed nearly always early. Controversy often raged and the man who had come determined to have his say and to wipe the floor with our incompetence would in his fury blow his top before the meeting started and that would be that – not always, but sometimes! During the meeting, one watched the eyes of Bertie Goldfinch. In the front row, he would rant on at question time: he was expected to have his say and nobody could see those laughing eyes but us. One waited for him to blow himself out: but if the eyes were angry and bloodshot, we had to watch our step!

There is a love-hate relationship between the public and ourselves and there always has been since the beginning of railways, when *Punch* was having ponderous fun at our expense. There will always be those who glory in the opportunity to pin a Divisional Manager to the wall, those whose criticism is constructive, those who will see no good in us. I have nearly always enjoyed taking on all comers and King's Cross, although I did not know it, was but a preparation for Liverpool where the Merseyside community had what almost amounted to a personal stake in the railway, more intimate than could possibly be so in London. At King's Cross our service was tolerably good, the commuters were reasonably satisfied with their lot (if one accepts the constant plea for the removal of the Craven DMUs) and we were blessed with a Public Relations Officer whose memory I shall always cherish.

Gerald Affleck-Graves had been with the railway for about five years since leaving the Navy. He joined us late in 1964 and I had no idea how best to blend an experienced naval officer into our comparatively homespun Division. I went to see the first and famous railway PRO Sir John Elliott, Chairman of Thomas Cook, and his advice was simple and invaluable. A PRO must know everything that is happening within a Division, and must be a complete member of the management team. But Gerald, until his untimely death, was more than a Public Relations Officer. He was the last but one link in our team and, with his totally different background, grew to be our counsellor, not on public affairs but on internal and personal matters. He died tragically and we missed him deeply.

Whereas in Liverpool we could entertain our customers at the Grand National, at King's Cross we made do with the use of the General Manager's Inspection Saloon. The joys of Liverpool tended to overshadow what we had done in London, but, on reflection, what enjoyment we were able to give those who joined us on a short tour of the Division! Half a dozen visitors, largely those whose business we handled, would join us at Great Northern House at 13.45. Cars would take us to Finsbury Park Depot where John Butt would do the honours, with 'Deltics' on parade and the 75-ton breakdown crane giving a brief demonstration. Then on to Finsbury Park station, where we would await the arrival of the Scotch Goods, 15.05 from King's Cross Goods. It would run in slowly, fully fitted and fully loaded: and there behind the Guard's brakevan would be our Saloon.

(Why, oh why, did I never arrange for this unique assemblage to be photographed, particularly as we nearly always had the wonderful old Great Northern coach, now in use on the Bluebell Railway?) We would run quite sharply down to Peterborough where we were unhooked and taken by 350hp shunter through the yards and over to the goods and parcels depot, then relatively new. Our visitors, absorbed by their privileged journey, still had enthusiasm for the intricacies of sorting parcels and freight but it would soon be time to move off for an early and excellent dinner at the Great Northern Hotel. There was always time to discuss our mutual problems in a relaxed manner but all too soon we were on our way across to the station. Our coach would be attached to the 17.00 ex-Newcastle, two of our visitors would have the opportunity to ride on the footplate, and we would be back in London by 22.00.

On one occasion, however, two people who had accepted the invitation dropped out less than a fortnight beforehand and I thought we would have a change from those who did business with us. One place went to my great friend Basil de Iongh, at that time Wing Commander RAF, although we had built up our friendship as apprentices at Doncaster. He retained his interest and appreciation of railways but who was to be the other guest? A man, maybe, who also shared a love of railways? I had the chance to ponder on this and it so happened that my thoughts happened to be inspired by music and good brandy. I started to think pleasantly of old Marlburians whom we might consider and I had got as far as 13'. Ah, Betjeman! Why not, who better, for he loved railways and loved people although he had not loved Marlborough. My invitation had an immediate and typical reply and on the appointed day there were the famous squashed hat and string bag, spotted by my secretary, Rhoda, and John Betjeman to our great joy came into my office.

It was the perfect start to a golden September afternoon. We slipped down to Peterborough in John's 'Victorian drawing room' as he chose to call the old GN Saloon. With Basil, he left us for Peterborough Cathedral and evensong, rejoining us for a memorable and heart-warming dinner, and then the future Poet Laureate to his great and obvious pleasure, accompanied by Chief Inspector Fred Hart, travelled on a 'Deltic' footplate back to King's Cross station.

In my last month at King's Cross, when Basil de Iongh was on the Directing Staff of the Joint Services Staff College at Latimer, we entertained him and his colleagues, who were all senior Service Officers. We took them to Hitchin in the 'Victorian drawing room', and allowed them the freedom of the footplate where they had some very clear instruction in locomotive handling on a beautifully turned-out Type 31. They visited the good old power box at the height of the evening peak and witnessed the precision of the signalmen, again with practical instruction thrown in, and we underwent a searching examination before, during and after dinner: an education to us railwaymen, an education to our guests who realised that we had so much to show, so much of which to be proud, and so many men of confidence who delighted in demonstrating their skill and ability. We knew that there were many outside the industry who regarded us as mindless bureaucrats, incompetent managers. We knew how wrong they were and so it became my policy as a Divisional Manager to open our doors so that people could see the good and the bad, the old and the new, and we knew that they would come away with the realisation that they had had a brief and satisfying look at the workings of a great industry.

In 1966 Gerard Fiennes became our General Manager, first at Liverpool Street and then at York. I had known him since 1947 and looked forward to serving under him for the first time. We had and still have much in common although our background and methods were very different. It was a tragedy, a great self-inflicted loss to the railway industry, when his railway career was abruptly terminated. We lost a great intellect and a fine leader when we could ill afford to do so and neither Gwenda nor I shall ever forget our feeling of bereavement when, on the eve of his departure, he had appeared on TV. But while he was with us life was full of interest and enjoyment, at any rate until he moved to York when the Eastern and North Eastern Regions were united and the busts and portraits of the great men of York stared down on us at our boardroom meetings. As one entered the hallowed portals, climbed the stairs, and walked along the high silent corridors, so different to homely plebian Liverpool Street, one sensed the awe-inspiring presence of early railway history, that one was back in the hard Victorian times, and felt a certain trepidation that even the gay and vital spirit of Gerry Fiennes could not dispel.

At that time it was not fashionable to talk of steam. Despite the fact that I was deeply involved in the Great Eastern revolution, I knew that there were folk who regarded me as an unrepentant steam man and who might well look askance at the man who, now and again, slipped across to France to try his hand on the Chapelon Pacifics at Calais, if he once more became involved. And so, when Gerry informed me that the Great Western *Clun Castle*, No 7029 was to come to our Division, 'because Dick Hardy's a steam man', I was none too pleased and, without the slightest success tried to wriggle out of the assignment. I knew, in my heart that I would love to see the old thing knocking around our railway and, in due course, Colin Morris and I decided on New England for No 7029; we could not have made a better choice. Steam had been gone from Peterborough a few years but the arrival of such a distinguished foreigner at a depot doomed to closure had an extraordinary effect on the maintenance staff, led by the dedicated old foreman fitter, Horace Botterill. Naturally, and in his opinion, there was a lot that needed putting right on a mere Great Western engine, both in his imagination and in truth, and so the owners benefited from his expertise. There is no doubt she returned to Tyseley the better for her Eastern excursion but all of us benefited as well. In the first place she had several profitable main line outings, and secondly we used her as motive power on one of the traders' outings from Finsbury Park, instead of going down on the rear of the 'Scotch Goods'. She was manned by Peterborough men when on the GN main line and they did splendid work.

Our men used good hard coal and there was none of this business of a huge fire and an even bigger shovel. The GW shovels were stacked away (nobody would be likely to pinch them for *railway* work!) and replaced by our long-handled 'teaspoons'. The Western men from Tyseley, who were the very best and who had come along to see fair play, were sceptical and then amazed when 'Clun' had come down from London to stand at Peterborough, ready to tackle Stoke, with the front firebars showing, a gently sloping red hot fire, ready to accept 'little and often' firing.

But there was one thing that I wanted to prove. A few years previously, I had travelled to Newport on a double chimney 'Castle'. I had not touched the shovel but had watched, fascinated by a firing technique of which I had heard but the like of which I had never seen. In all honesty I had never seen a fireman work so hard and so continuously with a moderately timed and loaded train. It was not the amount of coal actually placed on the fire that constituted the hard work but the opening and closing of the firehole flap for each shovelful of coal, together with that enormous Great Western spade, the contents of which were thrown on to the huge 'hump' in the centre of the firebox. I had thought that there must be an easier way than this, with what seemed to be a free steaming engine.

No 7029 gave me the opportunity to do the work myself, for as a Divisional Manager, in those halcyon days, one could quite simply organise one's own trial. The previous day and evening I had business in Peterborough with the Development Corporation. Early next morning on my way back to London, I was ready to do battle with No 7029 and 13 coaches on a 'clearance' test, to Hitchin. We went as hard as we could up the Ripton bank to vindicate my LNER method of firing quickly with a small shovel whilst leaving the door open and so cutting out that extra mileage necessary to open and close the flap. We had an ex-LMS Spital driver, Whitehead, along with Inspector Bill Buxton and by the time we passed Holme near the foot of the 1 in 200 gradient to Stukeley, she had begun to hang her head to the tune of 190lb/sq in instead of 225, but I realised my mistake in letting the front end get a bit bare. There was time to rectify this quickly and three successive rounds with plenty up the front brought the smoke to the chimney and the pressure round to the simmering point where it stayed for the rest of the trip. This method of firing was very different to that normally used on Great Western engines and it proved to me that we could make No 7029 go just as well our way without what seemed to me the extra toil of the time-honoured Great Western way of doing things.

My many Western friends both on and off the railway must realise that I have a deep respect for their methods and creations; but we have always had to have our bit of fun at the expense of the 'Broad Gauge' men and what we call their 'dung spreading' methods.

Although I did not know it, time was running out for me on the Region on which I had spent much of my railway life. We were on the fringe of modernisation but I was to leave before a start was made with those things on which we had pinned so much faith, and my successors reaped not only the rewards, but also experienced the same problems and difficulties that I had dealt with at Liverpool Street. Men were retiring, not only those that joined the railway in 1919, when the eight-hour day was introduced, but the signalmen who had stayed on over their time in the smaller signalboxes. I used to see every man on retirement with 47 years of service or more as, of course, did his departmental chief, and it was a joy to listen to the reminiscences of these old-timers and then to look at their papers to find the original testimonials from the vicar or the squire or the schoolmaster that they needed to join the service in those far-off days. Every retirement interview, over a cup of tea, provided some interesting aspect of railway work or a look into the social conditions of the past. Rarely were any of these men disappointed with their lot, rarely had they not enjoyed their railway life. It was fascinating to know that the Great Northern signalman's uniform cap was shaped like a rather distinguished cloth cap and that the majority of young GN signalmen who were recruited in Hertfordshire or Huntingdonshire were trained at Retford in the Signalling School and then despatched for their first posting to the outposts of the West Riding or perhaps the Vale of Belvoir. While they were waiting with their wives to see me, they would be entertained by Rhoda Powell, my Secretary, whom I have left to the end. I have been the luckiest of men with my Secretaries – to have a good one is a joy, to have a poor one must be a penance. They know one's moods, strengths, weaknesses, one's family, one's colleagues. A Divisional Manager's Secretary must handle the public who thunder at her on the telephone with the utmost discretion: she must know exactly what her Chief is thinking and, if possible, doing. She must work hard, she will meet all manner of people, and most certainly Rhoda had to cope with many unusual visitors and difficult situations with charm and tact. She too was a member of the management team, a vital link in the chain, and I have left her until last so that I can conclude this chapter by acknowledging not only what she did for me when she was my secretary but also the fact that she has typed this book (in her retirement, of course!), corrected my spelling, sorted out my grammar and, on occasions, called me to order as she was wont to do when we were working together at King's Cross in great harmony.

In March 1968 I was interviewed for the posts of Divisional Manager, Liverpool and Doncaster, both of which were more highly graded than King's Cross. Derek Barrie, by now my General Manager, rang me to say that I would be supporting Liverpool or Everton in May and, despite a long held disinclination to go to the London Midland Region, I was excited at the prospect, although I had no idea of what lay ahead for Gwenda and myself.

The Liverpool Division

A few figures may not come amiss in a story that has been predominantly about people. The Liverpool Division, in common with much of British Railways, was going through a difficult time financially in the years 1968-1973. In May 1968 there were 8,500 staff in the Division: by September 1973, this figure had shrunk to 5,618. On the other hand, pay awards always outstripped the economies made and the increase in receipts – whether from additional traffic or from rates and fares increases – was not sufficient to offset the worsening situation.

1972 was a particularly bad year. Expenditure increased no less than in other years but receipts actually decreased because of a series of industrial disputes, not all within the industry, and because of the depressed state of the national economy.

Liverpool Division

	1969	1970	1971	1972	1973*	1974†
Passenger Receipts (£m)	7.1	7.9	8.7	8.8	9.7	10.3
Freight Receipts (£m)	11.5	11.4	10.9	10.1	11.5	12.6
Total: (£m)	*18.6*	*19.3*	*19.6*	*18.9*	*21.2*	*22.9*
Staff Costs (£m)	9.0	9.7	10.1	10.8	11.5	11.8
Other costs (£m)	1.7	1.7	1.9	2.0	2.1	2.3
Total: (£m)	*10.7*	*11.4*	*12.0*	*12.8*	*13.6*	*14.1*
Excess of Receipts over Costs (£m)	7.9	7.9	7.6	6.1	7.6	8.8
Excess of Receipts over Costs as a percentage of Staff Costs	87	81	75	57	66	76

* Outturn at 36 weeks; † Budget forecast

The last line of figures gives a measure of the productivity – in the broadest sense – of the Division. Although the Division was not a self-contained accounting unit, it shows the relationship between staff costs and the contribution made by the Division to the profitability (or otherwise) of British Railways. The passenger receipts over the period showed that Liverpool-Euston accounted for 69% of the total receipts from InterCity services and for 73% of the budgeted increase in receipts between 1971 and 1974. This service was very profitable and in 1973 showed an operating ratio of 31%. The local services, however, were loss makers and therefore grant aided. In 1973, 14 services earned a total of £4,504,000 and cost £8,922,000.

Total freight receipts declined from 1969 to a low point in 1972 and then improved considerably in 1973. But within this total there were marked variations in individual commodities. Those which provided the increases in revenue were largely those particularly suited to rail transport, in our case chemicals, iron and steel and motor cars. Economies in bulk handling can be reflected in competitive rates and by the exploitation of new business.

Over the period the figures for iron and steel had highlighted the sensitivity of the forwardings of basic materials to the fluctuations in the national economy, whilst those for motor cars reflected the then notoriously uncertain labour position in that industry. The earths and stones group consisted largely of rock salt from Winsford to be spread on roads in the winter. The severity of the weather was paramount in determining the level of this traffic and I prayed in vain for hard winters! Coal and coke had declined in the Liverpool Division and there were but six collieries in all.

By 1973 63% of the freight business was being consigned in train loads and company trains were to account for almost half the freight trains run in the Division. 1,073 against a total of 2,215, including Freightliner, Network and Merry-go-round services.

The expenditure figures covering 1970 to 1973 showed that the most significant economies had been made at freight terminals, marshalling yards, signals operation and wagon maintenance and examinations, whereas train crew and traction maintenance costs had increased.

Whilst the London services were going from strength to strength, the journey times available on many InterCity routes were unbelievably bad, but not for want of trying on our part.

Old stock, and poor refreshment facilities. We knew that the M1/M6/M5 link-up and the opening of the M62 Trans-Pennine link had to be challenged.

In 1973, the Newton-le-Willows Motorail, showing a deficit, was withdrawn; 330 excursion trains produced £234,000; and in 1972, Mersey Bookstalls showed a net profit of £20,770.

These are just a few figures that happen to be available of a business going through a very difficult time. Constantly one searched for economies, for reasons to rationalise, for new and profitable business. We were accountable and, on figures alone, it cannot be said that we would appear to have distinguished ourselves. Read the story and judge for yourselves.

XV

'This is the place for me'

PRIOR TO MY APPOINTMENT to Liverpool, I was interviewed by W. G. Thorpe. Willie Thorpe had been my Line Manager at Liverpool Street and one knew his methods, if not always the direction that they would lead one. He had been a super governor who had taken a personal interest in my career. His ebullient temperament and affability which had tempted the unwary towards familiarity hid a great dedication: he was very serious indeed.

Willie opened fire in his customary manner by saying he was looking for 'a large man of aldermanic appearance' to go to Liverpool. He talked briefly of the role of the Division, its work and its future and questioned me on our performance at King's Cross. Bobbie Lawrence, the London Midland General Manager, had one point to make – 'In Liverpool, you will find you are next in line to the Lord Lieutenant and the Lord Mayor. It's no good keeping a home in the south, are you prepared to upsticks and move at once?' Obviously, my answer was yes. His point was vital and, in any case, neither Gwenda nor I could bear to be parted for long. So I went to Liverpool for 5+ years: a long term appointment, the job changed our lives and gave us lifelong friends. Had we stayed in London, we would have been thousands of pounds better off. When we moved back to the south in 1973, we had to double our mortgage against the explosion in house prices but we would never have experienced those years which enriched our lives.

As I write of the Merseyside, and the Lancashire and Cheshire railway scene, I am already aware that so much happened, so many people crossed my path, so many amusing and heart-warming incidents arose, that to my readers it may seem that we skipped across wave after wave, that there was no depth, no hard grind. I would not blame them for there were those within the railway who knew neither Liverpool nor Dick Hardy, who made the same mistake. Our splendid management team knew the meaning of hard grind, that nothing came easily, that business had to be fought for and was readily lost if we could not operate efficiently and at the right price. But it may be as well to set out a little of my own philosophy of management, not in depth but to set the scene. Some of the values may seem old fashioned but they were my beliefs and they still are, although I freely acknowledge that, in retrospect, there were many things I could have done better.

In March 1973, not long before I left Liverpool, I gave a talk to a group of staff from all grades in which I spoke of my work and the methods that I believed to be basically right. From biographies I had absorbed the experiences of some great men. Of Stonewall Jackson, the great

Confederate General in the American Civil War, it was said that 'his supervision was constant but his interference rare'. I had studied and then followed to advantage the views on leadership expressed by Montgomery, when I took over, against the tide, at Liverpool Street in 1959. 'Monty' had written that a third of his working hours were devoted to the consideration of personalities: that, in dealing with subordinates, justice and a keen sense of fairness was essential: that merit, leadership and ability to do the job were the criteria for appointment, and men, once chosen, were to be backed to the limit. Any commander is entitled to help and support from his immediate superior and he must have qualities of leadership, initiative, drive, character and, above all, moral courage. A commander must decide how we will fight the battle before it begins. He must then assemble his officers down to a specific level and explain the problem, the plan and the fight. The spoken word is vital. Command must be direct and personal.

These principles are right. They held good in railway management and much of my responsibility could be settled within that simple framework. I was expected to lead and co-ordinate all activities in the Liverpool Division and I was as financially accountable as the railway structure would allow. Thus in the management team the operator, the mechanical and electrical engineer, the marketing men, the planner, the accountant, the personnel man, the PRO, each had a function but I did not. I was there to lead and inspire but I was responsible for the major decisions that were taken. One took the lead at Planning Group meetings, at probing visits with Area Managers, at communication meetings, at important meetings with outside bodies, and one headed the Divisional team when our general and financial performances were under scrutiny. I was in touch with the management team – always available, in a position to encourage them, sometimes to criticise, to fight for them when necessary but never, never to interfere with their detailed conduct of day to day affairs. To spot something very right or very wrong, to ask the question, yes – but not to interfere.

I had learned from Harold Few at Liverpool Street how to keep calm, never to appear to lose heart whatever the turmoil inside me, I had learned to be where the going was hard and never to shirk the limelight when its glare was likely to be unfavourable. Just as one must never be too big to ask for or take advice on a line of action, so credit must be given to he who has conceived it. If one accepts advice and the decision is the wrong one, only one man is responsible: the Divisional Manager. One must have the courage to do the hard things to loyal men if the need is there, to take the hard but right decision in the interests of the business and not to compromise over vital matters of discipline. I had to break men's hearts at Liverpool Street, King's Cross and Liverpool, by taking them out of their position of responsibility. But it had to be done, straight and face to face, and after previous warnings had failed to bring about the required improvement. It had to be done quietly, not a matter for discussion afterwards, and the man's career had, if possible, to be rebuilt and his self-esteem re-established.

Our financial control was becoming more sophisticated and we had come a long way in a few years: each Area Manager now prepared his budget and his performance was regularly checked. I remembered my days as a Shedmaster where the word 'budget' did not exist but where we knew what was going on and how to maximise the effort of our staff. I fully understood the wisdom of accountability and yet, as I spoke that night, I wondered if we were going too far the other way and whether the financial constraints I was imposing were denying the managers in the front line their freedom to lead and inspire. Could they now achieve the impossible as my generation had done? These and many other points I made that evening. I realised, as I spoke, that I was concentrating on the human factor rather than on the harsh realities of marketing, operations, the search for and retention of business, public affairs, and industrial relations. When, with hindsight, I examine my performance as a Divisional Manager, there are many things I could have done better: certainly I did my best to observe the principles of leadership, but should I have been less of a railwayman, more of a businessman, less of a railway manager, more of a professional manager? Should I have curbed my independent and unorthodox streak and followed convention and the party line more rigorously? Instead of getting things done quietly and effectively – and our Division had an excellent record – maybe I should have made figures talk to my own advantage? I couldn't be bothered with railway politics but I enjoyed my work and I'll wager those who worked with me in Liverpool did likewise.

Early in April, six weeks or so before we were given a wonderful send off from King's Cross, Gwenda and I paid our first visit to Liverpool. We were to be met by John Shone, a stranger and a Liverpudlian. Our adventure had begun.

Reginald Jennings had been a famous Housemaster at Marlborough where a love of railways had brought us together. I had written to him for advice about preparatory schools on Merseyside for our son Peter who was all but 10 years old. Our choice of day prep school would govern where we lived, either in the Wirral or in Lancashire. Reginald put me in touch with John Shone who was an old Marlburian, at that time barely 36, a miller and a man who knew little of railways but much of life in Liverpool at all levels of society. He helped to run a thriving milling business, and he became a great friend and a marvellous unofficial PR man for British Railways, Liverpool. That evening he engulfed Gwenda and me with the warmth, friendliness and humour of Liverpool. He and Ibbs, his wife, took us out to dinner in West Kirby, talked schools and villages and then took us round the Wirral to show us villages and prep Schools – *in the dark.*

Only later did I find out they were going away on holiday next day. One thing was certain, I knew I was going to love this place and when we made our home in the beautiful village of Burton-in-Wirral, I knew that we would want to stay for ever.

On 20 May 1968 I joined my predecessor Hugh Mugliston for a few days prior to his retirement. I knew but three men out of the 8,000 in the Division and I knew nothing of the Liverpool scene. Hugh was held in high esteem amongst the top people in Liverpool and with the firms with whom we did business, such as ICI, the British Steel Corporation, Shotton (known to us as JS&S) and the Mersey Docks & Harbour Board. I was to do my best to maintain these links, to hold the pulse commercially. Hugh, and Arthur Williams, his Assistant, with whom I worked closely without discord for over five years, had made a great contribution to the plan to develop the Loop and Link extensions to the electric services, working closely with the City Council officials so that part of my first week was spent with the Town Clerk, Stanley Holmes, the personification of dedication and Liverpudlian wit. When the Merseyside Passenger Transport Authority and Executive were set up and the work was finally authorised, it had become too easy to forget the part played by my predecessor.

Tired at the end of that first week and even more at the end of my second which had included a crisis at our Isle of Man Office, I had nevertheless taken stock of our team and I realised that, at 44, I was the youngest and that there was a wealth of experience every bit as sound as at King's Cross. I had a personal bodyguard in Muriel Brownbill, my Secretary and Albert Bennett, Chauffeur, who looked after me wonderfully well, both true Liverpudlians. Albert kept a beautiful car, first of all a Humber Hawk and later on a new Ford Consul.

He took me on a tour of Liverpool, the good and the seamy side, the back streets, the docks, the Dingle and Liverpool 8. He seemed to be known by everybody and knew every street, every building and its history. Later on he would take our visitors, railway and non-railway, on a tour, inevitably the Mersey Tunnel, the Wirral (at Meols on the shore he would take great gulps of fresh air, 'The best air in Britain'); but it was in Liverpool that he excelled himself. Always humorous about the 'Wigwam', he fancied himself as an historical expert on the Anglican Cathedral and I'm quite sure what he didn't know he made up. But this was one of the delights of the Liverpool personality. In that first week, however, he was to advise me that if I wanted, say a new sink or a washing machine, to leave the purchase to him. Naive as I was I had not realised that, in Liverpool, discounts could be obtained on just about anything, but not if you spoke like Gwenda and me! Albert gloried in methods which he would never disclose.

Hugh had suggested a small hotel at Hoylake and this became my base until we moved our home to Burton on 8 August – three days before the last train hauled by a BR steam engine left Lime Street. I travelled on the 'Black 5' No 45110, and much to the surprise of all concerned, worked my passage some of the way to Manchester. My first days were spent concentrating on my new task, evenings either on social or railway engagements, 'learning the road' to Southport, to Chester, to New Brighton, to Wigan, to Crewe, hours in the Control, unexpected visits to depots and yards. This is the time of day to see and learn. Usually I was on my own, but sometimes Danny Whelan, the Divisional Operating Superintendent, came with me. Danny was ultimately promoted to become Divisional Manager at Preston but his delight was short-lived for he died on Good Friday, 1980. There must have been 300 people at his funeral, a much loved and respected railwayman. He had started at the bottom, his father was a shunter in Edge Hill when he was killed, and Danny had worked his way up the ladder, step by step. He visited France with me three times and nothing escaped the attention of that little white-haired, stiff-built figure.

I had soon come to the conclusion that Liverpool was a great village where everybody knew everybody else. I had been invited to join the Merseyside Industries Dining Club; I was a member of the Atheneum, where my subscription was paid by BR because I should be fighting the railway cause on every visit; I later joined a private shipping dining club, had visited the headquarters of or lunched with the heads of a number of Merseyside businesses, and had taken my first step towards becoming an unpaid member of the Liverpool establishment. At every level of society in which I moved there was a personal interest in my affairs and I realised that the position that I held counted for a great deal and that I had better live up to it.

But I had not really begun to understand my Liverpool until a small incident occurred on the night that we moved up from Surbiton, on Thursday 8 August 1968. We were looking forward to our venture in the north but we had left a lovely home and many friends. By the time we reached Hoylake about dinner time, we were tired and I was in no mood to take the sleeper to London for a 9 o'clock meeting with the General Manager and the Mersey Docks & Harbour Board. So when I walked to Hoylake station at about 22.30 I felt jaded and rather than doze in the train to Liverpool Central Low Level – I decided to join the driver.

The train arrived, three coaches of the 1956 Wirral electric stock that I was to know so well and, opening the cab door, I showed my pass to the driver asking him if I might travel with him. 'Of course you can', he said, 'Come in and you can drive the bloody thing'. 'Well no', I said, 'I could but I'm not going to, and maybe you had better have a look at that pass'. By this time, the bell had rung and we were on our way. Jack Fletcher, for this was his name, turned on the light and looked at my pass and then at me, by no means abashed but appraisingly and with twinkling eye. 'So you're Misterardy, our new Divisional Manager'! 'And I hear you're living in Burton'. 'Yes, that's right', said I, wondering how this could be common knowledge seeing we had only arrived that day, and the move would take place on the morrow! 'Yes and you've got three children and your interests are...' This floored me and I interrupted to ask how on earth he knew all this. But his reply was humorous and non-committal.

In fact my arrival, thanks to Frank Allen, our Public Relations Officer, had been reported in the Liverpool and Birkenhead newspapers. Frank, and his Assistant, Ernest Lawton, who complemented him so completely, had made a thorough investigation into my history which was spread across the papers, an interesting news item read and no doubt discussed by many of my own staff. Incidentally, Jack Fletcher had come to Birkenhead North from the GC Depot at Bidston – yet another GC man and, of course, our village of Burton was very near to the GC line over which he would have worked a thousand times as fireman and driver on his way to Manchester and points east.

Now one has to begin a story somewhere and as my journey with Jack Fletcher terminated at Liverpool Central (Low Level), let us begin the Merseyside tale with that Liverpudlian institution, the Mersey Railway. To me, the Mersey Railway means the whole of the third rail 650 dc network to West Kirby, New Brighton and Rock Ferry, but a Mersey man would have rebuked me severely as the MR proper ran from the Low Level to Rock Ferry on the one hand and Birkenhead Park on the other, no more and no less, and this *was* the universe to a true Mersey man.

Bobbie Lawrence told me that the Mersey Railway was a species of private club and he was not far wrong. In 1968 first class accommodation with carpets and comfort was available in the centre coach of the three-car sets. When we abolished the first class, there was a fair amount of indignation amongst not only the Wirral commuters but those who came first class from Bebington, Bromborough, Hooton and Chester. With the abolition of the post of Travelling Ticket Collector there was a rush for the first class seats and individual effort by any of us who felt inclined to turn second class travellers out of the first class had little lasting effect. Throughout my time in Liverpool, the 1956 stock worked the Liverpool-West Kirby service and the 1938 stock covered the New Brighton and Rock Ferry trains. The newer stock was cleaned and lovingly wax polished under the eagle eye of George Lee and Charlie Newton at Birkenhead North and, by 1972, the 1938 stock at Central had been raised to the same standard.

Drivers for the West Kirby services signed on duty at Birkenhead North and at Central for the Rock Ferry and New Brightons. Quite a few of the North men had been steam firemen or cleaners when the Wirral services were steam hauled, whereas at Central were based the remaining genuine Mersey man who had never worked for the LMSR and had been independent up to nationalisation. Some of the Central men were from Mollington Street, so could be either from LMS or GW lineage and, of course, the LNER (ex-GC) men were everywhere, so at that small depot one could meet men from four constituent companies. Furthermore, the original Mersey drivers were not in the normal line of promotion, having been drawn alternately from the traffic and workshop grades which prevented them from transferring elsewhere.

Maintenance was carried out largely at Birkenhead North Shops, a sizeable set-up run effectively and efficiently. The foreman in charge was Frank Holt and he waged total war on the drivers. In fact the Wirral stock was very prone to flats on the wheels and Frank's long nose would quiver with indignation when yet another set rumbled by with seemingly octagonal wheels.

And so a private club, certainly for the passengers and especially so for the staff. The men's diagrams for both depots were prepared at Rail House, Crewe. No doubt they were masterpieces in their own right but the Mersey thought otherwise. Crews were apt to change at peculiar places like Hoylake and some quiet observation convinced me that the actual operation of this extremely efficient little railway bore little resemblance to the workings laid down by HQ. Sound principles of train working were nevertheless observed. A West Kirby into Low Level was a West Kirby out and the only time that changes were made in this pattern was on Sundays when men were changed over to keep in touch with the other routes.

During my first week when I was travelling to Hoylake with Hugh Mugliston we just missed a West Kirby so I witnessed the departure from the Low Level of a Rock, a Brighton and then another Rock whilst listening to the splendid loudspeaker announcement encouraging people to go to 'Street, Squr, Park, North, Village and Brighton'. Hugh and I stood halfway down the island platform and I noticed that each train was driven by a man with a Jimmy Edwards moustache. Vastly interested in this phenomena, I had not at first understood the very smart turnround at the terminus. A train would arrive one side of the island platform, the driver would leave his cab and walk down the platform to the departure point. Meanwhile, a turnround man would get into the rear cab, propel up to the dead end, the automatic points would change, and in a matter of seconds he would draw forward into the other platform to be relieved by the driver who had brought the train into Low Level. The turnround man would then walk across to the next train and repeat the cycle. Thus Percy Ablard with the handlebar moustache on every train!

It can be seen that the turnround driver was a key man and particularly so if there had been a hold-up. I had often seen a tall, saturnine driver wearing the old soft-topped motorman's cap. I had not yet travelled with him and one evening I found that things were out of sorts. I was in no hurry and watched this driver at work on the turnround, with joy in my heart. His name was Tommy Lawrence. Now, in 1969, I was aware that there were two Tommy Lawrences around the place. One was based on Anfield, the Liverpool goalkeeper, and loyally as I followed the fortunes of that remarkable club, Tommy Lawrence of the Mersey and British Railways was the man for me. I watched him at his work, while without fuss, without excitement, unsmilingly, he pulled the minutes back until we were right time again. I had witnessed the Mersey miracle. The written workings bore no resemblance to what men were actually doing and yet no train was delayed waiting for relief; possibly Chief Inspector Swindells, a Mersey man of nearly 50 years' service, might have known who should be where, but when things had got behind the genius of the Merseyside temperament came to the fore – genius for improvisation, for the unorthodox, for the humorous. There were no puzzled men asking for orders, the trains ran and ran and ran, and in time order was restored, the turn-round man the centre of the scene.

Years later, after the Loop had been opened, Tommy Lawrence was joined at James Street by the then Chief S&T Engineer of the London Midland Region who opened the conversation by asking the archetypal Mersey man what he thought of his new railway. 'It's rubbish', replied Tommy, following up with a general castigation of engineers and management, with the merest twinkle of the eye. The train thumped noisily round the Loop until, suddenly released once more on to true Mersey territory, the driver remarked triumphantly: 'Now you're on a real railway, Mister!'

This unusual railway was a source of interest to railway visitors who knew not Liverpool, and who were fascinated by the operation of the junction at Hamilton Square, the working at Low Level, and the cavernous and rather shabby, gloomy stations echoing to the squeak of the Ferrodo brake blocks; and there was one more idiosyncrasy to be noted – the bright little bookstalls. Even these had a history and they were the property of the Divisional Manager. The original bookstalls were established by the Liverpool Daily Post Co as retail outlets for their newspaper, which local newsagents were refusing to sell. A year later, in 1911, the bookstalls passed into Railway ownership and now, in 1968, Jimmy Trainer – the Passenger Manager – and I, with the help of the Property Chief, John Phillips, were trying to avert their proposed closure. They began to thrive once more, and new stalls were opened, one of them as far away as Wigan. But the manning of the bookstalls was a typical but effective Mersey arrangement. Most of the staff were from railway families, recruited by Ken Houghton, the Manager, into no known grade and no known rate of pay for the simple reason that this little outfit was unique on BR and nobody had got around to creating either a grade, a rate of pay or a promotional structure!

The West Kirby trains called at the draughty island platform of Bidston and then headed west across the top of the Wirral peninsula. Away to the left curved the line to Wrexham, the Great Central, over which the Stoke Division worked the DMU service to New Brighton and later to Birkenhead North and we worked an iron ore service of quality and punctuality. That first week Hugh Mugliston introduced me to Philip Clarke, the Managing Director of Reas at Birkenhead, stevedores and tug owners, and to George Tatterson, 27 years a railwayman and 20 years with the BSC Shotton as Transport Manager and later Director. They wanted a first class service, they got it, and they paid for it. Sometimes a couple of ore boats came over the horizon together and all hell was let loose; sometimes things were quiet as there was no boat in the dock. It was a joy to serve these firms: in the first place, there were some splendid people dedicated to their task – sometimes we used to jangle but that was neither here nor there. Secondly, the service we gave created an interest in the working amongst our own staff, and punctuality was exemplary. I maintained the links with the firms at top level, living in the next village to Philip Clarke and lunching every few weeks at Shotton. But it was our Freight Managers who did the hard work. Harry Hardy had agreed, and later on Geoffrey Fell maintained, a very good rate for the job to which we had always stuck, maintaining that a good service was worth paying for. JS&S would talk about our 'Gold Plated Railway' but they got their trains of 11 bogie hoppers of ore from Bidston as regularly as shipping schedules allowed.

The working of an ore train over the road from Bidston Dock to Shotwick Sidings just over the Welsh border was interesting. The motive power might be a Type 40 or a couple of Class 25s thumping away up the 1 in 100 gradient to Heswall Hills. Up this long gradient the speed would be held at about 15mph; the controller would be eased near Heswall Hills, and the driver would feel the loose coupled train stretch out behind him when the guard applied his handbrake, for there was no continuous brake: an engine, 11 big ore wagons and a guard's brake in the rear. The driver would keep the speed down through Barnston and when they passed over the Burton road beyond Neston, he exceeded 15mph at his peril, for the line descended through Burton Point, near our village, at 1 in 100 and across into the yards at Shotwick where we handed the train over to the Summers pilots. Not very much in this, you might say: and yet it fascinated my mécanicien friends from Calais who had always used the Westinghouse continuously fitted brake and it fascinated visitors to our Division and seemingly it fascinated the Burton WI when I gave them a talk on the railways on their doorstep. But just now and again, there were problems.

One evening in January 1971, the telephone rang at about 10 o'clock. I had had a day's leave, and a strenuous time out hunting. I had got home eventually about 7 o'clock, and later Gwenda and I sat by the fire, thinking of bed – and then that phone had rung. The news was bad. An ore train had run away down Shotwick bank, gone through the yard and finished up with the locomotive derailed after a collision with a Summers shunting engine. Wagons, ore and engines all over the place. Naturally, I rang George Tatterson at once and took our old Riley car straight down across the marsh road to Shotwick Yard, only a few miles away. By and by, when the cranes were in position and the job in hand, I left the experts to it and took the Birkenhead driver and secondman, who had been shaken up a bit, back to Bidston. Halfway across the Wirral, in the pitch dark of 01.00, I smelt hot metal. But my dear friend, W. O. Bentley, had told me it was surprising how far one could go without water in a motorcar engine if one took it easy. We coasted downhill on the last lap and the engine, in neutral, stopped, but with no damage done. But no bed either and by now I was very tired and very cold: however, my chauffeur, Albert, came to the rescue although we had to leave the old Riley in the rougher streets of Birkenhead. (And that reminds me of the train of coal that would pass along the dock railway during the night past a housing estate on its way to the Concentration Depot near Birkenhead North. Somehow the wagon doors would open, with the inevitable result. But the coal never stayed long on the ground!)

From Dee Marsh Junction, the Great Central curved off towards Chester, becoming the Cheshire Lines in the process, and if we too followed the route, we could land up in Chester General station, outside the Liverpool Division. But we shared the DMU service from Chester General into Rock Ferry with the Stoke Division and the trains nearly all stopped at Hooton, the station nearest to my home. I had not used this station many days before I decided there was something wrong with the service, indeed there was something wrong with nearly every DMU service that we operated. Used as I was to a pretty detailed follow-up on both Eastern and Southern Region, I was amazed that the London Midland Region Headquarters was not giving me hell and deduced that it was up to us to put things right. On the Chester services, the DMUs were well maintained and reliable but the trouble was the passengers themselves. In those early days I kept quiet, and it was a week or so before my fellow commuters realised who I was. Then one morning Derek Bibby, Chairman of the Bibby Line – came after me unsmilingly – 'Are you in charge of the railway here? Well, you make a bloody poor job of it', he said, but permitting himself a twinkle. Something had to be done: day after day we missed our connection at Rock Ferry. Rightly, the Mersey trains could not wait and spoil the sequence over Hamilton Square Junction but, in fact, the trouble was largely at Bromborough. Each station was allowed 30 seconds or so to load its passengers but the Bromburians slummocked about as though they had all day. Whistling had no effect, but departure whilst some were still on the platform at least had them worried. Eventually, having fixed a gate, we locked them out of the platform as the train arrived. That did the trick. Little or no more trouble, but the gate stayed where it was as a deterrent to be used just now and again. Incidentally, the younger the passengers at that station, the slower they moved.

Our station, Hooton, had some characters on the staff; only little Frank Peacock remains, but perhaps the best known of all was Leslie Ames, the leading railman, always on the late turn, a much loved man. Nothing was too much trouble for Leslie, no passenger received even a hint of discourtesy and, years after retirement, he is always remembered as the passengers' friend!

XVI

The Great Village of Liverpool

A ND SO to Liverpool itself. There was Lime Street and there was Exchange. There was Central, the Cheshire Line terminal, full of ghosts, serving only the Gateacre line which had not been closed in 1968 – nor would it be so until April 1972 when it had become inevitable and the last battles of the eight-year war had been fought. There was Riverside just: and before I left Liverpool this fascinating boat train terminal, steeped in history, was to close as was Central so that work on the new Loop line could be progressed. We reduced to two interchange points with the Mersey Docks railway, at Canada and Alexandra Dock, and those who had known Liverpool in the heyday of railways and at its peak as a port would not have comprehended the change. Later, the Mersey Docks & Harbour Co – the successor to the Board – closed its own railway although the door was left open for a railway extension to the Seaforth container terminal. After many months of debate and negotiation, Seaforth was opened and the pattern of Liverpool shipping irrevocably changed. I was in Liverpool when the last Canadian Pacific liner left, Gwenda and the family amongst the visitors on this occasion, which brought to a close the glorious days of the great ocean-going liners which had come majestically up the Mersey to tie up in the centre of Liverpool. The city itself was a most interesting place and the waterfront, even with so little traffic, an ever changing source of enjoyment.

Out of Exchange ran a halting DMU service to Wigan and Manchester. Early one morning in November 1968 I went to see what on earth was wrong. The units were maintained elsewhere, and badly at that. It is often difficult to put things that are not one's responsibility right without hurting a neighbouring colleague but, in time, there was a great improvement in reliability and punctuality. There was plenty wrong in our own Division, distants not being pulled off, signal lamps not alight, stations unattended, a general malaise and lack of interest, for it must be remembered that the staff felt that many of these lines were still under threat of closure, even in 1968. That early November morning I learned the meaning of the word 'Demic'. I had never heard of it – was it a DMU coupling code or a class of DMU? The driver with whom I had travelled to Wigan told me he'd already had a 'Demic'. Ignorant but careful, I held my peace until I returned to Rail House. Ken Lord put me right about the Lancashire for 'failure'.

Out of Exchange ran the Southport electrics which also served Ormskirk. Like the Wirral stock, they were very reliable and the punctuality of the whole of the third-rail electric network was extremely good. Whereas the Wirral trains were not particularly powerful nor fast they were nevertheless solidly built, but the bodywork of the Southport trains had seen better days. When a driver opened his controller it seemed as if the chassis of the driving unit would leave its coachwork behind, so breathtaking was the actual start and acceleration. Draughts in the cabs were an endless source of complaint but there was very little that could be done to improve bodywork that had, long since, gone down the slippery slope. And now Exchange has gone so Liverpool is left with Lime Street and its history; but what a history. The great hewn granite cliffs, through which trains passed; Edge Hill, Olive Mount cutting; it needed little imagination to bring one's mind to the very early days of the Liverpool & Manchester Railway.

Lime Street was the starting point of the InterCity services to Euston, Birmingham and the South-West: also of the Leeds and Hull services of which we were not proud. However, the London services were very good and were improved still further in 1974. Punctuality was first class, failures were rare; there was the close attention to detail so necessary to maintain a high class service. The Edge Hill men had a share of the London work but it was not enough for my liking: gradually we improved the quality and increased the number of the London workings as well as those to Birmingham and Leeds. We tried for Carlisle work but that did not quite come off! Both Edge Hill drivers and Lime Street guards were of excellent quality and it was just as well that I was able to stop, soon after my arrival, the plans that had been made to remove the post of Shedmaster at Edge Hill on the pretext that an Assistant Area Manager could look in twice a week. There were over 200 highly individualistic Liverpudlians at Edge Hill depot and there was no way that I was going to let them look after themselves: and so Jack Mitchinson, with whom I had served my apprenticeship at Doncaster, took charge and remained at Edge Hill and later at Lime Street until his retirement. Jack was a Motive Power man through and through, and his caustic and sometimes profane humour was generally a match – if no more – for his independent and sometimes volatile Scousers.

The electric locomotives of Class 86 were the mainstays of the London jobs. They were marvellous machines and here at last one felt the same tremendous urge that I had encountered in France on the '16000BB' locomotives between Amiens and Paris and Lille. Men reared in steam are not bigoted when they have a machine that would leave Runcorn and pass Halton Junction at the top of the bank at twice the speed of a 'Duchess' or a 'Lizzie' under normal conditions of working, and the Liverpool men were proud of these locomotives and glad to show off their paces. People say that there is no longer the skill or the artistry that there was with steam. But the total responsibility is still there and things happen very quickly at the higher speeds. And in the late 1960s many of the London Midland electric locomotives had been fitted with the new air brake driver's valve. Most of the trains were still vacuum-fitted and the actual business of stopping quickly and at exactly the right place had become more difficult to perfect than anything that I had previously experienced. However, the arrival of the air-braked stock revolutionised the whole business of braking express passenger trains and the steady old vacuum, slow in application and release, became a comparative rarity.

There are the neutral sections: prior to each of these, the controller must be closed and run off by the time that the locomotive reaches the magnet in the 4ft which operates the circuit breakers. If the controller has not been closed in time, power will not be immediately regained after the neutral section. If the train is running uphill, speed will fall and valuable seconds will be lost if power is shut off too soon. And so, when there was time to be regained, it was important to strike that neutral with the tap changing complete and at the highest possible speed. Nothing in this, you might say! But in good old Britain there were no illuminated signs to tell one that it was time to close the controller, as in France. Each driver had to learn exactly where he was going to close the controller at a given speed to meet these criteria. And so were built up in men's minds, a series of markers: overbridges, lengthmen's cabins, a certain number of signals beyond a given point, or a lineside guide such as the pink house on the up road, south of Berkhampsted for the Berko neutral. It was a revelation to see a really good Edge Hill man pulling back lost time in the French style but with the extra panache that springs from our rule of thumb, homespun methods.

With an '86', time could be steadily regained. With an '84', in the rare event of one being let loose in emergency, it was very difficult to keep time, but with an '81', far less powerful than the '86', there was a little dodge, not covered in the instruction manual and frowned on by the electrical engineers, that was useful in pulling back time. Instead of closing the controller, it could be left until the last moment and then dropped one notch out of weak field. After the neutral section, the controller could be immediately opened wide without any loss of speed.

Such was the quality of the work that it was my pleasure, on a number of occasions, to arrange for our critics and also for influential people to ride on the footplate to Euston and back. Our detractors often have little room for the men who do the job: enginemen, guards, maintenance staff, signalmen. All too often this stems from a lack of understanding, a belief that a man can have no pride for his work within a nationalised industry, or simply that men who strike or work to rule cannot be conscientious, normal people. But there is no more natural PRO than the Scouse engine driver. The welcome to the visitor was invariably friendly, a seat would be made available, the tea or coffee immediately shared, whilst demonstration and explanation was always forthcoming. An inspector was always present but they were tarred with the same brush and knew exactly what was wanted. The visitor would be regaled with interesting stories, would see a new countryside, new landmarks, would absorb the confidence, alertness, the lack of tension; and certainly they would be doing the ton by the M1 south of Watford Gap and watching those cars which were not breaking the law seemingly going backwards on the crowded motorway. He would be shown where the Great Train Robbery took place in 1963 and, on the way home, he would soak up culture, albeit inaccurately, when shown the 'Gates of Jerusalem', the folly above Shugborough Tunnel. He would see a job well done and come back to Liverpool with memories he would be unlikely to forget.

The 'Liverpool Pullman' was a vacuum-fitted eight-coach train, tightly timed, and there were one or two places, for example south of King's Langley on the up road, where, when double yellows were sighted at 100mph, only a full brake application would stop the train at the next but one signal, if it was red. But, one morning in the early 1970s, when bomb scares had just started and when nobody took them seriously, I took my seat, not on the locomotive but in the first class Pullman near the front of the train. The Pullman Conductor, Bert Lock, had welcomed me aboard, the train was well filled and, after a few minutes, I realised that we were going to be late away. The Liverpool businessmen, waiting for breakfast, were deep in their papers and, on looking up, I found myself beckoned by a police officer to join him on the platform. On my way out I passed Bert Lock in his office, who said that there was some sort of a bomb scare but nothing to signify. The police had received a telephone call – a very Irish voice – to the effect that there was a bomb on the 'Liverpool Pullman'. They intended to search the train but that wouldn't take long they said, and nor did it, a number of officers strolling in the dignified and time honoured manner through the train and looking under the odd seat. Meanwhile, Bert Lock and his loudspeaker had sprung into action and he welcomed the passengers in his customary cheerful manner. He then went on: 'Sorry to say we shall have a few minutes' late start but nothing serious, the police say there's a bomb on the train'. This, in his muted Liverpudlian accent, with the second 'b' in bomb clearly pronounced, produced some laughter amongst the younger, more dashing first class passengers whilst the older members of his audience noted the point and carried on with *The Times*. Within a matter of days, trains and stations were being cleared for every hoax; but not on that memorable morning, when the 'Pullman' left five minutes late and British humour and British solidity reigned supreme.

Lime Street station had its characters. The Chief Inspectors were a race apart for more than one reason. Whereas at any comparable station on the Eastern Region there would have been a Loco Supervisor, at Lime Street the Chiefs, all operating men, arranged changes of working, engine changes and so on. This practice was accepted and worked very well indeed although, in all, the Inspectors had too much on their plates; and so we decided to introduce Assistant Station Managers on early and late turns.

Things happened at Lime Street in the evening, with people missing who should have been working, and there was much to do to improve not just the working but the tidiness of the station and the smartness of the staff. The two men appointed were sound, experienced and quite dissimilar: Cliff Beck, a local man, serious and determined; and Eric Steward, imported from King's Cross. Lime Street needed some new blood, and when that new blood came from London and enforced its wishes in the richest 'gaw blimey' accent, it was bound to serve a useful purpose. A few peculiar manoeuvres were sorted out and the two Assistant Station Managers, so very different, made a splendid contribution to the running of the station. In later years Eric Steward became a JP on the Wirral bench where I am sure he enlivened the proceedings. Almost worth appearing before him in court to hear him say, 'Fined 40 quid, Guvnah'!

Liverpool railwaymen were artists, not necessarily in deception, but, shall we say, in making the best of a difficult situation when on the carpet. I made it my business to hear appeals against most of more severe punishments, such as suspensions, reduction in grade and dismissal: I also heard all appeals against severity of punishment of drivers who had passed signals at danger. At an appeal it is easy to take the line of least resistance against the vehemence of the advocate, but to take the easy line undermines the authority of local managers and, in general, lowers the regard of one's staff for management. Equally, however, the case must be fairly made and, if it is not, there can be no grounds for support.

I have already referred to Arthur Williams who was my Assistant: he was older, greyer and more analytical than I was. He could smell a ruse a mile off and he had taken many of these hearings, impassively and unemotionally coming to a conclusion. There was a period when some lady carriage cleaners were in trouble and when they arrived in my office to appeal against their punishment, they were dressed to kill, hair a la mode, made up to perfection. Their advocate was a Liverpudlian with charm and a certain turn of phrase. The ladies, softly spoken, wide-eyed and innocent, won the day and I decided that the far less impressionable Arthur Williams would, in future, hear all appeals against lady carriage cleaners!

Fog, the greatest enemy of railwaymen, though not so commonplace or widespread today, places great responsibility on the shoulders of trainmen working at high speeds. I had been to a Euston meeting and travelled back to Liverpool with Driver Douglas Jones of Edge Hill and once of Wrexham, GC. It was a dark and murky winter afternoon. At Euston he had asked me over which section I would like to have hold and I said Stafford to Runcorn would suit me fine. By the time we reached Crewe the murk had turned to thick fog and going down towards Winsford at 100mph the green signals were just visible as we flashed by: but the AWS and its friendly bell made 100mph acceptable if uncomfortable at nil visibility. Suddenly the warning horn sounded to be quickly cancelled and the speed killed immediately: we were down to 35 at the single yellow and then had to set about finding the next signal. We crawled up to the AWS magnet, it gave us the horn and we knew we were 200yd from the invisible signal. We moved up yard by yard and suddenly the red light was staring at us and we came to a stand. 'Wait for the signal' said the telephone and so we did at this and several others, reaching Runcorn 40 minutes late. I got up with a sigh of relief and changed over with Douglas. The weather improved as we approached Liverpool; at Lime Street the, passengers streamed by us, unseeing, except for one. A clerical collar, a purple stock and a voice called out, 'Thank you, driver, for doing your best' – the then Bishop of Warrington, whom I saw the next day in the Atheneum and was able to tell that we really had done our best, against the odds.

Main line work is full of surprises: returning from a London meeting before the air brake became universal, I found a Liverpool driver being passed out in the use of the air brake, accompanied by an Inspector, Ernest Hillyard. Passing through Rugby at the regulation 60mph we saw the road set for Birmingham. We stopped at the next signal, asked what the hell was going on and were told that there had been a derailment on the Trent Valley line and that we should have been advised of this before we left Euston. None of us knew the road to Birmingham but to wait for a pilotman would have meant an interminable delay. We were practical railwaymen, we had the air brake and its quick release, and with luck we would branch off at Stechford and bypass Brum altogether. However, at Stechford the road was set main line. Under normal circumstances our driver would have known Birmingham-Wolverhampton, but Jimmy Rimmer had only recently transferred to Edge Hill from Birkenhead and before that he had been an Aintree man. Never mind, we took New Street in our stride, all green lights. At Soho we turned off to the right, dropping down until we joined the line from Stechford and Aston, eventually coming out on to the main line once more, at Bushbury. The air brake was our friend; so was the weather, sunny and warm; although we were on a strange railway we could go gently along at the low line speeds involved, watch for speed restrictions, observe them exactly and read the signals as they came to us. We were clear of Birmingham rather than standing waiting for a pilotman: and we knew our collective experience would see us through.

By 1973 the excursion business was very much in vogue again in the Liverpool Division. 1972 saw us running 180 special trains, whilst in 1973 we ran 336 trains producing revenue of £234,000. On one occasion the Division took 1,000 people on an excursion to Paris and this was but a pipe opener to a developing business and even greater endeavour. However the days had gone when specials were ordered by the dozen, when power, stock and men were found from almost limitless resources. The ordering of special trains had to be related to what was theoretically available whilst the day of the Running Foreman's magic touch was seen to be over. I remember being told by the Operating Headquarters at Crewe that I could not run an inspection saloon on such-and-such a day because no power would be available. Enough to make Fred Pankhurst of Stewarts Lane turn in his grave! But we ran our inspection train with No 5206, the Wigan Class 25 always cleaned to perfection at that splendid little depot.

Nevertheless, the Passenger people usually got what they wanted and it was Ron Cotton, coming from Newcastle, who introduced the Merseyrail concept. I was delighted to support him in what was a revolutionary idea. BR HQ had one or two acid comments to make about non-conformity with the corporate image but, thanks to support at both Marylebone and Euston, the Merseyrail tag had come to stay. The PTE ultimately agreed and it was not long before others followed the example of Newcastle and Liverpool.

And talking of those two cities, I travelled on the footplate one morning to Leeds, en route to York, with Driver George Griffiths of Edge Hill, he of the flashing teeth and smile and the one-time curly black hair earning him the nickname of 'Pedro'. As we approached Leeds, I opined that Newcastle men, Geordies, would relieve George and his mate and this turned out to be correct. 'The trouble with Geordies is you need a bloody interpreter'. This from George Griffiths, broad-spoken Scouser. The resultant exchange of information between the four men when we arrived at Leeds was worth watching and it was an open question who understood least!

Mention of George Griffiths reminds me of the only serious local labour trouble during my years in Liverpool, allegedly the home of industrial dispute. Liverpool was in the Cup Final in 1971. The specials were going to be well filled and Edge Hill was looking forward to working them. Locomotives and men would be no problem under these circumstances but Operating Headquarters, in search of economy, decided to take the Liverpool men off at Crewe or Rugby in order to stick to single manning. As long as I'd been a railwayman, men from the cities concerned in the Cup Final worked the trains throughout if Wembley was on the line of route. But not this time and once the workings were issued, we were in trouble. I came into the picture in the last 48 hours but to little avail. The LDC, chaired by George Griffiths, was told that we should either cancel all the Specials – in which case he'd do well to board his windows up – or we should run the trains by other means. In fairness to George, he went to great lengths to effect a reconciliation but the temper of a Liverpudlian, once aroused on such matters, is within easy reach. It was not to be and the majority of the Edge Hill men struck work on Cup Final day. We ran the specials at the expense of many other trains: don't ask me how! I was up all night and did not go home until the last special was home and safely inside at Edge Hill. And Liverpool lost to Arsenal!

I was deeply involved with railways when off duty. Lunches and dinners were part of the normal way of business life as were the many social or staff gatherings to which Gwenda and I were invited. But, as previously mentioned, the public was interested in us! Sometimes critical, of course but interested and very different from the comparative impersonality of London. Certainly, people would ring one's home to play pop; certainly James J. O'Brien, then our Operating Manager, spent hours one Bank Holiday in successfully finding a carriage from which the window could be removed to accommodate a stretcher, carrying the friend of one of our best customers who expected us to do this without hesitation, but who was equally grateful. Certainly, our village butcher, Charlie Swift, would take me to task over the state of the railways while I collected the joint and, without doubt, if Gwenda and I went out to dinner at friends, another of the guests within five minutes of our arrival would have me pinned in a corner on the subject of car parks at Lime Street or, worse still, the expense of rail travel. Never mind, I loved it, I was proud of our railway and I missed it all dreadfully when I came south to Amersham and was no longer the central figure that I had been on Merseyside.

Arthur Behrend was retired and lived in a beautiful home on the fringe of Burton. He had been connected with the Port of Liverpool all his life and his firm of Bahr Behrend was widely respected. He was an author of standing and we met through *Steam in the Blood.* We only knew each other a few years and yet I felt I had known him all my life. In his youth, rather than the family firm, he joined the LNWR, but soon afterwards came World War 1 and when he returned in 1919 the railways offered little prospect, and so he ultimately became the senior partner of the firm, to finish for ever with railways; and yet not quite for ever: Arthur had a delightful and whimsical sense of humour and the recognition of his services to the LMSR during the General Strike caused him much amusement. When he died, his widow Chris gave me a treasured medallion presented to her husband 'For service in National Emergency May 1926', and how he loved to tell the story. He and Chris had not been married all that long when the General Strike was called, but some days passed before he felt the urge to volunteer for railway service. He decided to go to Chester, only 10 miles away, to report for work. At Chester General he found the Stationmaster who took his name and address, and promised to send for him, although not very much was moving in the Northwest. As Arthur was leaving, the stationmaster asked him whether he would kindly take a luncheon basket to the engine shed and suggested he might well find work there. In due course, he handed over the basket and had been signed on when the foreman received a message to say that the strike was over. A few months later Arthur received his treasured medallion for services rendered in the General Strike. Never was the carriage of a luncheon basket over 300yd more handsomely rewarded!

Members of the Parliament were fairly thick on the ground in the Liverpool Division. There was a very stern critic of British Railways who represented Ormskirk, presenter of the unanswerable case: there were Conservative and Labour MPs who went out of their way to help us while at the same time strongly pressing their ideas for change and improvement and there was Bessie Braddock. When I took her out to lunch at the Adelphi to learn all I could from her about the life and people of Liverpool, she had not long to live, but I was not disappointed.

The Adelphi had seen many past glories. The older members of the staff loved to relate stories of Grand National nights and of the many great personalities of a bygone Liverpool. But the Adelphi, the great Railway Hotel, still had atmosphere and very personal and friendly service. The Exchange Hotel, over the L&Y station and closed during my time in Liverpool, was smaller and less famous but still had a certain atmosphere that was very Liverpudlian.

Every year it was our delight to entertain Mirabel Topham of Aintree. Jimmy Trainer and I would be waiting on the steps of the Adelphi for the elderly Rolls to arrive. The Head Porter would open the door and as the marvellous lady swept up the steps, usually in purple and wearing a huge hat, one savoured something of the great days. She was always accompanied by her nephew and niece who lived with her at Paddock Lodge near the Aintree gates of the famous racecourse. Every so often, Jimmy Topham would ring to say that 'Aunt would like you to come to lunch at Paddock Lodge'. You could bet your boots that Aunt wanted to do business and this usually meant going through the wringer, but it was followed by a marvellous lunch, prepared by her niece Pat, which helped one to recover from the force of Mirabel's requirements. Lunch over, Jimmy would propose a run round the motor race track in his 1927 6½-litre Bentley!

Every Grand National day we entertained our customers and friends in a manner appropriate to the occasion. A train of seven or eight coaches would arrive during the morning in Fazakerley Yard, from whence one looked down over the boundary fence on to the course between Beecher's and the Canal turn. We would lunch our friends in splendid style, we had TV in every coach and our own railway bookie who came from Crewe. Gradually the time for the great steeplechase grew nearer, the excitement and atmosphere indescribable and then the thunder of hoofs, the fallers, horses running out or leading the field without their riders. Our guests will never forget the experience and nor shall I ever forget Red Rum's first victory a few months before we left Liverpool.

XVII

Cheshire and Lancashire

In 1968 the Cheshire Lines passenger services in the Liverpool Division ran from Lime Street to Manchester Central and, after a short while, to Oxford Road, via Warrington Central; and from Liverpool Central to Gateacre and, through deeper Cheshire, from Chester to Manchester via Northwich, the journey through the Delamere forest being rather lovely.

Northwich was the manufacturing centre of ICI activity in our Division, although the Mond Divisional Headquarters was at Runcorn and the huge Castner Kelner complex nearby. Northwich railwaymen were there very largely to service the ICI factories and I insisted on maintaining a separate Area Manager at Northwich, although on paper this was barely justifiable. However, the presence of an experienced manager and a young assistant not only served ICI well but ensured excellent practical experience for a succession of young men in both railway operation and in liaison with an extremely important customer. At Divisional level we regularly met our opposite numbers in ICI, and occasionally members of the Mond Board. We shifted their traffic pretty efficiently and if we failed to do so we heard about it very quickly, not only from ICI but from our own Northwich staff.

Many railwaymen play an important role in public life and we were fortunate to have such men at Northwich – one of whom was awarded the British Empire Medal shortly after I left Liverpool. Ron Carey was a Northwich Councillor and later Vale Royal Chairman: he was a County Councillor, he was on innumerable committees, governor of an excellent grammar school, ASLEF Branch Secretary and Chairman of the LDC. This would have seemed fairly easy had he not had to come to work but he rarely asked for special leave and frequently changed over with other men to take nights or early morning turns in order to attend to his public duties. His great gifts earned him deep respect, not only in Northwich and in Cheshire but in the industry that he served so loyally, and I never heard a word said against him.

The staff at Northwich were concerned that, quite frequently, they were short of locomotives to work ICI trains which had then to be delayed or cancelled. There was no depot allocation at Northwich, no fuelling facilities, indeed only trouble-shooting fitters on each shift who kept the passing DMUs going and attended largely to wagon maintenance, in what was the skeleton of the old CLC steam shed. Locomotives would arrive off their previous workings with fuel tanks nearly empty, usually late in the day.

Before working their next duty, the tank had to be filled, which involved running light to Birkenhead or Crewe for fuel, a scandalous waste of time, crews and money and a curse to Running Foremen with few men to spare for such a futile exercise. Obviously, a fuelling point was needed at Northwich but this would cost money: the London Midland Region had laid down where fuelling and examinations would take place and showed no willingness to amend the plan: the HQ operators seemed happy enough with the situation. We were not and we had to find our own solution.

J. K. Lord, Divisional Maintenance Engineer, loved the unorthodox and so did I. Denis O'Reilly, Area Manager, Northwich, no Liverpool Irishman but the real article, loved anything out of the ordinary and, in any case, he was there to run trains to time. The wall of opposition served only to sharpen my resolve to achieve what should have been patently obvious: the solution, in our hands, lay in a siding at Kirkdale, Liverpool, where we were about to close a small and homespun DMU fuelling point. One Sunday, two four-wheeled tenders of great age made the journey to Northwich. Pipes were connected, a tanker arrived and we had our fuelling point: furthermore, staff were there to do the fuelling at no extra cost and to carry out maintenance examinations that were not officially recognised – for a time. Availability and punctuality improved overnight, our staff were delighted and the venerable little tenders made their last and by no means least contribution in a career which had already lasted well over a hundred years.

Mouldsworth was a delectable spot and so was Delamere. The former was presided over by two 'old retainers', Eric and Albert, proud of their station which was an interesting place, with gardens, rhododendrons and a large yard in front of the station house and booking offices. The signalbox had two single line instruments to the west and ordinary block bells in the Delamere direction: tokenless block to Mickle Trafford, the first of its kind in our part of the world and normal single line equipment down to West Cheshire Junction and Helsby. It was in that yard that I met the Cheshire Forest Hounds for the first time in 1969. Needless to say I was heavily disguised and, for some extraordinary reason, the bush telegraph had not advised Eric or Albert that I would be sitting on a horse in their yard. And so I was not apprehended! On the other hand, long after I had left Liverpool, and when I was hunting at Alvanley, nearby, word reached the Delamere signalman now in retirement that I was on the premises and he came across from Norley to see me. An hour later, I was on my back in the mud but luckily he did not see his former Divisional Manager in so undignified a position!

Delamere had a white beard and a crisp turn of phrase. His first well-chosen words to me had been, 'So you're Mr Hardy, our Divisional Manager, well, allow me to congratulate you on your mismanagement of the Division.' Words said without passion, with precision, civility and with sincerity, for Reg was a man of independent mind and of intelligence who disagreed strongly with national and therefore Divisional policy on railways in general and on the Cheshire Lines in particular. I would always rise to this sort of challenge, by no means isolated, and, if possible, return it blow for blow and I looked forward to our regular visits to Delamere with the inspection saloon where we took lunch in the siding which still existed at that time. On one occasion, we were honoured with the presence of John Betjeman and I took him across to the signal box while lunch was being prepared. As one would expect, there was an immediate understanding between the two men and, much to my amusement, the future Poet Laureate was asked to pull off for the train which had just entered the section from Mouldsworth. John took two or three ineffectual little tugs at the lever controlling the home signal, just outside the door and turned round to roar with laughter, 'Come on, Sir John, you'll have to do better than that, I can't stop this train with my Divisional Manager here after all I've said to him!' More tugs and much to everybody's surprise, the home signal came off and Reg Holmes finished the job amidst a tumult of laughter.

Later that same day we dropped down from Edge Hill to what was left of the steps up to the Moorish Arch. I had never been so deeply affected by the historical significance of the remains of these great works where once had stood the original locomotives and the original coaches and the original men of the Liverpool & Manchester Railway. John walked slowly towards the original tunnel; hat in hand, and followed by a mixed but respectful group of railwaymen, he declaimed on the historical past of Liverpool and its railway. It was wonderful and memorable.

It may seem quite extraordinary to some when I say that a day's hunting holds the same fascination for me, the same almost primeval excitement as does a day's battle with a steam locomotive. As I write I shall in 10 days' time be firing the *Flying Scotsman* up the Long Drag from Appleby to Garsdale, a fair assignment for one not far off 60. In five months' time we shall be hunting again and I shall have that same feeling of excitement that is pervading me at present at the thought of the battle to be won, high up in the Pennines. Is not the iron horse almost flesh and blood, does not it depend for its performance on human skill and fortitude? There is the same challenge, the same excitement, the same partnership, the same comradeship as with a horse out hunting. One finds too the same type of personalities, the same willingness to help and sometimes, the same language.

The two horses on which I have most enjoyed my hunting were great characters. The first was 'Hereford', owned by a Cheshire farmer called Bob Peacock. Both he and Hereford have passed on but neither will be forgotten. In later years came 'Osberton Lazarus', a thoroughbred Cleveland Bay, owned by Cathy Clarke of Arley in Cheshire. He was young and had much to learn: I was older but also had a fair bit to learn and between us we got along wonderfully well. He was enormous, a personality, the years stretching out ahead of him, no day too long, no fence too high (except the little ones to remind me of his independence), no load too heavy, a one-speed job who goes anywhere and does anything. 'Big Os', all 17.3 hands of him, is the living manifestation of the greatest of all British freight locomotives, the Great Central ROD or 'Tiny'.

The Garston area was tough, though not geographically large, for it covered South Liverpool, Widnes and Runcorn. It was presided over by Ted Merryweather, a Midlander but Liverpool by adoption, well equipped to deal with the motley collection of Scousers in the vicinity of Speke. There was plenty of unemployment in Liverpool in the early 1970s and although we were constantly reducing our staff, we were nevertheless very short of guards in this part of Liverpool. From the Job Centres came folk who took one long look at our conditions and left us to it and we did our best to fill these vacancies by advertising in the press.

Driver Maurice Wilkins sometimes acted as a relief Running Foreman and he was unique. In 1969 a party of Calais SNCF enginemen were about to visit the country and I needed two interpreters. One was easy, Ben Hervey Bathurst of Eastnor Castle who had worked with the Resistance and who loved to meet the same type of Frenchman, in peacetime. Who could be the other? Muriel Brownbill, my secretary, had heard there was a Frenchman at Garston, a driver. And so I met Maurice Wilkins, Parisian by birth, Liverpudlian by adoption, who spoke a mixture of Cockney, Scouse and French which, when he was aroused, grew in volume and complexity of sound. When he met the Calasians they thought he was an Englishman, especially as he wore his driver's uniform, but within an hour, the shackles had been cast off and his French came flowing back.

The SNCF party was overjoyed and could not believe it possible that a man born on French soil could be an engine driver in England. But then, anything was possible in Liverpool. Maurice had come across at the time of Dunkirk, little more than a boy.

He joined the Navy and somebody said, 'Call him Wilkins'. So Maurice Wilkins he became and after the war he joined the LNER at good old Stratford as an engine cleaner. He met a Liverpool girl and transferred to Brunswick CLC and later on to Garston, and lived happily for the rest of his life on Merseyside.

It is small wonder that I sometimes forsook the office for the front line of the Division where I could see for myself what was going on, and whether it could be improved: where one could learn some new facet of railway operation, or meet somebody who had not yet crossed my path, perhaps to the advantage of both of us. In the same way as Bobbie Lawrence, on his visits to Liverpool, would go through my pockets, so I would do the same with my own staff in the Division.

I have left the northern end of the Division until last. With the closure of Bamfurlong and some of the Crewe freight yards, Warrington assumed even greater importance, and when the closures were first made the congestion in Arpley Yard was considerable but this was a problem to be overcome by Danny Whelan and his people. The Warrington Power Box was opened in 1972 and was manned by a number of experienced signalmen. We had minimised the redundancy in this grade by leaving as many vacancies unfilled as possible in the surrounding signalboxes, particularly on the freight lines in the ' neighbourhood and we got properly caught when the NUR called a 'work to rule' over some National issue. Most of the boxes to be closed were covered on 12-hour shifts and even then by the skin of our teeth, so restriction to eight hours per shift left great gaps which were impossible to fill for the duration of the trouble.

We combined the Warrington, Wigan and St Helens areas and put them under the management of a young man, Paul Watkinson, destined, one hopes, for very high places. At Wigan North Western, we opened in 1972 a new main line station, replacing a mid-Victorian slum, whilst down below street level was the Wallgate station with Liverpool to the south, Manchester to the east and Southport to the west, a railway stretching across level farmland and memorable for the thousands of red-hot pokers growing by the side of the track. At Southport, there was a signalbox every bit as impressive as those huge North Western establishments, there was an indifferent DMU service to Manchester, replacing as one was frequently reminded, the steam hauled club trains of better days. Southport was the terminal of the dc electric railway from Liverpool: we had closed Meols Cop electric train depot, impressive in its triangle of lines and concentrated maintenance at Hall Road and Birkenhead North. An Area Manager presided over these railways, covering Liverpool Exchange station until years later when the Link was opened and the L&Y trains at last crossed Liverpool.

Between Southport and Birkdale there were two elderly signalboxes controlling the crossings at Duke Street and Portland Street and, in one of these, gloomy and gaslit, worked a certain William Preston, one of the Southport LDC representatives.

Profiting from the experience that I had gained at King's Cross, I decided soon after coming to Liverpool to set up the informal review or communication meetings with groups of staff on a six-monthly basis. It was natural that there would be suspicion for such activity was unheard of and so, at the very first meeting, my patience and temper was tested when Raymond Dickey, signalman at Hooton, rose both to his feet and to the occasion. Never had I had such a lambasting as the blood gradually rose up his neck and he became scarlet with indignation as he informed me that he had no love for Divisional Managers and little respect for them. I loved these meetings, I enjoyed the cut and thrust and so did some, if not all, of our team but I repeat that it is good for the senior members of a management team to be cross-examined by those whose destinies they control.

I never analysed what our staff, as a whole, made of these meetings. I knew that certain individuals looked forward to them for a variety of reasons: was it just a day out with an early finish, did they look forward to coming, did they think it a waste of time, did we really mean what we were saying, could we be trusted to keep our word? One or two of my fellow Divisional Managers maintained that the Area Manager should hold the meetings and deal with questions at Area level. That may have been all right in their Divisions but we were not ready for it and with us the lead had to come from the top.

Bill Preston, of Southport, always sat at the back, and delighted in his dry and humorous way, to throw a small but smoking bomb into the arena. For example, in very deliberate Lancashire tones, he would ask: 'What do you think of Worker Participation and how would the concept be applied in the Liverpool Division?' Such questions took some answering and almost invariably started a splendid, if controversial, discussion which had to be carefully and firmly controlled.

By 1972 we had reduced the number of areas to six from the 14 which existed in 1968: the areas were much larger and we felt it was time to allow the Area Managers to run their own meetings whilst insisting that we came along just the same, but under local chairmanship. Frank Worsdale, the Area Manager at Southport, decided he would allow written questions to give those who normally did not speak to have a chance to put their point of view. Bill Preston, when he spoke, was much to the point but on this, my last meeting with the Southport Area, he decided to put pen to paper. Such was his letter, with the sting in the tail, that it is reproduced in full. Such was his letter that I have kept in touch with him in his retirement these last 10 years. He has been a very sick man but if ever a sense of humour, a philosophical outlook on life and a devoted wife have seen a man through, Bill Preston is that man.

To Area Manager,
   Southport   12.4.73

Mr Worsdale,

     I wish to raise the undermentioned at your Review Meeting.

     No doubt, several representatives will remember the uncertain times that existed after Dr Beeching's proposals were published a few years ago.

     What would be the position on this Division? How many men would join their brothers, who had fashioned out of steel in railway workshops engines like 'Mallard', never to work again.

     The uncertain outlook and the morale of the men at that time was at its lowest and men who had military training during the war years likened it to the position facing Field Marshal Montgomery before the battle of Alamein.

He went amongst his men and explained the position facing them all, raised the morale of his men sometimes with a cigarette, but where was the leadership coming from on the Railway on this Division?

In my opinion it's worthy of consideration that leadership was given by Mr Hardy when he started his Review Meetings.

I recall being asked by men I represent 'was Mr Hardy just one of those old school tie wallahs that knew nothing about Railways and what was he like?' And I had to confess I didn't know but I did tell the men he was partial to French cigarettes and sucked 'a mint with a hole'.

Later after reading the review of his book, in 'Modern Railways', I told them he was a good railwayman. Since knowing him, I have added to it he is a 'Jolly Good Railwayman'.

It was noted with interest where a railwayman was polishing some brassware and Mr Hardy let it be known that if he wanted to do it, the materials would be made available to that man. It seemed to me this was not a lot of 'bullshine', if that is the correct expression but it meant that if you took an interest in your job, you would have all the encouragement from the 'Top Brass'.

Of course there will be differences of opinion between management and men. One has in mind the position of men at Portland Street and Duke Street cabins, working under the most dismal gas lighted conditions.

It would be wrong of me to use this occasion to bring this to Mr Hardy's attention, so I will not approach him but one cannot but remember an answer he once gave to a question when he said 'there are more ways than one of getting what you want'.

Maybe when Mr Hardy is tidying his desk to leave to his successor, he will leave the application for electric light for attention, what a wonderful way in later years Mr Hardy will be able to turn to his grandchildren and say 'I was instrumental in bringing light into some dark places'.

The men I represent charge me to wish Mr Hardy good luck in his new appointment and we will follow his career with interest.

The letter was read out by the Chairman, and once the laughter had subsided, I said I would see what could be done and, of course, Bill had his electric light within a couple of months! Yes, there are many ways of getting what you want. One day my grandchildren will read these words and they will laugh too but I hope that they will also come to understand the genuine joy that their grandfather experienced not just from railway work but from so close an association with real practical men.

When Bill Preston was writing that letter, the sands were running out for me. The Field Reorganisation was imminent, my job was to go but as I had done the same thing to many others, I had nothing to grumble about. On the other hand, I was bound for BR HQ in September 1973, to a job I did not want, from an area in which we could have lived for the rest of our lives. Had we been able to read the future we would have stayed where we were, for 18 months later the Field Reorganisation fell flat on its face and the Divisions carried on as before. A ludicrous waste of time and money that did nothing to support the contention that management could manage. But, for a few more months, there was much to do and I was still surrounded by the same splendid team and personal bodyguard.

For example, my secretary, Muriel Brownbill, who was so good with the public. She would take a battering from he who had lost a parcel and who wanted to take it out of the Divisional Manager. She knew who could find a lost parcel and it certainly wasn't me. In silvery tones, sitting at the table where our dictation had been disturbed by the call, she would charm the furious contestant who had worked himself up to lecture me on the evils of nationalisation. 'I'm afraid you can't speak to Mr Hardy at the moment, he's on his way to Wigan but I know that we can find your parcel and I'll put you through right away to Mr Alty.' Mr Alty was a genius at finding property that had gone astray and he was equally masterly with the soft answer that turneth away wrath. His greatest triumph was to locate Sir John Nicholson's daughter's trunk which was found in a corner of the LIFT Shed at Stratford E before it had even started on its railway journey. It was Jack Alty, not me, who should have had a splendid lunch with the Board of Martins Bank, on Sir John's right, for he had done the work.

On the personnel side, we were blessed with the persistent, quiet and unassuming Donald Walker. When I arrived at Liverpool, with King's Cross memories of Reg Clay – forceful and experienced – I wondered at this little man whose voice was sometimes little more than a whisper: could he really be effective? One must simply say that he was a step ahead, that he saw to it that we were the first LM Division to complete an Area reorganisation and that he was a tower of strength to us all.

It must have been about 1970 that we realised that we had a 'mole' at Edge Hill and that a newspaper was being produced and circulated amongst all grades of staff. Naturally, a copy came our way and we found that it was even more anti-NUR than anti-management. We were in a dilemma even though we thought we had found out who was behind it. The man whom we felt was responsible was young, well spoken, an exemplary timekeeper and a sound railwayman who co-operated with supervisors and colleagues alike. His name was never associated with the newspaper and yet we knew that, while he was with us, we should not have any peace and that his ultimate aims were far more sinister than the publication of a scurrilous news sheet. As one cannot dismiss a man on such a flimsy pretext, Donald applied himself to the problem. He found that, three years before, the man had given a false reference when he applied for a railway position; facing him across the table in his office, with no witnesses present, Donald succeeded in getting him to resign forthwith. What he said and how he did it he would never disclose, but the last we heard of our friend was that he was addressing a large gathering of Pilkington's strikers in St Helens, and telling them that the railway workers of Liverpool were solidly behind their cause!

References in the old days were essential before a man could start on the railway, even as a labourer: the references were followed up but this practice lapsed with the war and had not been reinstated by 1948 when a certain Lime Street guard, Taffy Evans, applied to join the railway at Liverpool Central. His application was examined, he was asked one or two simple but searching questions and then it was time for an eyesight test.

'What's the colour of that wall', raps out the Stationmaster.

'Maroon and cream', says Taffy.

'Right, start on Monday', says the Stationmaster!

Fred Comeskey was our Accountant and he mothered me particularly when the time approached for our budget inquisitions. It was one thing to ask the questions, quite another to account for our financial activities to a General Manager. In retrospect, I should have taken the detail more seriously, studied it more deeply, asked more questions, accepted less on its face value. The more I asked, the more I should have learned of such matters. My style of management was steeped in the old fashioned principles of leadership, of getting things done, of raising the performance of a large body of people to the highest possible standard whilst still reaching our targets. The next generations of managers and engineers were to have a far better grasp of accountancy, of statistics and of business requirements, and in the last nine years I have done everything I could to encourage this attitude of mind in younger men, whilst stressing, with all my heart, the ever present need to inspire and to lead with courage and fortitude.

The Liverpool Division had become part of my life: there was much to be done and a change in the top manager after 5+ years is no bad thing. Bill Bradshaw, my successor, had arrived to take charge for about nine months until the Field Reorganisation was completed. In fact he stayed two years, and upheld to the full the traditions of service and involvement that had meant so much to me and my predecessors. He built on what was good and changed what needed changing: above all, in the months after I left, he sometimes sought my advice on personalities and that I shall never forget. I am glad that I made our changeover friendly, enjoyable and comprehensive and that the same splendid people with whom I had worked served him so well.

During the last week we were given a fantastic send-off, but this did not prevent me taking what I knew would be my last disciplinary hearing at which the verdict went against Driver Jack 'Robbo' who revelled in controversy. When the last day came, I was uplifted by so much kindness and generosity and by the little gifts pressed on me by the most unlikely people. At the end of the day, when I walked out of Hooton station, Leading Railman Leslie Ames, proud of his station and his passengers, was waiting to hand me an electric clock which hangs in our kitchen to this day; every year until he retired we rang Leslie at Hooton to tell him the old clock was still going strong.

Our garden in Burton had been of over two acres. Billy Ollerhead, Driver, Birkenhead North, had tended it for over five years: he too had become part of the establishment and whilst we rarely talked shop I suppose it occurred to me that there was some fairly flexible Mersey Railway rostering even in those days. Seriously though, running an intensive underground electric service with such short peaks as existed in Liverpool inevitably produced a surplus of men when the service thinned out later in the morning and evening.

And so we sold our house and bought a house of character at Amersham in Bucks. With Peter still away at school, at Marlborough, it was a dreadful move for us financially but night follows day, life is for living and the Provident Mutual saw us over the hill and eventually, years after, into calmer waters. Railway Officers of my generation were, in general, not particularly good at handling their own affairs but the PM Agency Manager in London, Peter Whipp, saw me through this crisis by personal service and guidance of the highest quality.

We were on our way on 18 September and Albert brought the 'Green Goddess', our old 24-litre Riley, down the motorway next day. We have returned many times to Merseyside, and as long as my bones will allow it I shall have my confrontations, along with 'Big Os', with the fields, fences and hedges of Cheshire. Merseyside is our second home for we made many friends, and whenever the train passes over the Wavertree arches, and I see the familiar panorama, the two Cathedrals and the backcloth of the Welsh mountains, as they say in Liverpool 'I'm made up'.

'Une figure legendaire'

France! How little I really know of that great country and yet how much have I enjoyed myself in that very different world across the Channel. What happiness I have had from my association with the French railwaymen, what happy homes I have visited and what memorable journeys I made in the days of steam which came to an end on the Nord Region in May 1971, when I was invited by the SNCF to travel on the last steam hauled train from Paris to Calais.

The debt that I owe to James Colyer-Ferguson is enormous. It was he who persuaded me to go to Paris with him on the footplate in April 1958, it was he who was the interpreter, who had done all the necessary preparation. I simply stood fascinated as the legendary Chapelon Pacific No 231E16 backed on to the train at Calais Maritime before James made the introductions to a man who was to become a close friend to us both, Mécanicien Henri Dutertre, a mere 35 years of age. But later in the journey, although I had lost most of my school French, my successful use of the shovel from Amiens to Paris was to create the first bonds of friendship, strengthened that same evening by our introduction to Monsieur Philippe Leroy, the Locomotive Superintendent of the Nord who was to do so much for us over the years. Next morning, a Sunday, he was at the Gare du Nord to see us off on the great No 232S002. Here we met another man who was to become a great friend and frequent visitor to this country, Mécanicien André Duteil, just over 5ft tall, a little ball of fire, who retired at the age of 50 in 1964. Henri and Philippe Leroy, the quiet, determined, serious Frenchmen; André, voluble, excitable and yet professionally calm and competent, and a master of the Gallic gesture, beautiful in his use of the Westinghouse brake. And then from Lille to Calais, on No 231E36, Emile Lefebvre and Jacques Deseigne. Emile, who finished his career on E36 and who retired when she was withdrawn, visited England with a group of Calais men in 1969. He sat in the cab of the Romney *Green Goddess* quietly and unashamedly weeping with the joy of being reunited to the form of traction that had been his life. Jacques Deseigne, Chauffeur de Route, black, smiling, gruff of speech who later became a Communist Councillor in Calais, produced a nice bottle of red wine from his corner. And it was Jacques who, in later years, whilst watching my LNER use of the shovel on a wide Chapelon firebox, shouted to my friend Bill Thomas, 'Quelle panache, quelle panache'!

Since that first weekend I have been across 88 times, nearly always with a railwayman or a personal friend to share my pleasure. Steam has gone these 13 years, all my friends have retired and, of course, the magic is no longer there. But I now have a love of France and the French people.

In *Steam in the Blood* I took my readers with me on those footplate trips. Except by the men we already knew, we were rarely if ever expected when we climbed up into the cab, showed our passes and were immediately welcomed as members of the brotherhood of 'guelles noires'. No special arrangements had been made and thereon lay one of the greatest pleasures. We were simply English railwaymen; I was able to share the work and atmosphere of the footplate whilst holding no official position in France; I saw things as they really were, and, with my experience of men and machines, could understand those with whom I worked.

Towards the end of my railway career I was paid two great compliments by practical railwaymen, and one was in France. I was travelling on a 'BB67000' from Calais to Boulogne. I had not previously met the Mécanicien who had recently transferred to Calais from Amiens. After a few minutes of conversation, he suddenly looked at me and said: 'So you are Monsieur Hardy! Do you know you are a legendary figure, "une figure legendaire?"'

.

XVIII
French Interlude

The Chapelon Pacific was the most remarkable, brilliant locomotive on which I have ever worked. It was delicate, complicated, incredibly economical when working hard, so very powerful and yet almost slender and graceful in appearance. Its brilliance became apparent when there was real work to do.

The last journey that I made on a PO Pacific was from Calais to Boulogne on the 'Flèche', Train 82. The engine was E23 and we had well over 600 tons, nearer 650 indeed. Our driver was Lucien Ducrocq, a young 'mécanicien de remplacement' or in our language, a 'spare man'; the fireman was the venerable Auguste Beauchamps, nearly 54 years of age, known as 'Le Vieux'. Auguste simply gave me a packet of Gauloises, a plastic tumbler of red wine, and the shovel, telling me at the same time to get on with the job and bemoaning the fact that I was getting off at Boulogne.

We mounted the long grind at 1 in 125 up to Caffiers as though it were 1 in 300. My fire was blindingly white and yet my rate of firing seemed no more than with a light train. Steam was being made available at a tremendous degree of superheat, it was being used to great advantage, and yet the fire was not being torn by the blast so that I was not unduly exerted. Lucien stood in his corner, gradually increasing the high pressure cut off to meet the low pressure figure of about 60%. He was a big man, obviously delighted with the turn of events. 'Le Vieux', one hand on an injector wheel, the other holding a tumbler of wine, his back against the direction of travel, smoked, drank and, as befitted a PO fireman of 20 years' experience, offered his young driver a stream of outspoken advice! Once over the top of the grade, he paused to whistle to the paralysed lady in the cottage on the low embankment and at Boulogne, with regret, we left them to it, the 36-year old driver, and the fireman who, despite his age and his constant stream of repartee and his apparently slow and untidy methods of working, always had steam when it was needed.

The last time that I travelled on an SNCF Pacific, the K82, was on 26 May 1971. This engine was of PLM design and the class survived both the Chapelon and Nord locomotives on account not only of their large numbers, but because of their comparative simplicity. With the lighter trains they were said to be more economical than the PO locomotives but, good though they were with the very heavy trains, they were not in the same league. However, they were easy to maintain under the conditions that existed towards the end of steam when both cleanliness and maintenance standards fell from the very high level that had existed for many years.

On 25 May I caught the night ferry from Victoria to Paris and together with a very dear French friend, Oscar Pardo, made my way to Joncherolles depot where K82 was being made ready to work the special train to Calais, commemorating the end of steam on the Nord Region.

The Mécanicien was Jean Guelton who, along with Gilbert Sueur, was the last of the Calais 'Senateurs', the men who had their own Chapelon Pacifics. Guelton had driven E32 and then E22: Sueur E41 and then E14. The Fireman, Michel Lacroix, had worked most of the special trains in recent years and had fired E5 to Pierre Beghin. Both men were just right for the big occasion, Guelton huge, smiling and phlegmatic.

With them, and in complete charge of affairs, was Edmond Godry, the Chief Locomotive Inspector at Calais: a great personal friend, a disciplinarian and a real charmer. He could have been the Chief Inspector of the Nord had he been able to move to Paris. Edmond had spent much of the war in Germany, in forced labour camps before returning to complete his apprenticeship to become fireman, driver and inspector by the age of 29. Guelton was 48, Lacroix 41, Godry 49.

There were others standing near the K82 at Joncherolles that morning. An Englishman, George Carpenter, then of Westinghouse, was deep in conversation with Edmond Godry; George, whose profound technical knowledge of the steam locomotive had won him the friendship of M. André Chapelon. Nearby, but not altogether to my surprise, stood the legendary Pierre Leseigneur, retired some years from his position as Chief Locomotive Inspector of the Nord, but nevertheless ready for a day's work in overalls and beret. And so, as we stood beside K82, we were eight and we enjoyed our talk and then it was time to join the train in Paris. At the Gare du Nord, the platform was thronged with French railway personalities, past and present. It was a mild, sunny day with a gentle breeze and I looked forward to a splendid lunch and a relaxing journey to Calais in first class splendour. I was wearing an excellent suit as befitted the occasion, and promptly at 12.07 we slid almost imperceptibly and silently away from the Gare du Nord, for the last time behind a steam locomotive. Lunch was served: Jambon de Prague Braise Bourbonnaise, Entrecôte Maitre de Chais, Salade de Saison, Plateau de Fromages and Bombe Brasilia; and Vin Listrac. You may wonder why I quote the menu – we shall see!

Exactly on time at 13.30 we arrived at Amiens: I finished my cognac and walked up to the engine where all was well. I sauntered across the track to where the pilot engine, No 231K8, was standing, to talk to Gilbert Sueur and his chauffeur, but on regaining the platform in front of K82 I was apprehended by a smartly dressed and authoritative little Frenchman, M. Ravenet no less, the Chief Mechanical Engineer of the Region. 'Monsieur Hardy, you will travel on the engine to Boulogne'. Monsieur

Hardy immediately thought of M. Bernard Weatherill's beautiful suit and what it had cost him but tactfully replied: 'Ah no, Monsieur, there are many French people here today who deserve that honour far more than I'. 'Monsieur Hardy, you are our guest and you will travel to Boulogne on the K82'. M. Ravenet was clearly a second L. P. Parker and I was able to scrounge a dustcoat from Oscar Pardo, who had come prepared, but so far had not been singled out. For once I had no intention of working and intended to get up into the fireman's corner and stay there, keeping tolerably clean in the process. My hopes were quickly dispelled by the arrival in the cab of Oscar Pardo to whom I handed back the dustcoat, and when Pierre Leseigneur climbed the steps to the footplate it was clear we were not in only for a tight squeeze but an exciting time.

We had begun to sort things out when there was a shout from the platform. Opening the door, we saw two restaurant attendants carrying trays with plates heaped with entrecôte Maitre de Chais and Cheese, together with a basket holding bottles of wine with the corks drawn. Delightfully and fairly, the crew had not been forgotten but as we were about to get the right away, it was difficult to see how the three principal contestants were going to do their work, whilst eating an enormous steak on a plate with a knife and fork, in working a hand-fired Pacific locomotive at high speed, along the valley of the River Somme.

Naturally, the problem was solved. Pierre Leseigneur, already well dined and lubricated, took charge of the regulator whilst the rather well dressed Englishman, M. Hardy, spent the first 15 minutes of the journey in vigorous action, whilst the three gourmets dodged his flying shovel and pronounced not only the steak but the wine uncommonly good. Not to be outdone and obviously moved by the spirit of the occasion, Leseigneur seized one of the bottles and scorning the tumbler provided, lowered several noble draughts which produced a delightful smile and a tremendous gesture which, to a French engineman, means more speed and more coal. We ran steadily across the northern plain, leaned to the huge curve at Etaples and climbed Neufchatel bank in the usual seemingly gentle manner. In no time we could see the change of gradient near the Port de Soupirs and drifted down to Boulogne where, very far from clean, I left my friends to their own devices: I had shared a marvellous, emotional experience.

Soon the train was climbing gently but purposefully to Caffiers, to pass under the last Bridge of Sighs for the last time but Michel Lacroix would have put his shovel away near Marquise so he had no need to sigh with relief. The regulator would be all but closed for the final dash down to Calais: as Jean Guelton brought the speed down to 70kph, and sounded the whistle as they passed the Cité des Cheminots where many of the retired and serving railwaymen lived, the lineside was alive with railway families who had turned out to see the last steam train arriving at Calais: and so through the Ville station, on by the docks and harbour and gently into the Maritime station, the end of the great steam era.

John Shone, my Liverpool friend, knew little of railways when we met. Secretly he felt he might be a little bored by a railway weekend in France but he was a fantastic success and, if his French was sketchy, his gift for improvisation and descriptive humour was enormous. The Calaisians had never met his like before and they loved him.

On one occasion, there was a reception committee on the quay when we arrived at the Maritime. One could see Edmond Godry and the tall, angular and comical figure of Maurice Vasseur, one time driver of No 231E7 and a perennial visitor to England with all the groups for which I had been responsible. Jean Ringot, a mécanicien and deputy mayor, was there and as we were swept off in a species of autocade, I gathered we were bound for the Town Hall. I began to realise that there was going to be some sort of reception but when the Maire of Calais arrived, I realised this was a reception in our honour, and that I should have to rise to the occasion in no uncertain manner. By half past four or so we were half railwaymen, half local dignitaries, but there was also a table with a large number of bottles of Vin d'Alsace. Wine is good for the production of words as well as steam and my reply to those kind people was from the heart.

The local papers reported us on the front page, in the process promoting me to be Chief Executive of British Railway. According to the papers we drank toasts to the vitality of BR and SNCF, and after shaking hands with everybody half a dozen times John and I made our way dazedly into the evening sunshine before setting off for yet another gastronomic evening. Early the next morning, we all arrived at the home of M. Paul Bomy, mécanicien. His mate, Michel Robillard, was sent upstairs to get Madame out of bed to join us for champagne and cake. John said to me in English (which nobody could understand) 'I wonder what Gwenda would say if she was got out of bed at 2 o'clock by another man to drink champagne'! Five hours later we were on our way to Etaples with Edmond, and later on John, for the first and only time in his life, drove a steam locomotive from Hazebrouck to St Omer.

Edmond Godry kept a little book locked in his desk. In it were many fascinating revelations about the ability and character of the Calais enginemen. As Chief Inspector, Edmond was rarely immersed in paper and rode regularly with the drivers for whom he and his colleague, Jean Querlin, were responsible. Whatever the form of traction, punctuality and performance suffers if inspectors do not spend a fair amount of their time in this very basic duty. His men were categorised simply:

TB (Très Bien) (very good); B (Bien) (good); AB (Assez Bien) (so-so).

Surprises in the AB column: and it was fascinating to see the Flaman speed recorder cards which Edmond had kept in order to deal severely with men who had transgressed the maximum permitted speed by a few kilometres an hour. Amongst the TBs there was always the name of Lucien Fasquelle, youthful in both spirit and appearance. He had preferred to spend the earlier part of his career as an examining fitter but a change in personal circumstances enabled him to transfer to the footplate where he was quickly promoted to take charge of No 231E17. And let me digress for a few moments: BR men would be mystified by such a transition, quite impossible in this country. In France, remembering the age of retirement for drivers and firemen is 50, or 25 years which is the pensionable length of service for those who are over 25 years of age when they start work, any man from any background may seek promotion to the footplate and may become a 'mécanicien de route' provided he passes the relevant examinations. By no means all have the workshop training of Lucien Fasquelle.

Edmond and Henri Dutertre were Pacific drivers in their mid 20s and Joseph Six and Paul Bomy were turners made redundant who then chose the footplate as a career. Emilon Delattre, an experienced and competent chauffeur on the PO and PLM Pacifics, decided at the age of 48 to have a stab at the driver's examinations. For a time, he was an 'éléve mécanicien' – passed fireman – and then a regular driver for his last year in activity. Very flexible indeed compared with our practices which, on the other hand, are scrupulously fair in that men pass (or fail) for driving in seniority order.

Lucien was perpetually smiling and laughing but very determined. He speaks a little English and he makes up what he doesn't know. He also inherited a house, whilst his wife, Edith, is a very businesslike tailoress. One day he made the following remark to his friend Frank Mayes, laughingly but candidly. 'Fronk (always pronounced thus), I am only petit cheminot and when I am work, I, Socialiste. But when I am à ma maison (in my home), I Capitaliste!!' And so are most of us if we get the chance. Lucien has shaken hands with me after a visit to England with tears pouring down his cheeks as they were when he got down from his '72000BB' for the last time. But, as he said in a letter to me soon after, 'When I got off my locomotive, I was very moved. But that was the end of the life of a railwayman and now I am going to enjoy the rest of my life with my wife and family, many other interests and no lodging, and no shift work'.

Now, Maurice Saison ought to have been at Stewarts Lane or Stratford. He had been driving PO Pacifics since he was about 30 but he looked English, sharp of feature, gruff of voice, and he loved a good row with the boss, although his 'patron', Eugéne Lavieville, did not share his enthusiasm in this respect. Eugéne and I sat listening to the SNCF version of the 'Quatre Saisons' in a Calais concert hall. With a glint in his eye, he murmured to me that 'one Saison' was enough for him!

But whilst there could have been a dozen Maurice Saisons at Stewarts Lane and I should hardly have known the difference, the uproar, to which I had been reared, simply did not take place in France. Consequently, Maurice Saison was well known as an 'Enfant Terrible'. He was the Communist délége of the CGT on the Calais Commité Mixte (the LDC in our language) and it was he who refused to clean his PO No 231E40, because he regarded the bonus as inadequate. This had never happened before! E40 got steadily dirtier, Maurice was accordingly barred from working the 'Flèche d'Or', and, whilst news of this extraordinary phenomenon reached the Paris Headquarters, there was deadlock in Calais which was never satisfactorily resolved until the 'machine titulaire' arrangements were broken up at Calais in 1965. Two crews were then required to share one locomotive and as Francois Joly and his mate, Jacques Deseigne, had shown signs of following the example of E40, Eugene had the last word, triumphantly plonked the four of them on E19 and let them get on with it.

One day Maurice was talking about Oscar Pardo. Oscar was the youngest son of a large family, the father of which was a deposed President of Peru. Brought up in France and very much a Frenchman, he was a wealthy man who lived with Jacqueline, his wife, at Bidart, near Biarritz, in a beautiful Basque house, in the hall of which were many railway treasures. Oscar had worked in this country for a couple of years in the middle 1930s and had had a spell with the Southern Railway but, apart from the organisation of his family commitments in Peru, he had no need to earn a living. He made the best possible use of a footplate pass which covered all Regions on the SNCF. He was not required to say where and when he was travelling but such was his charm and personality that no chef, no ingénieur begrudged him this remarkable facility. He was what the French call 'simple', a word we should interpret as 'a gentleman'. He treated all with the same courtesy which was inevitably reciprocated, whilst his knowledge of railways and the men that worked them was profound. But he was extremely wealthy, had no need to work and here was Maurice Saison, the hard man, saying: 'Ah, oui, oui, charmant garçon, Monsieur Pardo, charmant garçon, Oscar. Un garçon simple!' That was perhaps the greatest compliment I heard paid to Oscar, who I last saw in 1976. We corresponded regularly but I could not get him and his wife to come to England. When Edmond Godry of Calais retired in 1979, his English friends gave him a wonderful send off in London. Oscar set a most generous contribution to a man that he revered. Ten days later, he and Jacqueline were dead.

There is no place in this book for that final tragic event, at the hand of an insane ex-Legionnaire, a triple murder. Let us simply say that, at their funeral in Bidart, there were present méchaniciens, chauffeurs and inspectors from Calais, La Chapelle, Bordeaux and Bayonne to pay their respects to 'un homme simple'.

XIX
My Good Fortune

WHEN VISITING this country, French railwaymen have been surprised at the closeness of a relationship between staff and management that may not be so obvious to us. It was apparently a revelation to them that we should know so many of our staff personally and appreciate their capabilities. They also noticed that there was a freedom of expression, neither too formal nor too familiar, that rarely existed in France.

Now, the grass is always greener, but there is nevertheless some truth in this. On the other hand, Philippe Leroy, the Locomotive Superintendent of the Nord who retired in 1969, was known and respected throughout the Region. But he had done it all, from chauffeur to superintendent. Henry de Fumichon commanded a similar respect but his background was quite different in that he had been a Regular Army officer, a 'Chasseur Alpin' before he studied engineering and entered the SNCF.

Jacques Vidal of Troyes was a driver and more than that for he had had a remarkable career and when I met him in 1961, he was starting his second stint. He had been a Troyes apprentice, leaving school at 14 and starting on the Est Railway in 1932. When the war came, he had become an under foreman on the maintenance side at Troyes and he joined the Resistance Fer. His cousin was one of the leaders of the Resistance in the City of Troyes and he was tortured by the Germans – but he was indestructible. In the early 1950s Jacques accepted a secondment to Abyssinia to take charge of maintenance, first of steam locomotives and then to convert the artisan staff to diesel traction. After a few years they had to return home owing to Madame's health and, as there was no suitable supervisory post at Troyes to which he could return, he decided to capitalise on his maintenance experience and become a mécanicien. He had a spell of two years as a fireman before passing for driving and immediately taking charge of his own '231G'. By this time he was 46 but as he had in a sense broken his service he was able to carry on beyond the normal age of retirement. Four years later he decided to take the examinations to become a chef mécanicien but a sudden heart attack prevented him continuing his career as a 'roulant'. But in the meantime, the Est line to Belfort and Mulhouse had been dieselised and there was a supervisory niche for Jacques in the depot at Troyes. The wheel had gone a complete circle!

Probably because of his Abyssinian experience, Jacques has taken a wide interest in world affairs and history. When he visited this country for the first time in 1981, it was a tour de force for he had a remarkable grasp of English and Scottish history and an insatiable appetite for knowledge.

His visit to Eastnor Castle was memorable. I have previously mentioned Ben Hervey-Bathurst and his connections with the French Resistance during the war. He knew and worked with men such as Jacques in those appallingly difficult days, and he and Elizabeth love to welcome such men to Eastnor. It did not do to tell Jacques that he was going to a chateau for a marvellous lunch, indeed I told him little about our day, except that it would be a good one. We drove easily down and I had plenty of time, just short of Woodstock, to turn off to Bladon. Again I said nothing until we stopped by the church and walked into the churchyard. His expression, when he realised that he was looking at Winston Churchill's grave, was unforgettable, for he had listened to the great broadcasts to France in Churchill's own particular brand of the French language and drawn inspiration from the great leader across the Channel. He marvelled at the simplicity of the grave.

As we drove down the hill to Eastnor we could see the castle on the left and Jacques cried out when he saw the French flag on the tower. After we met James Colyer-Fergusson off the train at Ledbury there was the sound of a steam whistle and a staccato exhaust. Yet again Jacques's face was a study, to be wreathed in smiles as the Harvey Bathurst's 1927 Foden steam tractor shot under the bridge and into the station yard. Our progress through Ledbury was hilarious, cramped as we were, the three of us behind the horizontal engine, with its whirling cranks.

The modern French railway is, in many ways, remarkable. Money has been poured into what is rightly regarded as a national asset and it has been well used, particularly in the field of electrification where great strides have been made. However, the French have a habit of getting their technical troubles sorted out in some comparative byway and not in the front line as we have tended to do. I have not yet travelled in the cab of the TGV but I think I shall delay a little longer. I would far rather wait until the excitement has died down, when Raymond Garde and I can make our journeys quietly and with great enjoyment. Raymond has worked immensely hard to make the TGV the success that it is and he deserves much of the credit. I have known him since he was Ingénieur en Chef d'Arrondissement at Amiens and watched him go on to greater things. The responsibility for the TGV project is a far cry from his days at the Ouest depot of Batingnolles, in Paris when as an 'attaché' of 24, training after he had qualified at the Politechnique, he was given a Pacific to drive.

Leaving Calais with a couple of '67000 BBs' on what now passes for the 'Fléche d'Or', one settles down to avoid certain things happening on the journey to Amiens. Almost habitually, one stands at one's work. It is perfectly possible to sit down with one's foot wedged on the dead man's pedal but many prefer to do it a different way. If one relaxes one's pressure on the dead man, a horn sounds, fit to waken the dead, and after a very short time the brakes go on, very hard: naturally, this must not happen but there is an alternative dead man's control available to the driver. The controller is in the form of a wheel and under the wheel there is a spring-loaded ring of the same diameter as the controller itself. This is the alternative dead man's control which I prefer to use. It is very positive, very comfortable but there is another reason, far more important.

When approaching a signal the train passes over the AWS lamp or 'crocodile', and if it is at green the driver hears a single melodious 'ting' of the bell, but if it is not, another horn, of a different tone to that previously mentioned but still capable of waking the dead, will sound and, of course, the brakes will apply if the driver does not immediately press the vigilance button. The same thing happens at temporary speed restrictions. An adverse signal must be acknowledged when it is sighted by pressing the vigilance button three times, which marks the speed recording band – somewhat like a barograph in appearance – to be checked at the depot in due course. But another sound will impose itself in this medley of horns and bells and this is the noise to avoid if only to prove one's vigilance.

Unless one makes a movement of the controls within 55 seconds a bell will ring, an attractive silvery toned affair like the luncheon call from the *Wagon Lit.* But ignore it at your peril for it is the vigilance device and it has the power to stop you, dead. On the other hand, a movement of a control, dead man's ring, pedal, controller or brake will immediately set things right but it is a point of honour not to let that little bell ring from Calais to Amiens and so, as one whizzes across the northern plain with the speed hovering near the permitted maximum, I have always stood with my hand on the controller and the dead man's ring beneath it, relaxing one's grasp for a fraction of a second, sufficient to satisfy the vigilance and quick enough to avoid the bellow of the dead man's device. These days, the journey to Amiens is scholarly and precise. No longer can wine be served, no longer is there a bucket of cold water suspended from the cab windows in which were placed the wine or citron bottles. It is the modern railway and all the better for it, but how glad and honoured am I to have known and worked with a few of the giants of the past.

My first journey in France, on No 231E16 from Calais to Paris, is still photographed in my mind. Everything was new: whether it was the complexity of the machinery, the performance when moving away from a stand, the rhythm of that locomotive oh so smooth and gentle – or the crew in their far from clean overalls, wearing goggles and standard railway issue boots. Everything was novel: the signalling, the track, the stations, the train, and yet we understood one another. The silent but courteous Henri smiled quizzically when I started to fire and in later years he told me he expected this mad Englishman to give up after a few miles but I had no need of rest on that locomotive. But the French are no more perfect than we are.

Why did Henri Dutertre and his mate, having polished the cab fittings to perfection, stand at Boulogne waiting for the time to depart, with smoke and flames seeping from the firehole door, clinging to the firebox front and spoiling their burnished fittings? A small touch of the blower would have altered that but no! They have their fads the same as we do, devoid of rational explanation. One tends to think that the French enginemen, so well trained and methodical, can never err from what is right, and I was delighted to hear that Henri, back in 1951, had had a tremendous conflagration with none other than Pierre Leseigneur on the subject of driving E16. Henri was young and proud and Leseigneur climbing up at Creil instructed him to use the wide open regulator on the rise to Survilliers. L. P. Parker would have approved. Henri, however, said that his father had only used half regulator on E47 and what was good enough for his father was good enough for him. No doubt the great Leseigneur consigned Henri to the fireman's corner but I never heard that part of the story. But how refreshing it is to know that these arguments occur in France, and as I write I think of old George Hawes of Ipswich, who drove a 'B1'. George did not have Henri's technical knowledge but he was a fierce old man with decided opinions. In 1950 the Inspectors, not to mention the District Officers and Shedmasters, were all full regulator men but not so George who would defy anybody to make him get the regulator handle right across to the stop!

Henri Dutertre gave me a serious lecture when he first saw a regulator fully closed on a 'Britannia' at 80mph. He gave Bert Hooker, a Nine Elms driver, a lecture when the latter had spoken on the telephone and received instructions to pass a signal at danger and obey all others. Where was Bebert's written authority? Which goes to show that there are differing ways of successfully achieving the desired result.

Henri had been taught never to close a regulator until the speed had fallen to 30kph whether he was on a Chapelon or even an American '141R'. His last move would have been to close the regulator at 30kph and place the reverser in full gear. L. P. Parker would have given Henri an even more serious lecture had they ever met on a locomotive footplate because his instruction was that, on piston valve locomotives, one closed the regulator, leaving the reversing gear at 15%. A glance at the steam chest pressure gauge would show about 50lb/sq in compression which would keep everything just so and minimise knock and wear. Try it on a 'B1' or a 'Black 5' and the difference in the ride between a coasting position of 40% or even 25% and 15% is quite noticeable.

One more memory of France: I have described this journey before but it was an amazing experience and repetition will do no harm. We were up against an engine that was no longer in the pink of condition for it was July 1961, and the end of steam – and of the great De Caso Baltics – between Paris and Jeumont was not far away. I met André Duteil of La Chappelle, my little gold-toothed friend, when he had backed his enormous No 232S003 on to our train at Jeumont. He told me that our engine, now common user, was not in good form but, as he had his own splendid fireman, René de Jonghe, we should no doubt get along very well. The De Caso Baltic is fitted with a mechanical stoker and, although the physical act of shovelling is largely removed, skill is required to direct the coal from the stoker in the right direction and at the correct rate to maintain the hottest possible fire when one is up against a heavy load and an indifferent engine. Our train was made up to 740 tons at Aulnoye, André told me that I was the mécanicien to Paris and that I was to run right up to the maximum permitted speed of 120kph uphill and down dale and never to deviate. I was advised to use the brake, rather than ease the locomotive, should l'Espion' (the Flaman recorder) move on to 122kph.

Renéde Jonghe left Aulnoye with about half an inch of water in the gauge glass. When we were on the move he raked the fire, applied coal continuously through the stoker and allowed the water to fall until it was barely visible: he had adjusted the feed pump very fine indeed. He wanted the dryest and hottest steam possible and the pyrometer would tell its own story; he wanted the maximum pressure on a shy steaming locomotive and for mile after mile this was maintained on the red mark. He had the confidence to run on the level stretches with the water at the bottom of the gauge glass, sometimes bobbing up but drawn now and again by the use of the top gauge column cock, to see that it was still with us! It was perhaps the most brilliant exposition of steam locomotive firing that I had ever seen.

Meanwhile, André was acting as chef mécanicien but not yet as wine waiter. However, we were checked by signal at Creil and when we had threaded the back platforms through the station and set to work to climb to Surveilliers, an excellent Bordeaux was served. With all his usual panache, André tasted the wine, pronounced it at the correct temperature, and filled our glasses as our great machine was opened out to climb 1 in 200 with 750 tons. As we listened to the beautiful clear exhaust of No 232S003, we drank to the days of steam and to our own good fortune. The engine shouldered her load, thundered up the gradient in the darkness, through the Forest of Chantilly, over the viaducts, gradually reaching 100kph before we reached the summit, passed under 'Le Pont de Soupirs' and spun silently down to Paris. We drew slowly into the Gare du Nord, stopped with the engine brake only at the buffer stops, and I looked down from the footplate of a steam locomotive at that most historic terminus almost for the last time. We had left Aulnoye 12 minutes late, we lost a further 15 minutes due to a speed restriction at Le Cateau and the deviation at Creil but we ourselves had gained 23 minutes on schedule. We shall never forget that journey.

Had it not been for James Colyer-Ferguson I should never have travelled on a French steam locomotive. Had it not been for Philippe Leroy's generosity, we would never have been able to come again and again. Had it not been for the friendliness and welcome of all those that we met and who befriended us, we might never have crossed over more than a few times but now we have built a remarkable comradeship. I may no longer go to France four times a year and maybe only once in 18 months, but with good friends this does not matter for the bonds are always there.

XX

Count your Blessings

In October 1973 I began what at first seemed a new life at BR HQ Marylebone. Financially, it had been a very bad move for Gwenda and myself and it was to be another seven years before I was promoted by which time I had grown to love my work, and not only the increase in salary but the added responsibility and pressure added the finishing touches to a job where one could exert an influence for good throughout the engineering departments and indeed, in general management.

In Liverpool, as a Divisional Manager, I was responsible for the management of 5,600 souls; at Melbury Terrace I had a staff of two. By the time I retired at the end of 1982, this had grown to three and my workload had increased enormously. Mavis, my clerk, Janet, the secretary and I were responsible for the recruitment, training and career development of what were called Designated Engineers throughout the railway; in 1973 these were confined to the Mechanical, Electrical, Signals & Telecommunication and Research & Development Departments as well as in BREL. A compact organisation had been set up in 1969 by my predecessor, C. F. Rose, an experienced engineering manager, to carry out this work on a very personal basis and, for the next nine years, I had no need to change Eric Rose's basic structure which withstood all the stresses of an ever increasing pressure.

At BR HQ the life was very different but I had the one position that could keep me in touch with the running of the railway and through which I could get things done and influence affairs at all levels, and where I was dealing with the lives of people, their careers, their hopes and disappointments, their successes and failures. To many people my translation, which would never have taken place other than under the shadow of redundancy, seemed a scandalous waste of a man whose strengths were clearly in the front line of management and the handling of men, whereas to one or two discerning friends, I was the man for the job, and I derived encouragement from this, because in my heart, I knew that it was extremely unlikely I should return to general management at a higher level.

Life at BR HQ required a sense of humour, an understanding of the ridiculous as well as the ability to command the respect of one's colleagues in Regions and Divisions. One no longer had the status of a Divisional Manager and for a short while I was not myself. But I soon realised that status is an imposter and that it was standing, prestige within the industry that mattered, that I should count my blessings, that I had a job with endless possibilities even if I had to remain there for 10 years, and above all that I would be my own boss, free from interference, reporting to and working with splendid people and, if I used my own personality and

361

strengths, a very considerable influence right across the railway scene. One learned to use the system but also when appropriate to beat it!

By 1980 I covered the recruitment and training of young engineers and personal career development of every chartered or potentially chartered engineer in the management ranges and below, in all engineering departments, including civil engineering. I was then given the additional responsibility for the preparation of plans for the career development and appointment of all senior engineering officers so that I could now play my part in achieving the best possible balance at the top which, in a constantly changing scene, was to prove very difficult indeed.

When Eric Rose set up his organisation in 1969, he became the Executive Member of the Central Engineering Training Group which was chaired by the Chief Mechanical & Electrical Engineer of BR, assisted by his Traction Engineer, two Regional CM&EEs, the Engineering Director of BREL, a senior Works Manager and the Head of Research & Development. The Group met every three to four months, and policy was agreed, whilst the Executive Member reported on his work: clear decisions were taken and he was then free to convert policy into action. Eventually the Executive Member served three Groups, the S&T in 1971 and, from 1975, the Civil Engineering CETG.

Yearly, one had to calculate the intake of engineering students and graduates and differentiate between the two. Graduates with good engineering degrees are recruited from universities and then given a sound and comprehensive training to fit them not only for their career but also to enable them to become chartered members of their particular professional institution. But for two reasons I preferred to select and sponsor young people through university on their way to becoming engineers. Firstly, at graduate level with our particular salary structure we were by no means always competitive; secondly, our sponsorship was first class, our selection generally successful, and matched by a standard of training always under scrutiny. We could thus select young people of 18 and by the time they had completed both their degree course at university and their training, they had been with us five years and had the makings not only of engineers or of managers but of *railwaymen*. Some of our own engineering apprentices were able to enter university after obtaining either Ordinary or Higher National Certificates in engineering. They competed against school leavers with good 'A' levels in vying for a place, because the Council of Engineering Institutions laid down that an engineering degree is essential for chartered membership of certain professional institutions.

Until Eric arrived on the scene in 1969, selection had been haphazard with little or no attempt to apply either academic or personal yardsticks but, with the help of the BR HQ Recruitment Department and, of course, the engineering chiefs, he brought a degree of order on which I was able, step by step, to build and improve. Ultimately I was involved in the final selection of all engineering sponsored students, concentrated early in the year. Try interviewing for some 27 full days over a two-monthly period without letting the normal work slide. Hard, sometimes exhausting work but for those of us deeply involved, remarkably rewarding. My civil engineering colleague at these interviews for nearly nine years was Maurice Sowden. We went to endless trouble to put each young person at ease at the beginning of what must have been a gruelling interview. Yet again, one had to learn beforehand something of each one so that there could be a point of contact, some mutual interest, some piece of background in common. Maurice and I between us seemed to know an awful lot of places and people and the joy and relief at a happy start of an interview was often our reward.

If anybody referred to me as a training officer, I usually had something to say. I was not one, but I had to work very closely with those responsible for the actual physical act, and such as J. D. Forster of BREL who masterminded all M&E engineers through works and depots and university had an extremely testing if interesting task. I was responsible for standards of training and, of course, liaison with universities and professional bodies. If the Mechanicals, for example, introduced a new training and development scheme, I had to interpret it and then make very sure that it worked properly. And this could be as demanding as the introduction of engine locks and keys at Stewarts Lane had been!

But the most fascinating, rewarding and challenging part of the whole business is yet to come. By 1980 the number of Designated Engineers, in training and right up to the fop of the management range and under the age of 42, had grown to about 1,200, across five engineering departments and including some in general management and operations for I foresaw engineers spread across the whole scene. My ideal was to know everything about each person with whose career I was involved. To do this, one must have a deep interest in the career of others and one had to come to terms with one's own ambitions in order to become dedicated to the service of others. But additionally, one must have constant sources of reliable information to support the judgement that one must make. To think that this information will only come through the formal channels of appraisal is to cheapen the task. It is here that my 33 years of front line railway experience helped me to assess so many of these people. First of all, one must understand the work they do and the circumstances of that work: one must know what should happen next, but to reach this decision, one must know how well they are doing in their present position. One must see them on the job, in their own environment, one must assess those for whom they work and for whom they may work in the future, and one will absorb whenever possible the reliable opinions of those men whose judgement one trusts but whose views would never reach the formal appraisal document. For example, when I heard from three different sources, without any prompting, that a certain young man commanded the respect of enginemen at Stratford, I knew for sure that, whatever else might be said, he had gifts of leadership and management that should be developed. When all the facts and all the knowledge available are put together, one has a picture and then one is in a position to say with determination and conviction what the individual can do. As I could ultimately be involved in every engineering appointment throughout BR and BREL it was essential that the views I expressed so forcefully – but sometimes tactfully – on who should be appointed to a particular job, carried total conviction. If one did not make one's presence felt, one could so easily become a cypher and if one did, one had to be right. I loved that part of it. It was a challenge for, although I had executive responsibility to implement the policy of the CETG, I did not have dictatorial power for never can one deny the manager or engineer, the man who actually makes the appointment, his authority, because it was he, not I, who carried the ultimate technical responsibility for failure. As a Divisional Manager, I would not have welcomed direction from some 'nark' at the Board as to whom I should have appointed, for example, as Depot Engineer, Wigan. Eric was then Executive Member and his views never reached me until long after the interview, and here was a lesson quickly to be learned in the art of cutting red tape to get to the root of the matter and then tying it again to avoid bruised feelings. Nevertheless, it was my job to influence and charm (and threaten just now and again) my own colleagues in Regions and Divisions and Headquarters to do what I wanted – not necessarily what they might think was right.

It was not only the seeking and sometimes finding an ideal to get the right man in the right place at the right time, it was the belief that one's own colleagues would use and trust the service that was offered, that gave me great pleasure. If they believed and trusted in me and used me as much as they could, then I was happy. Thanks to those colleagues all over the railway, my opinions and advice came to be accepted at every level. I became known amongst the younger engineers because I sought to know more than they knew about themselves and my position was therefore unique in BR HQ in that I was not only involved with policy, not only with senior people but with men and women who were running the railway and on the threshold of their careers.

Once I became involved in the career planning of senior engineers, the span was remarkable. One could be dealing with the appointment of a very senior engineer one minute and at the next the problems of a 19-year old sponsored student in his first year at university. Although I felt that I had considerable influence on the senior appointments, it was at the lower levels that I knew that my talents, such as they were, would be fully rewarded. It was my responsibility to influence, to develop, to give guidance, to give very firm words of advice and the necessary kick up the backside. Certainly I have helped many of my senior colleagues but when a person is between 22 and 40, there is so much that can be done. It is relatively easy to develop the career of the high-flyer or of the average person but what of the one who is in real trouble. I remembered only too well the feeling of helplessness when my own career was at the crossroads in 1973. Strange as it may seem I felt, at that time, that I had few to whom I could turn for advice about my own career, although I was a Divisional Manager and all but 50 years of age. I was, in truth, a doughty fighter for others but not for myself and I had a strong feeling that my usefulness in the years ahead would be limited. However, I was absolutely wrong in that assumption and after 14 years of riding very high, a hard knock was the best thing that could have happened to me. It knocked self-seeking and selfishness out of the window and I became determined that no Designated Engineer should feel as I had done and that I would fight for those in difficulties, once I knew the background and provided they would join me in that fight.

Every six months I would set out for each Chief Engineer details of any person about whom I was particularly concerned. I would sum up his personality, point out what was wrong and what had to be done to put matters right. I was outspoken and I never let up. It could be a clash of personalities, somebody in the wrong job, a front line man in a specialist post, an unaccountable failure to succeed, or equally somebody whose career I had directed elsewhere to gain wider experience for a limited period and who had to be brought back at the right time. Often, one had to give time and encouragement for months on end. I would never tell my colleagues that a man in difficulties was the man for a job when I knew, in my heart, that there was the slightest doubt. But when it all fitted together and a person, in difficulties for so long, went on his way once more, it made my day.

On one occasion, during my time as a Divisional Manager, my future career was discussed with me and I was gullible enough to believe what was said. A future in General Management was predicted although I found out months afterwards that there were inadequate grounds for such a prophesy. Therefore, when I came to Marylebone, I was more than ever determined that those with whose careers I was involved should know the truth. Whether it was unpalatable or not, they would get the soundest practical guidance and advice but no promises; no suggestions that could be taken too optimistically. Sometimes, this led to men seeking employment elsewhere but this had to be their choice in the long run and once a person, however promising, had decided to leave us, I had nothing more to say, though generally the best of our young men stayed with us – but one could never take this for granted.

I had always applied this policy whether to a fireman, a graduate or an up-and-coming manager. Only once did I deviate and the circumstances are worth mentioning because they highlight the constraints that, occasionally, foiled one in achieving a target. We had a brilliant young mechanical engineer whom we sponsored through Oxford whence he emerged with first class honours in Engineering. After completing his training and making a brief but successful start as a manager of men, he was allowed to return to Oxford to read for a doctorate. He returned after a further three years at Oxford and was appointed Personal Assistant to the Board Member for Engineering & Research. This was a 12-14 month stint and he did an excellent job. Because I had taken him out of his normal progression to a specialist task, it was my responsibility to help him back into the main stream in the right job at the right time. Here I ran up against a difficulty. Because of the critical financial position, the advertising and filling of vacancies had been held up throughout BR HQ and the Regions. Both the young man and I knew that he particularly wanted to run an area and that this type of post must be the next one. I knew that a London Midland area in the Home Counties was likely to be vacant but, for various reasons, I could not say exactly when this would be. I hoped that he would wait until I could see the way ahead clearly, but after a comparatively short time his patience began to wear thinner. A position became available at an engineering headquarters, more lucrative and indeed senior to the area position but quite the wrong type of job at that stage in his career. However, he was prevailed upon and he very soon found he had a job where there were constant changes of policy and direction, where he had no total responsibility or authority and where he became unhappy at his inability to push ahead. He had a brilliant intellect and I knew that his computer of a brain would produce a thousand questions, and a thousand solutions, none of which he could implement. By no means experienced in the political scene in which he found himself, he decided to leave the railway. When I heard of this, I sent for him to dissuade him, the only time in my life that I had done such a thing, once a man had said he was on his way. But he was committed too far, despite the fact that he knew the railway was his metier.

Meanwhile, the Area Maintenance Engineer in the Home Counties moved on, the embargo on promotion having been lifted, his vacancy was advertised and filled. I knew this would happen sooner or later but I could neither say exactly when nor could I have promised my young man that the job would have been his. I could, and did, tell him that he would be a very strong contender when the time came but I could say no more than that. Could he but have waited, his outstanding qualities of leadership and intellect would have commended themselves to the Chief Engineer of the London Midland Region and would have seen him through as Area Maintenance Engineer to greater things, for he had the potential to go to the very top. He was a great loss but no man is indispensable and we are fortunate enough to have others of his calibre.

And so I spent nine years in a position in which I consciously and willingly took on more and more responsibility. I never tired of the task and when I decided to retire in December 1982 I was both fresh and happy for I knew that we had built a depth of engineering talent across the railway, the like of which had never previously existed. There would be weak patches, but there were enough first class brains and enough managerial and practical talent to see to the future in all the departments.

No manager at BR HQ has had more help and cooperation than I, in these last few years. Only once, at any rate to my face, has anybody talked disparagingly about 'You people at the Board don't know what it's all about' and he got the bollocking that he deserved for he could not come within a mile of those 33 years of front line experience. But I say advisedly that we, not I, built up a great depth of talent because I could not work in a vacuum. I reported to a Board Member for whom I had great respect, to splendid Heads of Departments, and my office, one of the best in London, was next door to the Director of Mechanical & Electrical Engineering. Whatever their position and title, they were all super people. We went our various ways but saw just the right amount of each other and so worked closely together.

The proud and independent Department of the Chief Civil Engineer wanted no truck with me until 1975 when the defences were lowered. Once a CETG had been formed and I began to get to know a large number of the younger civil engineers, I was amazed at the help I received and the cooperation afforded me in improving or even changing some of the notions so firmly held on training and the career development of civil engineers. I had worked with civil engineers since I was first a Shedmaster, and by the time I had become a Divisional Manager the bond with the Divisional Civil Engineer was particularly close both at King's Cross with Ken Haysom and in Liverpool with Hubert Roberts. But we were then in separate departments, whereas now I regarded myself as working within the Civil Engineering Department. This gave me enormous satisfaction, and the knowledge of the progress that we made between us in the development of the careers of engineers throughout the management ranges will always remain with me. With Hubert in Liverpool, for example, our working relationship had been close, enjoyable and full of challenge but once I was able to strike a professional association with him and his staff from within the department, we reached a splendid feeling of achievement and understanding. I was able to work once again with Civil Engineering colleagues, notably Brian Davis who had been an Assistant to the DCE at Stratford. Being the men we were, with a mutual turn of phrase, we had always got on well and now here, in later years, was Brian as Chief Civil Engineer of the Eastern Region, with an onerous and demanding job, more so than mine, and yet with the time to know his men, his engineers and to command their respect. Brian had said back in 1975 that the Civils wanted something out of me. They got it and with pleasure.

I enjoyed talking to and meeting those on the threshold of their career for they were so full of hope, enthusiasm and interest. And then one would see how responsibility would mature and improve them and one would see them gradually mounting the ladder and achieving, in many cases, successes of which my generation would have been very proud. I have no fear for the future of the railway in the hands of our young engineers.

By 1977 we were recruiting, not without the odd qualm, girls to be trained as engineers. Not many but enough. They had to be very good for they were entering a man's world. In fact, the first girl to receive an appointment after training was a graduate, Carolyn Griffiths, recruited from university; and Carolyn, as I write, is doing well as a Supervisor on shifts at Stratford. At Stratford, mark you, in the footsteps of the Cock Allum's, the Jack Taylor's, the Casselton's, the Tommy Newman's. Impossible?

Not these days, quite the reverse. I shall never forget when Carolyn came to me to discuss her first appointment. There was fire in her belly and a laser beam of Thatcherian intensity in her eye. We were diametrically opposed as to what she should do.

Nothing less than supervision at Old Oak Common would do for her but she had to go to Paddington Headquarters, away from the rough and tumble. She was the first girl we had trained and I had a responsibility for her career. In time, I saw her again and the laser beam was not quite so fierce. Life was interesting but there was still the passionate wish to become a Supervisor, which still had to be denied her. Then again not long before I retired, I saw her at Cardiff. She was a Senior Technical Officer at Canton depot and had begun to gain the experience of depot life and supervision that she had sought so avidly. She had been dealing with failures of locomotives and I asked her to what extent she had travelled on the footplate, remembering, of course, that footplate and maintenance staff are no longer in the same department as in my day. But she had been out quite a few times. Now, when I first got on to Benny Faux's footplate back in 1943, he had said to me: 'Alta had howd afore, Dick'. Of course I had, so he told me to 'Get thi'sen up in't corner and get cracking.' To have hold of an engine' to drive it, is a loco man's phrase I've used all my life and so I asked Carolyn, 'Have you ever had hold.' Now, what she thought I meant, I don't know, but the laser beam appeared, Force 9, before I hastily explained what 'having hold' really meant in my language! Now she has achieved her first milestone, to be a supervisor in a locomotive depot: she has had the basic experience and she has the strength and keenness to succeed. Meanwhile, in retirement, I shall hear from time to time of the fortunes of all those Designated Engineers whose lives became so much a part of mine. My work was very personal. Our little team, Mavis from the Rhondda, Linzi and my Irish secretary, Margaret, did a wonderful and highly confidential job. We held a thousand secrets and so there is much that must be left unsaid. In some ways, it was the most rewarding task of all.

XXI

Steam in the Blood

S TEAM IN THE BLOOD has never stopped me seeing the way ahead. Events have proved that. We could never again tolerate steam as the prime mover of our network but it still has a magic touch. One would think that enough is enough in retirement but, on the contrary, there is time now and again to thrill once more to the challenge of steam. More or less the same muscles come in handy for firing and riding (which does not mean to say that a horseman can fire an engine) and, provided people don't go mad, the Long Drag is not yet my master although, in the words of Driver Lew Bell of South Lynn, there is a particular necessity 'Always to be ready' these days. The best advice to a fireman, it had been drilled into me from my earliest days on the railway as had the need to avoid smoke and blowing off when so doing.

These lessons served me well in the summer of 1981 when I fired the *Flying Scotsman*, born the same year as me, up the Drag from Appleby. I had been able to make up my own fire in the sidings. There was plenty to bite at, though not as much as one would have had years ago, and one could adjust matters to avoid the smoke and blowing off which can so easily frighten the children. The driver was Jimmy Lister of Carlisle, once of St Margarets, Edinburgh and he knew his 'A3'. Before we started, I asked his advice on whether to push forward the fire from immediately under the firehole door as we departed. He said to leave well alone. I did nothing for the first minute or so by which time Jim had the engine set for a fast acceleration. Immediately, the boiler responded, the first firing produced sufficient smoke to tell me that all was well and the exhaust injector was set to work and was never touched again until we were within less than a mile from the summit. The pressure never strayed from the red mark and the live steam injector could be used to check blowing off and wasting steam. The regulator was wide open from start to finish and all variations to cut off were made gradually so that there was no sudden demand on the boiler and on my back and arms. Between the two of us, we served up the classic recipe for good results. The firehole trapdoor remained open throughout the journey thus cutting down the emission of smoke, and finally and to my delight it was cold and pouring with rain, for the classic recipe means little and often firing and a clean footplate with little time to pause for breath. We climbed to Ais Gill in 29min 38sec, and that was not too bad for a couple of old gentlemen although it was Jim's handling of the locomotive that set it up for me.

It was an unforgettable journey. George Gordon, our Inspector, had tripped and fallen, breaking his shoulder between Carlisle and Appleby.

None of us realised this, nor did George and he maintained a remarkable feat of endurance until he was much nearer home. He had to spend the night in hospital in Carlisle and was off work for several months! I realised, perhaps for the first time, how the fireman occupies the centre of the stage. By arrangement, visitors were allowed to come through the corridor tender to observe what was happening. They came briefly one by one, into an extraordinary world of heat, noise, driving rain, strength and movement. They would see the driver in his corner, with the inspector behind him, the fireman in almost constantly vigorous activity in front of the blinding, white hot fire in the centre of the footplate. As for me, I was oblivious to everything and everybody except the job in hand, to get to Ais Gill in perfect order with a heavy train. No time nor, in truth, enough breath to hold a conversation.

Jimmy Lister and I had never met until he stepped aboard at Appleby. But when, at Garsdale, he turned to me and said: 'Well, Mr Hardy, I'm not sure who you are but you are a credit to your profession', I felt 10ft tall. Here was I, nearly 58 years of age with over 40 years' experience behind me and yet I felt like that. But then steam has the power, the magic touch to do these things. The two engines, *Flying Scotsman* and *Sir Nigel Gresley*, are marvellous creations. Both are a credit to those that maintain them and, as for No 4498, beautifully kept, spotless as to cab and firebox front, it is an engine and a half. Never did I think that I should have had the honour of firing an 'A4' up the Drag nor seen the darkness of Birkett Tunnel brightened by a thousand sparks as the historical streamliner attacked the gradient with every confidence and a white hot fire gradually getting thinner – but not too thin to raise doubts in the mind of the elderly stoker as to his ability to reach Garsdale with flying colours.

Some time ago I gave a talk to the Bath Railway Society and started on the subject of 'Rough Trips' when everything goes wrong, through some mechanical or human derangement. I have had quite a few such journeys. Sometimes one is disgusted with one's ineptitude and yet, sometimes, one is elated with a limited triumph against extreme adversity. Long before I became a Director of the Festiniog Railway, I spent a day fighting *Linda* without the slightest success. It was her last day's work as a coal burner and my only consolation was that neither Allan Garraway, the driver, nor I could produce enough steam to keep out of trouble.

I was asked to join the Festiniog Railway Board in 1977. One of my tasks in the early days was to agree a negotiation, consultation and discipline procedure with the NUR North West Area Organiser. Home from home! Much has happened since then and much remains to be done on a railway that is now a sizeable business in the tourist industry with some 40 permanent staff. The railway has been brought back to Tanygrisiau and thence to Blaenau Ffestiniog, a great feat of professional and voluntary endeavour. It has been headed by Dick Wollan, once a Director of ICI Mond, now by David Pollock, and they have brought business professionalism and drive into an organisation which already has much skill and talent at its disposal. Whenever I visit that railway, I know I am amongst railwaymen, whether they be professional or volunteer. Recently, I spent a couple of days on *Prince* working heavily loaded trains. What William Hoole thought of the earlier edition of *Prince* when he was translated from King's Cross, I do not know, but I found a little engine, lively and energetic but one which could round corners in a series of very straight lines and after a couple of days I felt about the same age as the machine. Both the drivers and the rostered fireman were volunteers and our trains were well loaded. On the first day we had to fight every inch of the way, not only against the load but a bad rail and oil that was 58% waste in the fuel tank. We won through, but nevertheless on my firing turn we had to stop at the summit to fill the boiler. On the second day we cleaned the oil filters between the first and second trip and came through triumphantly with a well-filled boiler and 150Ib/sq in throughout. With oil burning, there can be no loss of concentration or anticipation with a little engine working a heavy train and with *Prince* working hard there is no time for conversation. On the second day, my driver was Paul Ingham, a technician at Leicester University, and my fireman and instructor Tony Rowlands, in charge of computers at Liverpool University. To give a measure of what a volunteer will do, Paul had made over 650 firing turns before becoming a driver and his handling of the engine reflected that experience. It was my eighth firing turn and so there might have been a great gulf between us. But, on the contrary, there was immediate understanding between the volunteer – with railways and steam in the blood – and the professional. Indeed, there are many ways that the experience of such as myself can be of service to those who run and man the smaller railways. And yet my experience of the Festiniog Railway has convinced me that I should not necessarily be the man to run a small railway nor indeed that men with a lifetime of BR experience are always the right men for that job. I would probably be too independent to take kindly to the multiplicity of ideas and instructions thrust upon the manager of a small railway, however much one would enjoy the challenge of running a railway business. But, provided one is not professionally involved, there is much that can be done and one's advice and experience is always at the service of those who care to use it.

Who would have thought that, after all those thousands and thousands of miles, I should delight in the attack of Freshfield bank on the Bluebell Railway. But, on the contrary, with a heavy load. and the ever youthful Wainwright 'C', one is back on the Southern of 30 years ago when our Chatham Goods link men worked the Hoo Goods from Hither Green and sent the rockets flying night after night on their old 'C' class whilst No 263, the 'H' class Tank, stood over at Victoria, or worked the 'White City Milk', or the 'Winkle' or 'Empties' to Eardley via Tulse Hill on a summer Saturday.

Again, how could a five-mile stroll from Loughborough to Rothley give me pleasure: and yet my day on *Butler-Henderson* rolled back the years. There was the same high firehole door, the same heavy beat in full gear, the same need to gather speed before too much notching up took place. I was last on a 'Director' in 1946 but that day we might just as well have been working a prewar stopping train to London whose gentle progress was in very direct contrast to the blood and thunder of the outward working because the Frances, the Simpsons, the Geo Parks of Neasden enjoyed their dawdle southwards through the countryside on their way back from Leicester after working the hectic 10.00 Marylebone Bradford.

For many years I have visited George Barlow at New Romney. For years he *was* the railway and until he retired in 1981, gave everything to its service. We have made many journeys together on his own *Green Goddess* and, with experience, one learns the road and also that the handling of such a small locomotive – so similar to its full size counterpart – holds pitfalls for the unwary or indeed for those brought up on the main line. Never shall I forget my first journey with George. We sat side by side, divided only by the requirement of drawing coal from the tender door. The shovel was domestic, and the firebox should have been chicken feed to one used to a Great Eastern 'B12'. But the *Goddess* was not herself that day; a superheater element was blowing but she must have known that the hand on the throttle was not that of her maestro.

I could not get that little to steam. 'Don't swear at my engine', said George as he picked lumps of coal out of his lap intended for the front corner of the firebox. I was furious with myself and there was no escaping the basilisk eye of my Lord Barlow as the pressure sank, for there is no respite on a level track and Hythe seemed a long, long way from Dymchurch. An eye opener, a struggle and the first of many days with George, to which I looked forward always with anticipation. The *Goddess* has just come back from a general overhaul as I write with a new and unusual looking tender but it is my fervent hope that the senior driver at Romney, Richard Batten, has now taken it as his own to continue in the Barlow tradition.

A few years ago, when George had stopped regular driving, I spent a day with Eric Copping, once of BR Colchester, a main line fireman who left the railway and found his way to Romney where he became not only a driver but a species of Sammy Gingell. I had been brought up by George on the firm principle of observance of timetable and economy in the use of steam and coal. Eric told me to have hold and we travelled steadily to Hythe arriving exactly on time. However, a friend of ours, who had been in the train, opined that the running was not up to Eric's usual standard, who warned me that his reputation was at stake. Now, I like a little sprint, as Sam would have said, as much as anybody and so we rolled up at New Romney well before time just as Mr Operating Manager Barlow arrived on the scene, looking at his watch and saying, 'Bloody BR men again'!

George, like me, settled down immediately to retirement. Like me, he has a wonderful partner. For years, his wage was that of a farm labourer, for years he had no prospect of a pension until the present management put that right, and yet he and Miriam are rich in the things that money cannot buy. They have friends all over the world.

The Severn Valley is a super railway. There one can experience the competence of the Great Western locomotives and one can also see how some of them are designed to make things as uncomfortable as possible for the crew. Compare that '51XX' monstrosity with a GC 'A5'. They used to stand side by side in Aylesbury depot and both were splendid engines on the road. And yet, for a bit of 'fluence, one can gently fire a 'Hall' sitting on the fireman's seat (using an Eastern shovel, of course): one could never do that on a Robinson four-cylinder. Perhaps the most enjoyable of all the GW engines at Bridgnorth is the Pannier Tank No 5764, strong and virile, full of personality and crude in many details. But the engine does a great job and so one forgives the inconvenience of coil springs in the cab, the double bending necessary to open the water valves, and the stick lever without a footrest against which to brace the legs when reversing or notching up.

Alun Rees of the SVR and I have a day together now and again, and one time we fell from grace. We had No 5764 and left Bewdley with what we thought was a full tank. At Bridgnorth, what with one thing and another, we were late out of the Shed for our return working, checking the tank level as we moved off, a daft thing to do as it gave a false indication. Approaching Highley, my injector struck work: the tank was empty and we had to fill up from a carriage washing pipe, a slow business. We took just enough to get to Bewdley where we were due away when we arrived. Alun would take water while I made up the fire for an attack on the timetable but I was stopped short by a shout to the effect that there was no water in the Bewdley tank. Calamity! One felt like blaming he who had shut the main stopcock to carry out repairs to the water tank without advice, but it was us two who had made the incorrect assumptions which landed us well and truly in the cart. Somehow we had to get out of Bewdley. We filled the boiler and eventually got about 250gal into our tank from the carriage pipe. We should have stayed longer but the risk was worth taking. We had to get beyond Hampton Loade before our tank was dry. L. P. Parker would have approved of Alun as an engineman for he worked with the shortest cut off and a wide open regulator to save water. I maintained steam 20lb below the blowing off point and kept the water well up throughout. Our opposite number, not surprisingly, was waiting for us when we crossed and so we did not hang about. The injector was still working when we reached Hampton Loade, the boiler was full and we were nearly home. Alun and I had worked as a team: for notching up, we lost no speed, I worked the regulator, he set the lever. Halfway up the bank, the feed blew off but we knew we were safe and that there would be water at Bridgnorth. We knew we had achieved something exceptional for we had used our skill and experience to win through, but it did not alter the fact that we were both culpable, and that we had caused unnecessary delay. I am still waiting to receive a disciplinary Form I!

The steam locomotive taught us all to pay attention, to neglect things at one's peril. There should never be conversation when running into a station, up to an adverse signal, up to or over a speed restriction. In France, there is no escape, the Flaman speed recorder sees to that but even the best slip up now and again. André Duteil once told me that when he had paid a fine of 600 old Francs for exceeding the limit by a few kilometres an hour, he was accosted by the boss, the Ingénieur, Chef de Depot. They shook hands as is the custom but then the Chef said:-

'Duteil, bon mécanicien, tres bon mécanicien mais faites attention.'

The wisest advice.

The use of the surname is still universal amongst older men in France whereas it can give offence in this country. When I was at school it was the standard method of address and for many years, I expected to be called Hardy by my boss. On the other hand, in my earlier years I could never bring myself to use men's surnames, for my own generation had a different outlook, but I well remember being keelhauled by my Superintendent at Cambridge and being told that my use of the Christian name would encourage undue familiarity. In fact, once one took charge, it did nothing of the sort for it did not undermine one's position as a Shedmaster if used correctly and with courtesy. It was clearly understood that there could be no reciprocal arrangement in the use of Christian names but again this came perfectly naturally. I can now see that it was actually welcomed provided that it was accompanied by the ability to command respect and maintain law and order.

Once or twice I slipped up and was neatly caught out. Not so long before I left Stewarts Lane, I wanted to see Fireman Fred Keene about something that had gone slightly wrong but did not warrant him being on the carpet. I came across him in the Shed. 'Keeny', I said, 'I want a word with you'. Very tall and beetle browed, he turned and faced me. 'Guv'nor', he said 'I'm Fireman Keene or Mr Keene to you'. Collapse of stout party; Fred Keene never got his bollocking and I had asked for it, for attitude will always be the art of gunnery.

Frank Mayes, now retired and once a King's Cross fireman and Stratford driver, had corresponded for many years with Edward H. Livesay who lived in Victoria BC and who wrote for the *Engineer*. An Englishman of the old school and a pioneer motorist, he had emigrated to Canada near the turn of the century and had travelled widely on locomotives, particularly in France. He was of a generation who would not use Christian names and encouraged Frank to address him simply as Livesey in correspondence, if not in person. So far, so good. Livesey came to England in 1960, called to see me and asked to meet Mayes at Stratford. I wrote a personal but formal letter to Frank, addressed to Driver F. Mayes but with 'Dear Mayes' handwritten by way of introduction. By return came the reply: delighted to meet Mr Livesey again and will report to the Shedmaster at Stratford as arranged but I was staggered to see that I was addressed as 'Dear Hardy' by one of my own drivers who had to be sent for very quickly and told very firmly that 'you can't do that there 'ere' which he smilingly accepted, but we laugh about it to this day. Nowadays, the use of the surname is regarded by some as a mortal insult but these last few years I found it very useful just now and again in dealing with young engineers who were getting a bit above themselves. It made them grind their teeth and how they loathed it!

A story must end somewhere. I have been lucky to choose a way of life that I have loved. I have met hundreds, nay thousands of people in all grades and levels, who have enriched my life; I have also met a few whose room I preferred to their company, but many of the events that stand out in my mind have been related to the deeds of others. Look what the grass roots railwaymen have achieved and one can be proud to have shared a working life with them. Of course, there were many with whom I crossed swords, who gave nothing but trouble, who had to go, but even with these there was a challenge, for it was either them or me!

One can think of George Pilch, boilermaker at South Lynn who went in a 'B12' firebox with the fire under the brick arch and 75lb of steam in the boiler, to stop the fusible leaking, as we had no substitute engine. He saved 30 minutes delay. To say he risked his life would be an insult. He knew what he could achieve. One thinks of the examining fitters at Ipswich: Wilfred Brown, Syd Lincoln, Russell Gooderham. At the old depot, almost devoid of decent pits, three middle-aged men examined locomotives for service, going over the top or underneath on the level or on the disposal pit. They knew, too, exactly what to look for and where. Or of Wally Mason, Stratford driver acting as an inspector, who pulled up the timekeeping of the Norwich services when the 'Britannias' were ailing before their transfer to Norfolk. He was a dying man but nothing deterred him from achieving his aim.

Think of the work of the great Denman at Copley Hill. Highly intelligent, a man of principle, with great charm and determination, Bill Denman was at home in any company. As an engine driver he was of the highest class. I have written of his feat, during the war, when with the GN Atlantic No 4433, blowing off at 155lb/sq in, he lifted 570 tons away from

Wakefield on the 1 in 100, without assistance. He moved forward at the sixth attempt but by judgement, not by luck. After the war, he was coming back from Grantham and his Pacific failed at Retford. He took the Retford pilot. It was facing up road and had to be turned: in that short time a fire could be partially made ready. The load was 14 coaches and Bill Denman worked through to Leeds with the Great Central Atlantic No 6089. All the way to Wakefield, he was improving slightly on sectional time. There he dropped the Bradford portion. An engine was waiting to help him but he refused assistance and took 10 bogies through to Leeds with a GC Atlantic on a Pacific timing. Time was kept precisely.

And now read this letter. It was written by a man who had, through ill health, failed to achieve his ambition to become a driver, in those days a crippling blow to the spirit, and financially. John Cook was a labourer in King's Lynn Loco when I first met him in 1946, finished with the footplate for ever, a couple of years before. And yet here is a letter written in 1974, describing an incident that took place 40 years before when he was in his prime with years of driving ahead of him. I had been invited to speak at King's Lynn. It was not a railway gathering and, to my amazement, I found the hall packed with South Lynn and King's Lynn staff, many of whom had retired and more of whom I had not seen for 26 years. It took me a few seconds to remember John but he would not have noticed any hesitation for one was adept at this sort of thing.

59 Creswell Street
18th Nov. 1974 King's Lynn

Dear Mr. Hardy,

It was very nice of you to write via Tom Stokes, he delivered your letter and we sat and had a long chat together about the old times on the M.&G.N.

One of the things was when I fired to his father who I was with for some time, the best driver I was with, he nursed an engine and got the very best from it. In the 1930's, we were booked to work an excursion train from South Lynn to Yarmouth, lodge and work same back to South Lynn leaving Yarmouth at 10pm. C Class engine No. 43 with 10 Midland coaches on. We were in Yarmouth on time. On our return trip, we were booked to stop at Melton Constable for water, if required, after passing Corpusty, Tom went on tender to check our tank and said John we are not stopping for water at Melton we have enough to get to South Lynn. Melton distant was on, we whistled for it and got the right away. Up to now it was the only signal we had against us and the first time I had passed Melton without taking water.

When we were booking on duty next day, Mr. Jack Neave (the then Shedmaster) met us saying what the hell were you doing last night running from Yarmouth to S. Lynn in 1 hour 26 minutes for 74 miles, it's a record time. Tom, in his quiet way, just said 'Good driving and firing' and I didn't mind where I went with him.

By the way, I have not mentioned your lecture: well, I thoroughly enjoyed it from start to finish, so had all I had spoken to, it was great and what a reunion.

Yours sincerely
John Cook

And this from a man who had finished with the footplate in 1944. When I come to think of it, it was a reunion and how it happened is difficult to say. I can only put it down to the brotherhood of practical railwaymen, for I had left that depot in June 1948 when I was not quite 25 years of age.

And now for nearly 42 years I have shared that brotherhood. I have had a wonderful life on the railway, I have no regrets and I envy no one. From 1949 to 1982, with the exception of four years, I have been my own boss, as a Shedmaster, a District Officer, Divisional Manager and in my final position at BR HQ. For the good of my career, I should have spent more time in the earlier years at the Regional level but it was remote in comparison and not for me: in any case, the question never arose.

Gwenda and I arranged a party at '222' when we retired. It was a gathering of the clans all right and a true cross section of those with whom I had worked. Stan Hodgson who introduced me to the West Riding that June evening in 1941 was there, so was Sir Peter Parker. It was a wonderful reunion and Michael Casey, friend and colleague, who presided, conducted it with dignity and humour. And now for retirement and we shall enjoy the years ahead just as much as those that have passed.

Surely I can use, once more, the words of Edmond Godry of Calais: 'Que de souvenirs imperissable'!

COMBINED INDEX

378

COMBINED INDEX

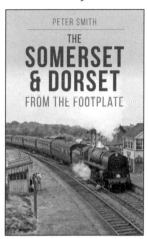

The Somerset & Dorset from the Footplate

Peter Smith

In 1987 the original Oxford Publishing Company produced an amalgam of two paperback books written by former Branksome fireman Peter Smith. *Mendips Engineman* and *Footplate over the Mendips* told the story of a young railway fireman and his driver Donald Beale. Enthralling the reader with stories of working trains over the old Somerset & Dorset line, the two books encompassed not just ordinary workings, but also early footplate experiences of his driver Donald Beale as well as the climax of Peter's own railway career, driving the very last northbound 'Pines Express in 1962.

The Somerset and Dorset route from Bath to Bournemouth across the Mendip Hills passed through magnificent scenery but had difficult gradients and tunnels which taxed the abilities of both S&D enginemen and the line's ever changing motive power. Each mile of the long closed line could tell a story, tales which the author recounts in this book with accuracy, clarity, and humour.

This edition contains the complete original text and also includes a new set of black and white images with which to illustrate what remains a still lamented cross-country railway.

The Somerset and Dorset from the Footplate is a book to be savoured, not just by those who remember this line, but by a whole new generation of railway enthusiasts.

288 pages, paperback

ISBN: 9781909328921 **£9.95**

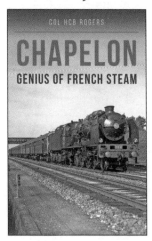

Chapelon
Genius of French Steam

Col HCB Rogers

The name Andre Chapelon is certainly known to every lover of the steam engine. Put simply, he was a genius, standing head and shoulders above all others in the field of steam locomotive design.

In his native France, Chapelon transformed the steam locomotives of various French railways from often mediocre machines into high performers surpassing the capabilities of similar machines used in other European countries at the time. His work was disrupted by World War II, but his importance was reasserted in 1946 with his superb 4-8-4 No 242 A 1 capable of producing a remarkable continuous output of 5,500hp.

Col Roger's book on Andre Chapelon originally produced in 1972 and difficult to obtain for many years remains the best and most detailed biography of the man and his work. Chapelon's influence was worldwide, Bulleid amongst others took note of his comments and this reprint will enhance the knowledge of the steam enthusiast who might otherwise have missed the opportunity to read about the true genius of French steam.

192 pages, paperback
ISBN: 9781910809730 **£8.99**

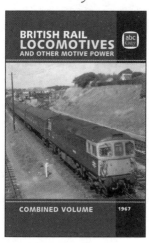

British Rail Locomotives and other Motive Power
Combined Volume 1967

Peter Smith

The Ian Allan 'abc' volumes have a lineage that can be traced back more than 70 years and have become sought-after collectables.

This latest 'abc' re-issue, sourced from original books in the Ian Allan archives, features all the steam locomotives, diesels, electrics, and multiple units that were in operation in 1967, including their allocated home depots.

Whilst it is no longer possible to see many of the locomotives described and illustrated in this facsimile copy, any railway modeller, historian or enthusiast will find much in here to reference and to research. The railways of the UK may have changed almost out of recognition since the 1960s, but nostalgia remains!

248 pages, hardback

ISBN: 9781800351448 **£13.50**

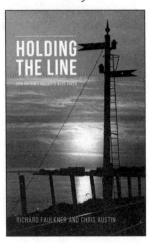

Holding the Line

How Britain's Railways Were Saved

Richard Faulkner and Chris Austin

At its zenith, there were 21,000 route miles of railway in Britain. Today the country's railways deliver more passenger miles than they did at their greatest extent despite a drastic reduction in the size of the network. Those cuts were the result of a campaign by a number of individuals who believed, erroneously as the passing of time has shown, that railways were a thing of the past and an impediment to progress.

Railway closures gained momentum in the 1960s in the harrowing years following the publication of the Beeching report. However, *Holding the Line* reveals, it could have been much worse and there were plans to reduce the size of the network even more drastically, to the point where only a few lines would have survived.

Now available in paperback for the first time, this book shows how close Britain's railways came to being eviscerated and how the dangers of closure by stealth still exist, even today.

344 pages, paperback
ISBN: 9780860936763 **£9.95**